Study Guide

AN INVITATION TO FLY

Basics for the Private Pilot

Second Edition

For further information about ordering or adopting this study guide or accompanying material, write or call

Wadsworth Publishing Company
10 Davis Drive
Belmont, California 94002
(415) 595-2350

Accompanying materials:

Textbook and partial San Francisco Sectional Chart
    (packaged together)

Instructor's Manual (complimentary on adoption of
    textbook for classroom use)

Videotapes (30 half-hour, color tapes)

Viewer's Guide

Flight Maneuvers Manual

Study Guide

AN INVITATION TO FLY

Basics for the Private Pilot

Second Edition

George B. Semb
University of Kansas

Wadsworth Publishing Company
Belmont, California
A Division of Wadsworth, Inc.

Aviation Education Editor: Carol Butterfield

Copyeditor: Joan C. Pendleton

Cover Photograph: Paul Bowen

Composition: Trilogy Systems, Inc.

All of the questions from the <u>Private Pilot Question</u> <u>Book</u> (FAA-T-8080-1) and the following figures from the <u>Private Pilot Question Book</u> (figure numbers are from this <u>Study Guide</u>) are FAA material and may be reproduced without permission: Figures 2.3, 2.4, 4.1 through 4.5, 5.1 through 5.3, 6.1 through 6.5, 6.9 through 6.15, 7.1 through 7.5, 9.1 through 9.11, 10.3, 10.4, 12.1 through 12.4, 13.3 through 13.9, 14.2, 14.3, Appendix A. The answers provided for each FAA question are copyrighted and may not be used without permission.

ISBN 0-534-04801-3

Printed in the United States of America

1 2 3 4 5 6 7 8 9 10 -- 89 88 87 86 85

To Pat, GT, and Ryan

CONTENTS

Preface   ix
To the Student   ix
Preparing for the FAA Private Pilot Written (Airplane) Exam   x
Unusable FAA Test Questions   xiv
Acknowledgments   xiv
About the Author   xiv

1/On Becoming a Pilot   1

2/The Practical Science of Flight   9

3/The Power Plant and Its Systems   31

4/Flight Instruments   53

5/Airplane Weight and Balance   75

6/Performance: Measuring an Airplane's Capabilities   93

7/Airports, Airspace, and Local Flying   127

8/Meteorology: A Pilot's View of Weather   171

9/Using Aviation Weather Services   205

10/Flight Information Publications   239

11/Federal Aviation Regulations   255

12/Basics of Air Navigation   283

13/Radio Navigation Aids   309

14/Composite Navigation: Going Cross-Country   341

15/The Physiology of Flight   357

16/Handling Airborne Emergencies   371

Appendix A/Private Pilot Question Book, Appendix 2   381

Appendix B/Private Pilot Exam Cross-Reference Guide   389

PREFACE

TO THE STUDENT

This study guide contains **all** of the items that may appear on the FAA Private Pilot Written (Airplane) Exam, together with answers and explanations. All FAA questions are fully integrated with the text and this study guide. Finally, FAA questions are cross-referenced to the appropriate chapter of both the text and this study guide.

This study guide is designed to accompany An Invitation to Fly: Basics for the Private Pilot by Dennis Glaeser, Sanford Gum, and Bruce Walters. It can also be used to review and prepare for the Federal Aviation Administration (FAA) Private Pilot Written (Airplane) Examination. The study guide is based on principles, regulations, and practices that were current at the time it was printed.

The study guide requires the use of the same supplemental materials as the text, including:

1. San Francisco Sectional Chart (partial)
2. Navigation plotter
3. Flight computer or flight navigation calculator
4. Calculator (optional)

The study guide is designed primarily to help you learn the basic information necessary to become a safe, competent private pilot. Each chapter includes: a summary of the main points in the text, vocabulary and technical exercises to make certain you know the "language" of piloting, discussion questions and exercises that emphasize the important concepts, and multiple-choice review questions, many of which are identical to questions you will encounter on the FAA Private Pilot Written (Airplane) Exam. Questions that actually appear on the FAA Private Pilot Written Exam are highlighted by having the **FAA question number** printed in **boldface**. Furthermore, each FAA question is cross-

referenced to the <u>Private Pilot Question Book</u> (FAA-T-8080-1).
This study guide contains all of the items that the FAA uses. It
also has several other items that reinforce concepts stressed on
the FAA exam. Finally, answers to the review questions, plus an
analysis of the questions and answers, appear at the end of each
chapter.

As a general guide to studying the text and using this study
guide, begin by thoroughly reading the chapter in the text.
Next, read the main points in the study guide. These summarize
and review material presented in the text. Once you are familiar
with the material, it is time to become actively engaged in
learning it. Try to do all of the vocabulary and technical
exercises without referring to the text for help. Once you have
attempted them all, refer to the text to check your answers and
to restudy those items you missed. Next, answer the discussion
questions and exercises, again without referring to the text
until you have attempted all of them. Finally, the review
questions are designed to see how much you have learned. Try all
of the review questions before you look up the answers at the end
of each chapter.

If you follow these guidelines you should be able to
correctly answer most of the review questions. Further, you
should have a good grasp of the material covered in the chapter.
One of the most important parts of this process is your attempt
to answer the questions before looking up the answers. Learning
specialists have argued for a long time that active involvement
in study-guide exercises promotes effective learning and
retention, and research has shown them to be correct. Therefore,
this study guide is designed to encourage you to actively
participate in the learning process.

PREPARING FOR THE FAA PRIVATE PILOT WRITTEN (AIRPLANE) EXAM

The FAA Private Pilot Written (Airplane) Exam is not an easy
test. Nor is it a terribly difficult one, if you are well
prepared. One purpose of the text and this study guide is to
help you prepare for it.

Beginning in March 1984, the FAA initiated a new testing
system. Under the new system, all questions that appear on the
Private Pilot Written (Airplane) Exam are published in the
<u>Private Pilot Question Book</u> (FAA-T-8080-1). Before you rush out
and purchase the <u>Private Pilot Question Book</u>, however, you should
be aware that all of the items that appear on the FAA written
exam for private pilots (airplane) are included in this study
guide, together with answers and explanations. Everything you
need to master the FAA Written Exam is contained right here!

Before you can take the test, Federal Aviation Regulations
(FARs) require that you: (a) have satisfactorily completed a
ground-instruction or home-study course signed off by a Certified
Flight Instructor (CFI) or by a General Aviation District

Officer, (b) present an airman certificate, driver's license, or
other official document as proof of identification, and (c)
present a birth certificate or other official document showing
that you meet the age requirement. For complete details, refer
to Part 61 of the Federal Aviation Regulations.

For written test purposes, you will be required to use the
Private Pilot Question Book provided by the testing center or the
designated written test examiner. The test administrator will
give you an "assignment" sheet listing the 60 questions you are
to answer. You will have four hours to complete the test. Each
question is to be answered by selecting a single best
alternative. You may take the following equipment with you to
the test center: a protractor or plotter and a flight navigation
computer. Textbooks and notes are forbidden, and the test
administrator will provide papers, answer forms, and special
pencils. You may also use a calculator; however, there are
several regulations concerning its use. First, the test
administrator will instruct you to turn the calculator on and off
before and after the test to ensure that any data stored in it
are destroyed. If your calculator is one that stores data even
when the switch is off, the administrator may ask you to
demonstrate that all memory registers are empty. Second, if your
calculator produces hard copy, all printouts must be surrendered
to the administrator at the end of the test. Third, you will not
be allowed to use any written instructions pertaining to the
calculator during the test.

The feedback you will receive will be in the form of an
Airman Written Test Report (AC Form 8080-2) that includes not
only the test score but also the actual question number of each
item answered incorrectly. You may then refer to this study
guide to locate and analyze the questions you missed.

Besides reading the text and completing the study guide, you
will fint it valuable to simulate examination conditions whenever
possible. One way you can do this is to consider the review
questions at the end of each chapter as a mini-examination. Set
a time limit of four minutes per question and answer all of them
before looking up the answers. Making the simulated exercises as
similar as you can to the actual exam conditions will greatly aid
in your preparation for the FAA written exam and may also help
you reduce your anxiety.

One of the best ways to prepare yourself for the actual exam
is to be well rested. Staying up late and "cramming" the night
before the test often leads to anxiety, fuzzy thinking, and
careless errors. So, get a good night's sleep the day before the
exam. Next, make a checklist of the things you will need for the
exam (plotter, calculator, and so on). Use the checklist, much
as you would before you started a flight, to be certain that you
take all of the required paraphernalia. It is easy to become
flustered when you discover that you have forgotten something
important, so use your checklist. Finally, arrive at the test
center early to acquaint yourself with the setting. Find the

restroom. Ask about the availability of beverages. Remember, too, that you are usually not allowed to bring food, so eat a good meal before your test.

The multiple-choice questions on the FAA exam have one answer that is considered **most** correct. Each item is independent of other items. This means that an answer to one item in no way depends upon or influences the answer to another item. However, as you read through a series of questions on the same subject, you may find answers to one question in the stem of another.

The minimum passing grade will be specified by the test administrator on the written test sheet you receive. Before you begin the exam, read the directions carefully. If you have questions, ask the administrator.

Many questions on the exam may appear tricky. Some of the questions may in fact be tricky, but careful analysis of most of them reveals that they are not tricky per se but rather that they require subtle discriminations. For example, a question that asks you to discriminate between 3,500 ft. MSL (above sea level) and 3,500 ft. AGL (above ground level) is attempting to make certain that you understand the difference between two ways of referring to altitude. Although such subtle discriminations are sometimes difficult to learn, they are extremely important in flying.

Read each question thoroughly and carefully before selecting an alternative. If the statement refers to a general rule, do not look for exceptions to the rule unless the question specifically asks you to do so. Once you have read and understood the question, carefully evaluate each of the alternatives. Treat each one as a true-false statement and eliminate those you know to be obviously incorrect. Then concentrate on those (if any) about which you are uncertain. When answering multiple-choice items like those that appear on the FAA written exam, I typically keep a list of the alternatives about which I am uncertain, so that when I return to the question later, it is easy to see exactly where ambiguities still exist. Finally, if time remains, I go through the test a second time.

In general, I do not spend a great deal of time on any single question. If, after you have spent more than two minutes on a question, you still do not understand what is being asked, go on to the next question. At some point later in the exam, you may discover an item that has a direct bearing on what you did not comprehend earlier or one that jogs your memory. When that happens, answer the question you are working on, then return immediately to the one you did not understand earlier. Keeping a "tally sheet" of questions about which you are uncertain while taking the exam makes retracing your steps an easy task.

Once you have attempted all of the questions, return to the items about which you were uncertain. Count how many items there are in the "uncertain" category and divide the number of unanswered questions into the time remaining. This will give you an idea of how much time to spend on each question. Attempt to answer each of the remaining questions, checking occasionally to be sure that you are not spending too much time on any one item.

When only 10 minutes remain, or if you are confident all of your answers are correct, stop and cross-check the answer sheet with your tally sheet. If there are any discrepancies, check to see which answer is the one you want and make certain the answer key is marked the way you want it. I cannot emphasize the rechecking process enough. Our research has shown that as many as 10 percent of the answers students record on the answer sheet do not correspond with their tally sheet, so cross-check your answers thoroughly. Finally, if you left any answers blank, make your best guess and record it. There is no penalty for guessing, so choose the alternative that you think best answers the question being asked.

If you do not pass the test the first time you try it, you were not adequately prepared. You will have to wait 30 days from the time you last took the test to reapply, unless an authorized instructor certifies in writing that appropriate instruction has been given and that he or she finds you competent to pass the test, in which case you can reapply before the 30-day limit has elapsed.

One final comment about the written test concerns cheating. If you cheat you are cheating yourself as well as others. It is something that simply cannot be tolerated. Federal Aviation Regulation 61.37 is presented here for you to read.

FAR 61.37 Written tests: cheating or other unauthorized conduct

(a) Except as authorized by the Administrator, no person may

(1) copy, or intentionally remove, a written test under this part;
(2) give to another, or receive from another, any part or copy of that test;
(3) give help on that test to, or receive help on that test from, any person during the period that test is being given;
(4) take any part of that test in behalf of another person;
(5) use any material or aid during the period that test is being given; or
(6) intentionally cause, assist, or participate in any act prohibited by this paragraph.

(b) No person whom the Administrator finds to have committed an act prohibited by paragraph (a) of this section is eligible for any airman or ground instructor certificate or rating, or to take any test thereof, under this chapter for a period of one year after the date of that act. In addition, the commission of that act is a basis for suspending or revoking any airman or ground instructor certificate or rating held by that person.

I hope the information provided here helps you achieve a high score on the FAA written exam. If you find other preparation techniques helpful, please send a description of them to me so I can incorporate them into future editions of this study guide. My address is: George B. Semb, Department of Human Development, University of Kansas, Lawrence, Kansas 66045.

UNUSABLE FAA TEST QUESTIONS

In August 1984, the FAA designated several questions as unusable for test purposes and removed them from the Question Selection Sheets. These items are noted by the letters DI in boldface next to the FAA item number. Because many of these items still have educational value, they have been included in this study guide.

ACKNOWLEDGMENTS

I would like to thank Greg Harper, Flight Training Director, Executive Beechcraft, Johnson County Industrial Airport, Kansas, for his critical analysis of the material.

George B. Semb

ABOUT THE AUTHOR

George B. Semb is professor of human development at the University of Kansas. He holds a B.S. degree from the University of Washington, an Sc.M. degree from Brown University, and a Ph.D. degree from the University of Kansas. All of his degrees are in the field of psychology, with particular emphasis on the design and evaluation of instructional systems. Dr. Semb received his private pilot's license in 1966 and has flown actively since that time. He has written study guides for textbooks in psychology, child development, and biology. His primary area of research is instructional systems, where he has published and presented over one hundred papers in the last decade.

1/ON BECOMING A PILOT

MAIN POINTS

1.   The aviation community consists of three sectors:
commercial airlines, military aviation, and general aviation.
Included under general aviation are student-pilot training, air
taxis, crop dusters, business jets, and other activities.

2.   Businesses that provide services for general aviation
at airports are called fixed base operators (FBOs).

3.   The Federal Aviation Administration (FAA) regulates
aviation in the United States. It issues and enforces Federal
Aviation Regulations (FARs), operates flight service stations
(FSS), and coordinates traffic flow at busy airports through air
traffic control towers.

4.   Aircraft are classified by category, class, and type.
Categories refer to the method of staying aloft and means of
propulsion. They include: (1) lighter-than-air craft; (2) rotorcraft; (3) gliders; and (4) airplanes. There are four classes of
airplanes: single-engine land, single-engine sea, multi-engine
land, and multi-engine sea. Common airplane types include the
Cessna 152, the Beechcraft Skipper, the Piper Tomahawk, and the
Boeing 767.

5.   This chapter explains how to obtain Student and Private
Pilot (airplane) Certificates as detailed in Part 61 of the
Federal Aviation Regulations (FARs).

6.   There are no minimum or maximum ages for taking flight
lessons. However, you must be at least 16 to fly solo and at
least 17 to receive the Private Pilot Certificate.

7.   Flight instruction (sometimes referred to as dual
instruction) must be performed by a Certified Flight Instructor
(CFI). Your training will also include ground school. Fees

1

associated with the various aspects of training will include the instructor's fee, airplane rental charges, and any fees associated with ground school such as enrollment and texts.

8.   To receive a **Student Pilot Certificate**, you must be at least 16; be able to read, speak, and understand English; and hold at least a **Third Class Medical Certificate**. You will also need to apply for a Federal Communications Commission (FCC) **Restricted Radiotelephone Operator Permit**. The Student Pilot Certificate and the FCC permit are required to fly an airplane solo and must be carried when flying solo. The Student Pilot Certificate is valid for 24 months; it will expire after two years at the **end of the month** in which it was issued.

9.   To fly solo as a student pilot (by yourself--no passengers allowed), a Certified Flight Instructor (CFI) must determine that you are familiar with visual flight rules (VFR)-- Part 91 of the FARs--and must enter an endorsement in your logbook and on your Student Pilot Certificate. The solo endorsement does not entitle you to fly cross-country; each cross-country flight must be endorsed separately.

10.   To apply for the Private Pilot (Airplane) Certificate, you must be at least 17; be able to read, speak, and understand English; hold at least a Third Class Medical Certificate; pass the Private Pilot Written (Airplane) Examination; have completed at least twenty hours of dual instruction, including three hours of cross-country and three hours in preparation for the flight test within sixty days of the test; and have completed twenty hours of solo time, including ten hours in airplanes and ten hours of cross-country flight. Flying at night requires separate training and endorsements, but it is not required to obtain a private pilot's certificate. Finally, FAR Part 141-approved schools can recommend students with a minimum of 35 hours (20 dual, 15 solo).

11.   Requirements for the **FAA Private Pilot (Written) Exam** are discussed in the **Preface** to this study guide, together with suggested study techniques.

12.   Once you satisfy the above requirements, you are ready for the flight test. It consists of three parts: an oral test, basic flying techniques, and a cross-country flight. Passing the test results in your being awarded the **Private Pilot Certificate**. This certificate is issued without an expiration date.

13.   There are two types of flight rules, **visual flight rules (VFR)** and **instrument flight rules (IFR)**. VFR pertain to flight free of clouds and in areas of adequate visibility. IFR pertain to aircraft operating in weather conditions below VFR minimums.

14.   Like other fields, aviation has many abbreviations. One of your tasks will be to learn many of them--CFI, FCC, FAA, VOR, TRSA, ADF, FBO, POH, IAS, CAS, MSL, AGL, and more.

KEY TERMS AND CONCEPTS

Match each term or concept (1-16) with the appropriate description (A-N) below. Each item has only one match.

___ 1. FCC
___ 2. sixteen
___ 3. VFR
___ 4. CFI
___ 5. type
___ 6. FBO
___ 7. IFR
___ 8. Third Class

___ 9. designated examiner
___ 10. FARs
___ 11. basic flying technique
___ 12. airplane class
___ 13. twenty
___ 14. pilot-in-command
___ 15 FAA
___ 16. cross-country flight

A. person who must endorse logbook for solo flight
B. agency that issues Restricted Radiotelephone Operator Permit
C. one part of the private pilot flight test
D. minimum age for solo flight
E. flight free of clouds and in areas of adequate visibility
F. type of medical certificate required of student and private pilots
G. minimum number of hours of solo time required for private pilot certification
H. rules that apply to weather conditions below VFR minimums
I. government branch responsible for civil aeronautics
J. refers to a specific make and model of aircraft
K. business that provides services at airports
L. civilian flight instructor authorized by the FAA to give flight tests
M. regulations governing all aspects of aviation
N. individual responsible for operation of an aircraft
O. single-engine land
P. a flight of more than 50 NM from your home base

DISCUSSION QUESTIONS AND EXERCISES

1. What three sectors constitute the aviation community? To which will you belong as a student pilot?

2. Distinguish among category, class, and type of aircraft. Identify four categories of aircraft, name four classes of airplanes, and identify what type you fly or intend to fly.

3.   What are FARs? Who issues and enforces them?

4.   What is a CFI? Name at least three activities he or she performs.

5.   What three requirements must you meet <u>before</u> you can obtain a **Student Pilot Certificate**?

6.   What documents must be in your possession whenever you fly solo or act as pilot-in-command?

7.   What is a **supervised solo flight**? What must a CFI do before <u>each</u> solo cross-country flight?

8.   What restrictions apply to flying at night?

9.   According to FAR Part 61, how much time must you accumulate, and in what categories, before you can take the private pilot check ride?

10. Briefly describe the three parts of the private pilot check ride.

11. Distinguish between **VFR** and **IFR**.

REVIEW QUESTIONS

1.   (FAA 1536) What document(s) must be in your personal
possession while operating as pilot-in-command of an aircraft?

1--A certificate showing accomplishment of a current flight
   review.
2--A certificate showing accomplishment of a checkout in the
   aircraft and a current flight review.
3--A pilot logbook with endorsements showing accomplishment of a
   current flight review and recency of experience.
4--An appropriate pilot certificate and a current medical
   certificate.

2.   (FAA 1538) A Third Class Medical Certificate was issued on
August 10, this year. To exercise the privileges of a private
pilot certificate the medical certificate will be valid through

1--August 10, 3 years later.
2--August 10, 2 years later.
3--August 31, 3 years later.
4--August 31, 2 years later.

3.   (FAA 1534) A private pilot acting as pilot-in-command, or
in any other capacity as a required pilot flight crewmember, must
have in their personal possession while aboard the aircraft

1--a current logbook endorsement to show that a flight review has
   been satisfactorily accomplished.
2--the current and appropriate pilot and medical certificates.
3--a current endorsement on the pilot certificate to show that a
   flight review has been satisfactorily accomplished.
4--the pilot logbook to show recent experience requirements to
   serve as pilot-in-command have been met.

4.   (FAA 1535) When must a current pilot certificate be in the
pilot's personal possession?

1--Only when operating in controlled airspace.
2--When acting as a crew chief during launch and recovery.
3--Only when passengers are carried.
4--Anytime when acting as pilot-in-command or a required
   crewmember.

5.   (FAA 1539) A Third Class Medical Certificate was issued on
May 3, this year. To exercise the privileges of a private pilot
certificate, the medical certificate will be valid through

1--May 31, 1 year later.
2--May 31, 2 years later.
3--May 3, 1 year later.
4--May 3, 2 years later.

6.   (FAA 1537) What is the duration, if any, of a private pilot
certificate?

1--It expires 24 calendar months after issuance.
2--As long as the flight review and the medical certificate are
    current.
3--It expires 90 days after the currency of the flight review has
    lapsed.
4--Indefinite.

7.   What age must you be to be eligible to receive a Private
Pilot (Airplane) Certificate?

1--15 years of age.
2--16 years of age.
3--17 years of age.
4--There is no minimum age.

8.   Of the following items (a, b, c, and d), which are you
required to carry with you when flying solo as a student pilot?

          (a) Student Pilot Certificate
          (b) FCC Restricted Radiotelephone Operator Permit
              or higher certificate
          (c) Social Security card
          (d) Driver's license

1--a, b.   3--a, b, c, d.
2--a, b, c.   4--a, d.

9.   According to FAR Part 61, how many total hours are required,
both with a flight instructor and solo time, before you can take
the private pilot flight test?

1--20 hr.   3--40 hr.
2--30 hr.   4--50 hr.

10.   A hot-air balloon is classified as a/an

1--airplane.   3--lighter-than-air craft.
2--glider.   4--rotorcraft.

11.   Which of the following is an example of a **class** of aircraft?

1--Boeing 747.   3--Glider.
2--Multi-engine sea.   4--Airplane.

ANSWERS

Key Terms and Concepts

1. B	5. J	9. L	13. G
2. D	6. K	10. M	14. N
3. E	7. H	11. C	15. I
4. A	8. F	12. O	16. P

Review Questions

1. 4--You must have your current pilot certificate in your personal possession to act as pilot-in-command. See FAR 61.3.
2. 4--A Third Class Medical Certificate expires at the end of the last day of the 24th month after issue. See FAR 61.23.
3. 2--You must have **both** your current pilot certificate and current medical certificate in your personal possession to act as pilot-in-command. See FAR 61.3.
4. 4--You must have your current pilot certificate in your personal possession to act as pilot-in-command. See FAR 61.3.
5. 2--The certificate expires at the **end** of the 24th month after it was issued. See FAR 61.23.
6. 4--Any pilot certificate, other than a student pilot's, issued under FAR Part 61 does not expire. See FAR 61.19. Although the certificate is good indefinitely, there are several things you must do to stay "current."
7. 3--You must be at least 17 to obtain a private pilot certificate.
8. 1--You are required to carry **both** your student pilot certificate and the FCC permit. Remember, your medical certificate also serves as the student pilot certificate.
9. 3--Under FAR Part 61, 40 is the required minimum, 20 hr. of which must be dual instruction and 20 hr. of which must be solo.
10. 3--Lighter-than-air is a **category**. A balloon is a **class** within that category.
11. 2--A Boeing 747 is a **type** of aircraft. Airplanes and gliders are **categories** of aircraft. Multi-engine sea is a **class** of airplanes.

2/THE PRACTICAL SCIENCE OF FLIGHT

MAIN POINTS

1.   You should be able to label an airplane, including the following common structures: **wings**, including ailerons and flaps; the **fuselage**; the **tail assembly** or **empennage**, including the vertical stabilizer, rudder, horizontal stabilizer, and elevator; **engine**, **propeller**, and **cowl**; and, **undercarriage**, including the nose wheel (if any) and landing gear. The angle at which the wing root leaves the fuselage forming a slight "V" is called the **dihedral**.

2.   The four forces of flight are **lift**, **weight**, **thrust**, and **drag**. The general rule is: Lift opposes weight and thrust opposes drag. Gravity acts on weight to cause airplanes to go down. Thrust, the force developed by the airplane's engine, acts in the same direction as the airplane's flight path. Lift and drag are forces produced by the motion of the airplane through the air. Lift acts perpendicular to the flight path and drag acts parallel to it.

3.   Airplanes have several **airfoils** such as the wings, elevator, and propeller. The airfoil that produces the greatest effect is the **wing**.   It has a rounded leading edge and a sharp trailing edge. The imaginary line passing through the leading and trailing edges is called the chord line. The angle that the chord line makes with the **relative wind** defines the **angle of attack**. The relative wind is exactly opposite the flight path of the airfoil.

4.   An airplane creates **lift** by redirecting passing air downward. Newton's third law of equal and opposite reaction applies. The action of redirecting air downward creates the reaction of lift. The wing causes this redirection of air because its airfoil shape causes air passing over the wing to accelerate while causing air passing below it to slow down. As Bernoulli observed, when you increase velocity, there is a lower

9

pressure. Thus there is a lower pressure area on top of the
wing, which draws passing air downward, creating downwash.
Higher pressure under the wing and the deflection of air striking
the lower surface of the wing also contribute to downwash.
Finally, lift consists of two primary forces--horizontal and
vertical. The differential interaction of these two forces
causes an airplane to turn.

5.   That proportion of lift parallel to the flight path is
called **induced drag**. Remember that forces perpendicular to the
flight path produce lift, while those parallel to it produce
drag. Another type of drag is called **parasite drag**. This refers
to the effect of air molecules that run into parts of the airplane such as the landing gear, wing struts, and the windshield.

6.   Four variables influence the aerodynamic features of an
airplane: (1) its size and shape, (2) air density, (3) the
wing's angle of attack, and (4) how fast it is moving through the
air. Increasing the angle of attack increases both lift and
induced drag. But there is an upper limit to how large an angle
the wing can support. It is similar to the limit set on how much
weight you can lift. As the angle of attack increases, the air
flowing over the top of the wing begins to separate from the
upper surface. The amount of separation increases as the angle
of attack increases, until lift begins to decrease. At this
point, the wing is **stalled**. The angle of attack at which the
stall begins is called the **critical angle of attack**. The only
way to recover from a stall is to reduce the angle of attack. We
put a great deal of emphasis on stalls because they are a leading
cause of accidents. **Stalls can occur at any airspeed and in any
attitude!** All you need to do is exceed the critical angle of
attack. Airplanes are usually designed to cause stalls to occur
close to the fuselage and then work their way out the wing.

7.   An airplane has three axes of rotation: **lateral,
longitudinal**, and **vertical**. Rotation about the lateral axis is
referred to as **pitch**, as when the nose of the airplane goes up or
down; it is controlled by the **elevator**. For example, when you
push the control column or yoke forward, the trailing edge of the
elevator (another airfoil) moves down, which in turn produces
lift on the elevator surface. The result is that the tail goes up
and the nose goes down. Rotation about the vertical axis is
called **yaw**, as when you turn right or left; it is controlled by
the **rudder**. For example, when you push the right rudder pedal,
the rudder moves to the right. Once again, the forces of lift
operate, this time to move the tail of the airplane to the **left**
and the nose of the airplane to the **right**. Rotation about the
longitudinal axis is called **roll**, as when the control wheel is
turned left or right (banking); it is controlled by the **ailerons**.
For example, when you turn the control wheel to the left, the
left aileron goes **up** and the right aileron goes **down**; this
produces more lift on the right wing (it goes up) and the plane
banks to the left. Go through these axes, their controlling
surfaces, and your actions several times until you can **see** what
is happening in your head.

8.   **Load factor** is the ratio of total lift to the weight of the airplane. As you bank the airplane, total lift increases; therefore, so does the load factor. In a 60° banked level turn, the wings must produce <u>twice</u> the lift (a load factor of 2). Load factors are expressed in terms of G units, which refer to gravitational force. Angle of bank also affects the minimum speed in level flight at which the wings will stall. In level, turning flight, **the steeper the bank, the higher the stall speed.**

9.   Two other important control surfaces are **wing flaps** and **trim tabs.**   When wing flaps are lowered, the wing produces more lift **and** more drag. Thus, they increase the stall angle of attack of the wing. During landing, flaps allow the fuselage to assume a more nose-down attitude during low speed flight, they allow a steeper glide path and lower approach and touchdown speeds. Trim tabs are fine-tuning control surfaces located on a primary control surface such as the elevator. Some airplanes also have trim tabs for the rudder and ailerons, again to fine-tune the aerodynamic features of the airplane and allow the airplane to fly "hands off."

10. **Stability** refers to the airplane's tendency to return to the trimmed condition following a disturbance such as turbulence or control movement. Stability may be **positive** (return to the original trim condition), **neutral** (stay in the same condition), or **negative** (move away from the original condition). Stability may also be classified as static or dynamic. Finally, it also applies to the three axes of rotation: longitudinal (pitch), directional (yaw), and lateral (roll).

11. Operating an airplane near the ground affects the airflow over the wings. This is referred to as **ground effect.** The effect is most pronounced when the airplane is less than half the wing span from the ground. Ground effect allows the airplane to start flying at lower than normal airspeeds and may cause the plane to travel farther (float) when landing.

12. **Wake turbulence** is caused by tornado-like **wingtip vortices.** It is strongest behind aircraft that are heavy and slow and that have landing gear and flaps retracted. As a general rule, stay above the flight path of larger aircraft! On takeoff, lift off well before the point the previous large aircraft did and stay upwind of its flight path. On landing, stay above the glide path and land farther down the runway. Wait **at least** two minutes when departing after a large aircraft.

13. Every airframe has a maximum load it can handle without damage. The **maneuvering speed** is the maximum speed to use during maneuvers or in turbulent air, so as not to exceed the maximum load limit of the airplane. This speed decreases when the plane is operated at light weights.

14. Several **propeller effects** lead most planes to display a left-turning tendency under certain conditions. First, the tendency of the airplane's propeller to produce unbalanced

thrust, thereby inducing a yaw force to the left, is called the **P-factor**. This force can be counteracted by applying the right rudder. Second, the propeller also produces **torque**, which induces a left-rolling movement. Third, raising or lowering the tail of the airplane when the propeller is spinning produces a gyroscopic effect called **precession**. Finally, the high-speed rotation of the propeller causes a spiraling motion to the air-flow behind it. This spiraling motion affects vertical control surfaces and is referred to as the **slipstream effect**.

15. Stalls were covered earlier. We return to them now because of their importance. Most airplanes have **stall warning devices** to help the pilot detect impending stalls. You will practice stalls in every possible flight configuration and flight condition. Stall recovery involves these steps: lower the angle of attack (control wheel forward); simultaneously add full power and level the wings. Return to level flight as soon as possible to minimize altitude loss.

16. If a yawing force is present when an airplane stalls, a steep corkscrew path will result. This is called a **spin**. There are several ways to produce yaw, such as uncoordinated use of the rudder, adverse yaw from the ailerons, and turbulence. In general, spin recovery involves several steps: close the throttle, neutralize the ailerons, add full rudder opposite to the direction of the spin, move the control column forward to break the stall, neutralize the rudder, level the wings, and recover smoothly from the resulting dive.

KEY TERMS AND CONCEPTS, PART 1

Match each term or concept (1-20) with the appropriate description (A-T) below. Each item has only one match.

__ 1.	neutral stability	__ 11.	up
__ 2.	angle of attack	__ 12.	left
__ 3.	flaps	__ 13.	down
__ 4.	gravity	__ 14.	Bernoulli's principle
__ 5.	airfoil	__ 15.	Newton's third law
__ 6.	lift	__ 16.	load factor
__ 7.	pitch	__ 17.	yaw
__ 8.	roll	__ 18.	thrust
__ 9.	right	__ 19.	maneuvering speed
__ 10.	wake turbulence	__ 20.	P-factor

A. force of attraction between the earth and the airplane
B. rotation about the airplane's longitudinal axis
C. direction the left aileron moves when the control wheel is turned left
D. force that opposes gravity
E. direction the elevator moves when the control column is pushed in
F. principle that explains the area of reduced pressure created above the surface of an airplane's wing

G.   rotation about the airplane's vertical axis
H.   tendency to remain in conditions produced by a disturbance
I.   direction the airplane rolls when the right aileron goes up
     and the left aileron goes down
J.   shape designed to obtain lift from air that passes over and
     under it
K.   direction the rudder moves when making a left turn
L.   force created by a power plant that gives the airplane
     forward motion
M.   allows slower approach speeds
N.   ratio of total lift produced to the airplane's weight
O.   principle that explains the force obtained from the creation
     of downwash
P.   rotation about the airplane's lateral axis
Q.   strongest behind heavy, slow aircraft
R.   angle between the airfoil's chord and the direction of the
     relative wind
S.   decreases as the airplane's weight decreases
T.   unbalanced thrust that produces yaw to the left

KEY TERMS AND CONCEPTS, PART 2

     Match each term or concept (1-20) with the appropriate
description (A-T) below. Each item has only one match.

___ 1. parasite drag   ___ 11. critical angle of attack
___ 2. tricycle   ___ 12. positive stability
___ 3. flaps   ___ 13. stall
___ 4. spin   ___ 14. camber
___ 5. induced drag   ___ 15. adverse yaw
___ 6. static flight   ___ 16. conventional landing gear
___ 7. relative wind   ___ 17. empennage
___ 8. trim tab   ___ 18. coordinated flight
___ 9. stall recovery   ___ 19. dihedral
___ 10. ground effect   ___ 20. precession

A.   motion of air **parallel** to an airfoil flight path, opposite
     direction
B.   speed that decreases as the angle of bank decreases
C.   type of landing gear with a nose wheel and two main gears on
     the wings or fuselage
D.   angle at which the wing root leaves the fuselage
E.   cushion created by downwash from the wing
F.   force that pushes the nose of the airplane opposite (away
     from) the direction of a turn
G.   allow a steeper glide path
H.   flight in which yawing and slipping motions are not produced
     by the pilot
I.   simultaneously applying forward pressure to the control
     column and adding full engine power
J.   tail assembly of an airplane
K.   force that slows the forward movement of an airplane through
     the air due to frontal areas and undercarriage resistance
L.   proper use of aileron and rudder

M. fine-tuning controls for primary control surfaces
N. type of landing gear with a tail wheel and two main gears on
    the wings or fuselage
O. curvature of an airfoil
P. occurs when yaw is induced in a stalled airplane
Q. tendency to return to the original trimmed condition
R. retardant force produced when lift is being produced
S. a stall will occur when this is exceeded
T. gyroscopic effect

DISCUSSION QUESTIONS AND EXERCISES

1. Label each part of the airplane shown in Figure 2.1.

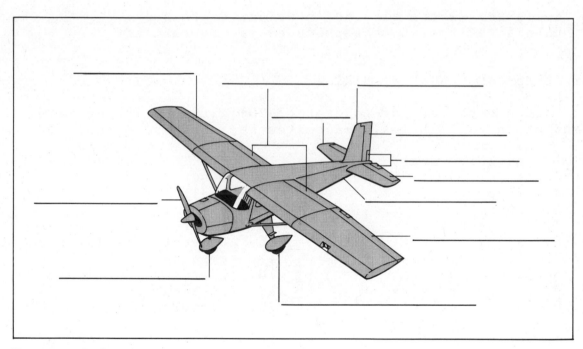

Figure 2.1

2. What are the four forces that affect flight? Explain the
relationship among these forces.

3. T F Drag opposes lift and thrust opposes gravity.

4.   Explain how lift works when you turn the control column of
the airplane to the right.

5.   Distinguish between **parasite drag** and **induced drag**.

6.   Describe in **your own words** what happens when the wing of an
airplane **stalls**. How is a stall related to the **angle of attack**
of the airfoil? How is it related to the airplane's airspeed and
attitude?

7.   Define **pitch, yaw, and roll**. Give an example of each, being
certain to explain how each relevant control surface moves and to
describe the resultant change in the airplane's attitude.

8.   T F   When you push the left rudder pedal, the rudder moves
              to the right and the airplane moves to the left.

9.   T F   When you push the control column forward, the elevator
              moves down and the tail of the airplane goes up.

10.   T F   When you turn the control column to the right, the
              right rudder moves up and the left rudder moves down,
              resulting in rolling the airplane to the right.

11. Explain how lowering the **flaps** affects each of the
following: takeoff distance, glide path, landing distance,
landing speed, and lift.

12. What are **trim tabs**? For what are they used? Give at least
one example of the use of trim tabs.

13. T F Elevator trim tabs control movement about the
          airplane's lateral axis.

14. Define **load factor** and explain how it is related to the
stalling speed of the airplane.

15. Explain the relationship among the following terms: **relative
wind, flight path, angle of attack,** and **stall.**

16. T F Relative wind is in the same direction as and parallel
          to the flight path of the airplane.

17. T F <u>Lift</u> is the term used to express the relationship
          between an airfoil's chord and its encounter with the
          relative wind.

18. T F The point along the trailing edge of an airfoil at
          which smooth airflow breaks away is called the
          stalling point.

19. T F The greater the angle of bank is, the lower the
          stalling speed will be.

20. T F Ground effect often causes an airplane to settle to
            the surface immediately after becoming airborne.

21. What is **slow flight**? Give at least one reason why it is an
important part of pilot training.

22. What is an **accelerated stall**? In normal flying situations,
when are you **most likely** to encounter one?

23. T F An airplane will enter a spin when the elevators lose
            their effectiveness due to a decrease in the velocity
            of the relative wind.

24. To answer questions a–d, refer to Figure 2.2.

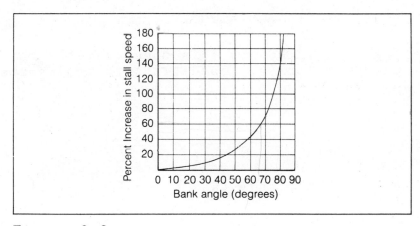

Figure 2.2

a. If your airplane normally (0° of bank) stalls at 40
kts., at what speed will it stall in a 70° bank?

b. If your airplane normally (0° of bank) stalls at 50
kts., but you encounter a stall at 60 kts., what is your
angle of bank?

      c. If your airplane is traveling at 100 kts. and stalls when your angle of bank reaches 80°, at what speed would it stall in level flight (0° of bank)?

25. Suppose that your airplane weighs 2,540 lb., its maximum takeoff weight. How many pounds are the wings supporting if you roll it into a 60° bank? At what speed will the airplane stall if it normally (0° of bank) stalls at 35 kts? At what speed will it stall in a 60° bank if it weighs only 2,000 lb. at takeoff?

26. What is **wake turbulence**? Under what conditions is it most likely to occur? Identify at least three things you should do as a pilot to avoid it.

27. Define **maneuvering speed**. How is this speed related to the airplane's weight? Under what flight conditions is it important?

28. What is the **P-factor**? What is **torque**? How do each of these forces affect the airplane?

29. Briefly describe how you enter a **spin**. Describe the general procedures you are to follow to recover from one.

REVIEW QUESTIONS

1.   (FAA 1150) The four aerodynamic forces acting on an
airplane are

1--power, velocity, gravity, drag.
2--power, velocity, weight, friction.
3--thrust, lift, gravity, weight.
4--thrust, lift, weight, drag

2.   (FAA 1151) When are the four aerodynamic forces that act on
an airplane in equilibrium?

1--When the aircraft is at rest on the ground.
2--When the aircraft is accelerating.
3--When the aircraft is decelerating.
4--During unaccelerated flight.

3.   (FAA 1152) What is the relationship of lift, drag, thrust,
and weight when an airplane is in straight-and-level flight?

1--Lift = drag and thrust = weight.
2--Lift, drag, and weight = thrust.
3--Lift and weight = thrust and drag.
4--Lift = weight and thrust = drag.

4.   (FAA 1153) What makes an airplane turn?

1--Centrifugal force.
2--Rudder and aileron.
3--Horizontal component of lift.
4--Rudder, aileron, and elevator.

5.   (FAA 1154) Refer to Figure 2.3. The actual angle A is the
angle of

1--dihedral.
2--attack.
3--camber.
4--incidence.

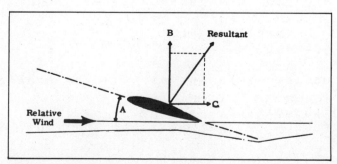

Figure 2.3

6.   (FAA 1155)   The term angle of attack is defined as the

1--angle between the wing chord line and the direction of the
    relative wind.
2--angle between the airplane's climb angle and the horizon.
3--angle formed by the longitudinal axis of the airplane and the
    chord line of the wing.
4--specific angle at which the ratio between lift and drag is the
    highest.

7.   (FAA 1161)   The left-turning tendency of an airplane caused
by P-factor is the result of the

1--clockwise rotation of the engine and the propeller turning the
    airplane counterclockwise.
2--propeller blade descending on the right, producing more thrust
    than the ascending blade on the left.
3--gyroscopic forces applied to the rotating propeller blades
    acting 90° in advance of the point the force was applied.
4--spiral characteristic of the slipstream air being forced
    rearward by the rotating propeller.

8.   (FAA 1162)   The purpose of the rudder on an airplane is to

1--control the yaw.
2--control the overbanking tendency.
3--maintain a crab angle to control drift.
4--maintain the turn after the airplane is banked.

9.   (FAA 1170)   What causes an airplane (except a T-tail) to pitch
nose down when power is reduced and controls are not adjusted?

1--The CG shifts forward when thrust and drag are reduced.
2--The downwash on the elevators from the propeller slipstream is
    reduced and elevator effectiveness is reduced.
3--When thrust is reduced to less than weight, lift is also
    reduced and the wings can no longer support the weight.
4--The upwash on the wings from the propeller slipstream is
    reduced and angle of attack is reduced.

10.   (FAA 1167)   Which basic flight maneuver increases the load
factor on an airplane as compared to straight-and-level flight?

1--Climbs.   3--Stalls.
2--Turns.   4--Spins.

11.   (FAA 1176)   Under what conditions can an airplane be stalled?

1--Only when nose is high and the airspeed is low.
2--Only when the airspeed decreases to the published stalling
    speed.
3--At any airspeed and in any flight attitude.
4--Only when the nose is too high in relation to the horizon.

12.   (FAA 1179) In what flight condition must an airplane be placed in order to spin?

1--Partially stalled with one wing low and the throttle closed.
2--Placed in a steep diving spiral.
3--Stalled.
4--Placed in a steep nose-high pitch attitude.

13.   (FAA 1159) Refer to Figure 2.4. If an airplane weighs 5,400 lb., about what weight would it be required to support during a 55° banked turn while maintaining altitude?

1--5,400 lb.   3--9,180 lb.
2--6,720 lb.   4--10,800 lb.

14.   (FAA 1160) Refer to Figure 2.4. The maximum bank that could be made during a level turn without exceeding the maximum positive load factor of a utility category airplane (+4.4 G units) is

1--71°   2--73°
3--77°   4--83°

15.   (FAA 1157) Refer to Figure 2.4 If an airplane weighs 2,300 lb., what approximate weight would the airplane structure be required to support during a 60° banked turn while maintaining altitude?

1--3,400 lb.   3--2,300 lb.
2--4,600 lb.   4--5,200 lb.

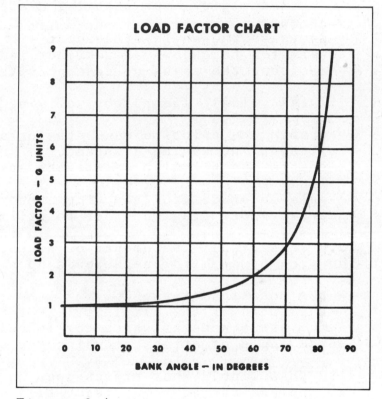

Figure 2.4

16.   (FAA 1158) Refer to Figure 2.4 If an airplane weighs 3,300 lb., what approximate weight would the airplane structure be required to support during a 30° banked turn while maintaining altitude?

1--3,100 lb.   3--1,200 lb.
2--3,960 lb.   4--7,220 lb.

17. (FAA 1054) One of the main functions of flaps during the approach to landing is to

1--decrease the angle of descent without increasing the airspeed.
2--permit a touchdown at a higher indicated airspeed.
3--increase the angle of descent without increasing airspeed.
4--decrease lift, thus enabling a steeper-than-normal approach to be made.

18. (FAA 1163) One of the main purposes of using flaps during the approach and landing is to

1--decrease lift, thus enabling a steeper-than-normal approach to be made.
2--increase the angle of descent without increasing airspeed.
3--permit a touchdown at a higher indicated airspeed.
4--decrease the angle of descent without increasing the airspeed.

19. (FAA 1164) The purpose of wing flaps is to

1--enable the pilot to make steeper approaches to a landing without increasing airspeed.
2--enable the pilot to reduce the speed for the approach to landing.
3--enlarge or control the wing area to vary the lift.
4--create more drag in order to utilize power on the approach.

20. (FAA 1165) The amount of excess load that can be imposed on the wing of an airplane depends upon

1--the position of the CG.
2--the speed of the airplane.
3--the abruptness at which the load is applied.
4--the angle of attack at which the airplane will stall.

21. (FAA 1166) What effect does an increased load factor have on an airplane during an approach to a stall?

1--The airplane will stall at a higher airspeed.
2--The airplane will have a tendency to spin.
3--The airplane will be more difficult to control.
4--The airplane will have a tendency to yaw and roll as the stall is encountered.

22. (FAA 1174) P-factor or asymmetric propeller loading causes the airplane to

1--be unstable around the lateral axis.
2--yaw to the left when at high angles of attack.
3--yaw to the left when at high speeds.
4--be unstable around the vertical and lateral axes.

23. (FAA 1175) In what airspeed and power condition is torque effect the greatest in a single-engine airplane?

1--Low airspeed, high power.   3--High airspeed, high power.
2--Low airspeed, low power.   4--High airspeed, low power.

24. (FAA 1177) The angle of attack at which an airplane wing stalls will

1--increase if the CG is moved forward.
2--change with an increase in gross weight.
3--remain the same regardless of gross weight.
4--decrease if the CG is moved aft.

25. (FAA 1178) As altitude increases, the indicated airspeed at which a given airplane stalls in a particular configuration will

1--decrease as the true airspeed decreases.
2--decrease as the true airspeed increases.
3--remain the same as at low altitude.
4--increase because the air density becomes less.

26. (FAA 1180) The phenomenon of ground effect is most likely to result in which problem in an airplane?

1--Settling back to the surface abruptly immediately after
   becoming airborne.
2--Becoming airborne before reaching recommended takeoff speed.
3--Inability to get airborne even though airspeed is sufficient
   for normal takeoff needs.
4--A rapid rate of sink and absence of normal cushioning during
   landings.

27. (FAA 1181) Which adverse effect must a pilot be aware of as a result of the phenomenon of ground effect during takeoff?

1--Difficulty in getting airborne even though airspeed is
   sufficient for normal takeoff.
2--Becoming airborne before reaching recommended takeoff speed.
3--Settling back to the surface abruptly immediately after
   becoming airborne.
4--Difficulty in climbing the first 20 ft. after takeoff.

28. (FAA 1182) The effect of floating caused by the phenomenon of ground effect will be most realized during an approach to land when

1--at less than the length of the airplane's wingspan above the
   surface.
2--at twice the length of the airplane's wingspan above the
   surface.
3--a higher-than-normal angle of attack is used.
4--at speeds approaching a stall.

29. (FAA 1183) An airplane is usually affected by ground effect
at what height above the surface?

1--Between 100 and 200 ft. above the surface in calm wind
    conditions.
2--Less than half of the airplane's wingspan above the surface.
3--Twice the length of the airplane's wingspan above the surface.
4--Three or four times the airplane's wingspan.

30. (FAA 1169) What determines the longitudinal stability of an
airplane?

1--The location of the CG with respect to the center of lift.
2--The effectiveness of the horizontal stabilizer, rudder, and
    rudder trim tab.
3--The relationship of thrust and lift to weight and drag.
4--The dihedral, angle of sweepback, and the keel effect.

31. (FAA 1425) Wingtip vortices created by large aircraft tend
to

1--sink below the aircraft generating the turbulence.
2--rise into the traffic pattern.
3--rise into the takeoff or landing path of a crossing runway.
4--accumulate at the beginning of the takeoff roll.

32. (FAA 1426) When taking off or landing at a busy airport where
large heavy aircraft are operating, one should be particularly
alert to the hazards of wingtip vortices because the turbulence
tends to

1--rise from a crossing runway into the takeoff or landing path.
2--rise into the traffic pattern area surrounding the airport.
3--sink into the flight path of aircraft operating below the
    aircraft generating the turbulence.
4--accumulate at the beginning of the takeoff roll.

33. (FAA 1424) Wingtip vortices, the dangerous turbulence that
might be encountered behind large aircraft, are created only when
that aircraft is

1--operating at high airspeeds.
2--heavily loaded.
3--developing lift.
4--using high-power settings.

34. The back edge of the wing is called the _____ edge.

1--elevated   3--leading
2--horizontal   4--trailing

35.   Refer to Figure 2.4. If an airplane has a maximum load factor of 4.4 G units, what is the maximum bank you could make without exceeding this load limit?

1--Roughly 42°.
2--Roughly 73°.
3--Roughly 77°.
4--Impossible to determine from this chart.

36.   Air resistance over the wings of an airfoil is called

1--drag.   2--gravity.   3--lift.   4--thrust.

37.   When the angle of attack increases, as in a climb or a turn, this leads to an increase in

1--induced drag.   3--induced thrust.
2--induced lift.   4--parasite drag.

38.   What airplane control surface is responsible for movement about the longitudinal axis of rotation?

1--Aileron.   2--Elevator.   3--Flap.   4--Rudder.

39.   What direction does the right aileron move when the control wheel is turned right?

1--Down.   2--Left.   3--Right.   4--Up.

40.   Elevator trim tabs influence movement about the airplane's _____ axis.

1--elevated   2--lateral   3--longitudinal   4--vertical

41.   As the angle of attack increases, induced drag _____ and power must be _____ to fly slower.

1--decreases; decreased   3--increases; decreased
2--decreases; increased   4--increases; increased

42.   Refer to Figure 2.2 If your airplane is traveling at 100 kts. and stalls during level flight when you are in an 80° bank, at approximately what airspeed would it stall at 0° of bank?

1--31 kts.
2--42 kts.
3--53 kts.
4--Impossible to determine since not enough data are provided.

43.   Recovering from a stall necessitates

1--increasing drag.
2--increasing lift.
3--reducing thrust.
4--reducing the angle of attack.

44. Suppose you are crossing the flight path of a large jet airplane that is ahead of you at the same altitude. To avoid wake turbulence, you should

1--descend and adjust speed to maneuvering speed.
2--descend and fly parallel to the jet's flight path.
3--descend below the jet's flight path.
4--fly above the jet's flight path.

ANSWERS AND EXPLANATIONS

Key Terms and Concepts, Part 1

1. H	2. R	3. M	4. A
5. J	6. D	7. P	8. B
9. I	10. Q	11. C	12. K
13. E	14. F	15. O	16. N
17. G	18. L	19. S	20. T

Key Terms and Concepts, Part 2

1. K	2. C	3. G	4. P
5. R	6. H	7. A	8. M
9. I	10. E	11. S	12. Q
13. B	14. O	15. F	16. N
17. J	18. L	19. D	20. T

Discussion Questions and Exercises (objective items only)

3. F--Drag opposes thrust and gravity opposes lift.
8. F--The rudder moves to the left.
9. T--The nose also goes down.
10. F--If you subsitute <u>aileron</u> for <u>rudder</u>, the answer is true.
13. T--Trim tabs on the elevator influence the airplane's <u>pitch</u> (movement about the lateral axis).
16. F--It is in the <u>opposite</u> direction of the flight path.
17. F--This defines the <u>angle of attack</u>.
18. F--It is called the <u>separation</u> point.
19. F--Stalling speed <u>increases</u> as the angle of bank increases.
20. F--The ground effect frequently causes an airplane to become airborne <u>before</u> sufficient airspeed is obtained to maintain flight above the ground effect level.
23. F--A spin occurs when <u>yaw</u> is induced in a stalled airplane.
24. a. 68 kts.; 40 kts. x .70 = 28 kts. additional; 40 + 28 = 68. Or 40 kts. x 1.70 = 68 kts.
     b. About 47°. To calculate this one you need to compute the percentage increase in speed above normal (0° bank) stalling speed. Since you are traveling 10 kts. faster, you divide 10 by 50 to obtain the proportion increase, .20. Multiply by 100 to get the percentage, 20 percent, and refer to the chart to obtain 47° of bank.

     c. 41.67 kts. This one is a bit more difficult: 80° is
     equal to a 140 percent increase, so 100 kts. equals a 240
     percent increase. Thus, speed = 100/2.4, or 41.67.

25.   5,080 lb. A 60° bank increases the load factor by 2 (you
     should remember that). Referring to the previous chart,
     a 60° bank produces a 40 percent increase in stalling
     speed, so the airplane will stall at 49-50 kts.

Review Questions

1. 4--Thrust opposes drag and lift opposes weight.
2. 4--In unaccelerated flight, lift = weight and thrust = drag.
3. 4--The opposing forces are offsetting.
4. 3--The vertical and horizontal components of lift interact to produce a turn.
5. 2--The angle of attack is the angle between the chord line and the relative wind.
6. 1--See answer 5.
7. 2--The higher angle of attack of the downward moving blade of the propeller produces a tendency for the airplane to yaw to the left.
8. 1--The rudder controls movement about the vertical axis (yaw).
9. 2--The slipstream effect is caused by the downwash produced by the propeller as it passes over the horizontal control surfaces. When power is reduced, the downwash over the horizontal control surfaces diminishes and the nose pitches down.
10. 2--The load the wings carry changes as the airplane turns. Load factors increase as the angle of bank increases.
11. 3--The only thing necessary to produce a stall is to exceed the critical angle of attack. This can occur at any airspeed and any flight attitude.
12. 3--The only necessary condition is that the airplane is in a stalled condition.
13. 3--Read up from the horizontal axis (55) to the curved line. Read across to the vertical axis. This produces a G load of 1.7. Multiply 1.7 x 5,400 to get 9,180 lb.
14. 3--Read across from 4.4 G units on the vertical axis until you intersect the curved line. Next, read down to the horizontal axis to obtain roughly 77°.
15. 2--Read up from 60° on the horizontal axis to the curved line. Read across to the vertical axis to find the load factor in G units (2). Multiply the aircraft weight (2,300 lb.) by 2 to arrive at 4,600 lb.
16. 2--Use the same logic as in the previous problem. The multiplier in this problem is roughly 1.2. Then, 1.2 x 3,300 = 3,960 lb.
17. 3--Flaps increase lift and drag, thus enabling the airplane to descend at a steeper glide slope without increasing the airplane's airspeed.
18. 2--See answer 17.
19. 1--See answer 17.

20. 2--The speed of the airplane determines the amount of excess
     load that can be imposed on the wings.
21. 1--Increasing the load factor decreases total lift; thus,
     the airplane will stall at higher airspeeds.
22. 2--The yawing tendency produced by the P-factor is most
     pronounced at slow airspeeds and high angles of attack.
23. 1--The yawing tendency produced by torque is most pronounced
     at slow airspeeds and high angles of attack.
24. 3--To repeat an old phrase, an airplane will stall at any
     airspeed and in any flight attitude. The only necessary
     condition is that it exceed the critical angle of attack.
25. 3--We haven't really covered indicated airspeed yet; but
     since this question relates to stalls, we'll include it
     here. We'll repeat it in the section on airspeed
     indicators. Indicated airspeed is the same regardless of
     altitude, so the airplane will stall at the same
     indicated airspeed, regardless of altitude.
26. 2--Ground effect is most pronounced within one-half the
     airplane's wingspan above the ground. It provides a
     "cushion" so to speak that may lead to the airplane's
     becoming airborne before reaching recommended takeoff
     speed. It also may increase the landing distance because
     the airplane may have a tendency to "float" down the
     runway.
27. 2--See answer 26.
28. 1--See answer 26.
29. 2--See answer 26.
30. 1--We'll cover this more thoroughly in the section on
     "weights and balances." In the meantime, it is the
     center of gravity (CG) that has the greatest effect on
     longitudinal stability.
31. 1--They sink below the aircraft and **may** be affected by any
     crosswinds that are present. Crosswinds below 10 kts.
     should be viewed with extreme caution because they do not
     move vortices away from the flight path. Vortices tend
     to sink at a rate of about 400-550 ft. per minute. That
     is why the FAA requires a minimum of at least two minutes
     before a small aircraft will be allowed to take off
     following the departure of a heavy aircraft. Finally,
     remember that heavy aircraft produce wingtip vortices
     when they are generating lift, as during takeoff and
     landing.
32. 3--See answer 31.
33. 3--See answer 31.
34. 4--The rounded, front edge is called the leading edge.
35. 3--Read across from 4.4 on the vertical axis (load factor in
     G units) to the curved line. Read down to the horizontal
     axis (bank angle--in degrees). The value is about 77°.
36. 1--Drag opposes thrust and weight opposes lift.
37. 1--When the angle of attack increases, this leads to an
     increase in induced drag due to greater air resistance.
38. 1--The ailerons control bank, or the amount of roll.
39. 4--The right wing goes down when the airplane rolls right;
     the aileron goes up to help produce this movement, while
     the left aileron goes down.

40. 2--Elevator trim tabs are the fine-tuning control surfaces
       that influence the airplane's pitch (movement about the
       lateral axis).
41. 4--This is the principle of **slow flight**. Induced drag
       increases; and to maintain slow flight with an increased
       angle of attack, more thrust (power) must be added.
42. 2--An 80° bank is equal to a 140 percent increase; 100 kts.
       equals a 240 percent increase. Thus, speed = 100/2.4, or
       41.67.
43. 4--The first thing to do when you encounter a stall is to
       reduce the angle of attack while simultaneously adding
       power (thrust).
44. 4--See answer 31.

3/THE POWER PLANT AND ITS SYSTEMS

MAIN POINTS

1. **Reciprocating engines**, the power plant of most general aviation airplanes, produce rotary motion, or **torque**, to do work, which is measured in horsepower. Fuel and air mixed in the **carburetor** flow into the engine's cylinders through intake valves (the intake stroke). Once inside the cylinder the mixture is compressed by the piston (the compression stroke) and ignited by electrical sparks provided by two separate ignition systems. The resulting force (the power stroke) forces the piston downward, producing torque. As the piston continues to move, this time upward (the exhaust stroke), the exhaust valve opens and exhaust fumes are expelled through the **exhaust manifold**. The four strokes of the piston give the engine its name: the **four-stroke cycle**. Piston movement is translated into torque by a series of **connecting rods** that attach to the **crankshaft**. A crankcase encloses the crankshaft and connecting rods. The **propeller** is at the end of the crankshaft. Although some aviation engines are liquid-cooled, most are air-cooled.

2. **Propellers** convert engine torque into forward thrust. Each propeller blade is airfoil-shaped with decreasing angles of attack from the hub (where the propeller attaches to the crankshaft) to the tip. There are two types of propellers, **fixed-pitch propellers**, whose blade angles cannot be changed, and **constant-speed propellers**, whose blade angles can be changed by the pilot during flight. The position of constant-speed blades is controlled by a governor that is connected to the propeller control in the cockpit.

3. Engine power depends upon the rotational speed (RPM) of the propeller (indicated on the **tachometer** in the cockpit) and the pressure of the air/fuel mixture (**manifold pressure**). Power changes are made by moving the **throttle control**. In fixed pitch airplanes, the amount of power is directly related to RPM and is indicated on the tachometer. Lower air density at higher

31

altitudes leads to a decrease in manifold pressure; this decrease means that if engine power is held constant, RPM must increase with altitude (fixed prop only). Airplanes with constant-speed props have both a tachometer and a manifold pressure gauge. When increasing power, one first increases RPM and then increases manifold pressure. When decreasing power, one first decreases manifold pressure and then decreases RPM.

4.   Some airplanes have liquid engine-cooling systems, but most are air-cooled. Some high-horsepower engines have movable **cowl flaps** to aid cooling during low-speed, high-power flight. A **cylinder head temperature** gauge registers engine temperature in the cockpit.

5.   Engine oil removes internal engine heat and coats the moving parts of the engine. There are two types of engine oil-- non-detergent, which is typically used during break-in, and detergent (ashless-dispersant) oil. Oils are also categorized by their viscosity. The oil system gauges include the **oil pressure gauge**, which should show movement within 30 seconds after a normal engine start and should stay within the green arc during flight, and the **oil temperature gauge**, which should also operate within the green arc during flight.

6.   Aviation gas (avgas) is identified by its **octane number**, which is a measure of how fast it burns. Higher octane fuel burns more slowly and more controlled, thus leading to less rapid, uncontrolled combustion (detonation). Various grades of fuel are color-coded for positive identification. The color dyes are designed to cancel each other out if different grades are mixed. You can use a **higher** grade of fuel if the proper grade is not available, but **never use a lower grade**. While detonation is one form of abnormal combustion, **preignition** is another. Preignition causes fuel to ignite **before** the spark from the spark plug and creates high pressures in the cylinder. It is often caused by operating at excessively high engine temperatures.

7.   You should make sure that an airplane is **grounded** before starting to fuel it to prevent fire caused by static electrical discharge. You should also make sure that fuel vents are clear and that the fuel is of the proper grade and not contaminated by sediment or water. To help avoid water accumulation in the gas tanks from condensation, it is best to keep the tanks full, particularly when the airplane is left overnight. Furthermore, it is the pilot's responsibility to visually check for the presence of water by draining fuel from the **sumps** located at the lowest point on each tank and from the **fuel strainer**, which is at the lowest point of the engine's fuel system. On low-wing aircraft, the tanks are lower than the engine, thus the need for fuel pumps.

8.   The purpose of the fuel system is to provide the engine with an uninterrupted flow of fuel. All airplanes have **fuel gauges** to indicate fuel levels in each tank, but you should also check visually to determine before a flight how much fuel you

have on board. There are two types of fuel systems in light airplanes: **gravity systems**, which are common in high-wing airplanes, and **pump systems**. Fuel pumps may be driven by either the accessory pad or the electrical system. Primary pumps are typically engine driven, while backup or boost pumps are typically electrical. A **fuel selector valve** allows the pilot to select from which tank gas is to flow. (Note: never run a tank dry before switching as this may cause vapor lock.) The **engine primer** allows the pilot to inject fuel directly into the cylinders before starting, a common practice when the engine is cold.

9.   Air and fuel are mixed in the **carburetor**. Opening and closing the throttle adjusts the **throttle valve** (butterfly valve) in the carburetor. As the liquid fuel evaporates into the intake air, it cools rapidly and may condense and freeze any moisture in the air. The butterfly (throttle) valve area of the carburetor is most susceptible to this type of icing. Carburetor ice, as it is called, is most likely to occur in high humidity when the temperature is between 20° and 80° F. The first sign of carburetor ice is a loss of RPM (fixed pitch prop) or a decrease in manifold pressure (constant speed prop). You may also suspect carburetor ice if airspeed decreases during level flight. The **carburetor heat** system removes ice from the carburetor by sending heated air into the carburetor and melting the ice that has accumulated. As the ice melts, the engine runs rough as the water goes through the engine. (Note: use of carburetor heat makes the mixture richer and engine power is less than normal.) Since warm air is necessary to melt ice, the engine must be developing power to provide warm air. Thus, if your airplane handbook recommends using carb heat before landing, apply it **before** reducing the throttle. Finally, remember that outside air entering the exhaust shroud is not filtered, so avoid using carb heat on the ground, particularly in dusty conditions.

10.   An alternate way to mix fuel and air without a carburetor is through a **fuel injection system**, which delivers fuel directly to the cylinders. Such systems are more common with high horsepower engines. They provide a more uniform flow of fuel, provide quicker acceleration, and are not subject to icing. However, they are most costly and they are subject to **vapor lock**.

11.   The **mixture control** allows the pilot to adjust the ratio of air to fuel. As altitude increases, the mixture becomes more fuel-rich. The mixture control allows you to reduce the fuel/air ratio through a process called **leaning**. You should change the mixture whenever you change altitude or power settings.

12.   **Ignition** occurs between the compression and power strokes. The **ignition system** includes a source of high voltage, a timing device, a distribution system to the cylinders, spark plugs, control switches, and shielded wiring. The airplane's ignition system is the **magneto**, which typically runs off the

engine accessory pad and produces a high-voltage electrical
pulse. Airplanes have **two** magnetos, and each cylinder has two
spark plugs (the concept of redundancy). Further, the entire
ignition system is independent of other systems and can be
stopped only by grounding the magneto or stopping engine
rotation. The chances of losing power because of ignition system
failure are remote since each magneto has independent spark
plugs. Ignition and starter operations are controlled by a
single switch: Off, R (right magneto), L (left magneto), Both,
and Start. An important part of the preflight runup is the
magneto check to verify that both magnetos are operating
properly. You should expect a small drop in RPM when each
magneto is switched off (within allowable limits). If the engine
begins to die or you lose more than the allowable RPM, you know
there is a problem. Similarly, if there is no drop in RPM, it
may mean that the magneto is not grounded and that **any** movement
of the propeller could start the engine. Finally, the proper way
to stop the engine is to pull the mixture control to full lean so
as to burn any remaining fuel out of the cylinders.

13. The airplane's electrical system operates on direct
current (DC). **Voltage** is initially provided by a DC **battery** to
get the engine started, and then by the alternating current (AC)
generator, called an **alternator**, which is driven by the accessory
drive pad. Diodes immediately transform the alternator's AC to
DC. The **ammeter** measures electrical current flowing into or out
of the battery. Some airplanes have **load meters** that measure the
electrical load being placed on the alternator or generator. The
electrical system has a **master switch** that must be on to activate
any of the electrical components. Further, each piece of
equipment has an independent **circuit breaker** or **fuse**, which turns
the equipment off in the event of a malfunction. Circuit
breakers can be reset in flight, but they should never be held in
place manually. Fuses must be replaced with a fuse of the **same
or lower** amperage.

KEY TERMS AND CONCEPTS, PART 1

Match each term or concept (1-24) with the appropriate
description (A-X) below. Each item has only one match.

___	1. cowl flaps	___	13. amps
___	2. oil	___	14. carb heat
___	3. fuel selector valve	___	15. carburetor
___	4. power stroke	___	16. mixture control
___	5. fuel strainer	___	17. manifold pressure gauge
___	6. fuel injection	___	18. preignition
___	7. master switch	___	19. volts
___	8. exhaust valve	___	20. AC
___	9. alternator	___	21. crankshaft
___	10. throttle valve	___	22. ammeter
___	11. magnetos	___	23. float chamber
___	12. pitch	___	24. constant-speed

A. these open and close to control air flowing over the engine
B. part of the carburetor most susceptible to icing
C. located at the lowest point of the engine's fuel system
D. system that counteracts carburetor icing
E. device that controls power to all electrical components
F. measures the amount of stored electrical potential
G. device that allows the pilot to control the air/fuel mixture
H. valve used to select fuel from one tank to another
I. hot carbon particles cause spark plugs to fire early
J. the air/fuel mixture is ignited during this stroke
K. this measures the flow of electrons
L. electron current that flows in cycles
M. an engine-driven, alternating-current generator
N. device that displays air/fuel mixture pressure in the intake
    air passages
O. system in which fuel goes directly into the cylinders
P. terminology describing propeller blade angle
Q. residual gases leave the cylinder through this
R. device where air and fuel are mixed together
S. heart of most airplane ignition systems
T. product that lubricates the internal parts of an engine
U. propellers whose blade angles can change
V. fuel enters the carburetor and is deposited here first
W. device that measures the alternator's electrical output
X. the propeller is attached directly to this part of the engine

KEY TERMS AND CONCEPTS, PART 2

     Match each term or concept (1-24) with the appropriate
description (A-X) below. Each item has only one match.

___ 1. connecting rods   ___ 13. detonation
___ 2. ground   ___ 14. oil pressure gauge
___ 3. circuit breaker   ___ 15. fixed-pitch
___ 4. exhaust manifold   ___ 16. primer
___ 5. accelerating system   ___ 17. ashless-dispersant
___ 6. intake valve   ___ 18. fuel boost pump
___ 7. octane number   ___ 19. RPM
___ 8. horsepower   ___ 20. intake stroke
___ 9. tachometer   ___ 21. piston
___ 10. accessory drive pad   ___ 22. leaning
___ 11. DC   ___ 23. two
___ 12. vapor lock   ___ 24. cylinder head gauge

A. capability of suppressing detonation (antiknock value)
B. gears where the crankshaft ends
C. device through which residual gases exit from the engine
D. propellers that cannot change their blade angle
E. bubbles of gas that block the fuel line
F. provides fuel pressure for starting some airplane engines
G. what you should do to prevent fire when refueling an airplane
H. gas burns too rapidly in the combustion chamber
I. free-swinging device attached to the bottom of a piston
J. the speed of the crankshaft's rotation is measured in _____

K.   type of oil used in most general aviation airplanes
L.   air-fuel mixture enters the cylinder through this
M.   air-fuel mixture enters the cylinder during this stroke
N.   capacity for doing work
O.   electron current that comes in a constant flow
P.   device that controls power to a particular airplane
     instrument
Q.   displays pressure of oil being sent to lubricate the engine
R.   carburetor system that compensates for sudden throttle
     movements
S.   device that directs raw fuel directly into the cylinder
T.   device that displays crankshaft revolutions per minute
U.   movable plunger that fits inside the cylinder
V.   procedure used to reduce the fuel/air ratio
W.   number of magnetos found on most airplanes
X.   displays engine temperature in the cockpit

DISCUSSION QUESTIONS AND EXERCISES

1.   Briefly explain how piston movement translates into
propeller movement.

2.   Distinguish between fixed-pitch and constant-speed propellers.
How does each operate?

3.   Briefly describe the functions of the tachometer and the
manifold pressure gauge. Describe the procedure you should
follow when increasing and decreasing power with a constant-speed
prop.

4.   Briefly describe the importance and proper use of the oil
pressure and oil temperature gauges.

5.   Distinguish among vapor lock, detonation, and preignition.

6.   Suppose your airplane uses 100-octane aviation fuel. Would
it be acceptable to use 80-octane fuel if 100-octane were not
available? Why or why not?

7.   What are the two general types of fuel systems? Why is it
important to check the fuel strainer and wing tank sumps before
each flight?

8.   What are the two primary functions of the carburetor?
Briefly describe how a fuel injection system differs from a
carburetor system.

9.   What is carburetor icing and where in the carburetor does it
typically occur? Specifically, how does it affect RPM and
manifold pressure? Describe the procedure you should follow if
you encounter carburetor icing.

10. What does the mixture control mechanism do? What is
leaning?

11. Briefly describe the ignition switch and its proper use for starting and stopping the engine.

12. Describe the preflight procedure for checking the operation of the magneto system. What are two indications of a magneto that is malfunctioning?

13. Briefly describe the components and functions of the engine's electrical system. Include a discussion of the master switch, ammeter, voltage regulator, circuit breakers, and alternator.

14. T F If the specified grade of aviation fuel is not available for your airplane, it is best to use a slightly lower octane, but never a higher octane.

15. T F Detonation occurs in a reciprocating aircraft engine when the unburned air/fuel mixture in the cylinder explodes rather than burning evenly.

16. T F It is generally considered good operating practice to check for water in the fuel system only when the aircraft is fueled, since this is the only time the system can collect moisture.

17. T F Float-type carburetor systems provide more even fuel distribution and faster throttle response than do fuel injection systems, but they are less efficient.

18. T F The carburetor's float mechanism seals the chamber and prevents further fuel from entering.

19. T F The engine ignition system of most general aviation airplanes is called dual because it operates independently of other airplane systems.

20. T F Central to the airplane's ignition is the alternator, which typically runs off the accessory drive pad and which produces a high-voltage electrical pulse for the spark plugs.

21. T F With a fixed-pitch propeller airplane, engine power is registered on the tachometer.

22. T F The red line on a tachometer indicates a maximum RPM reading that may be exceeded only in straight-and-level flight.

23. T F Carburetor icing almost always occurs simultaneously at the needle valve and in the float chamber.

24. T F On a warm, sunny day, if the oil pressure does not reach the green arc within 30 seconds after starting the engine, the engine should be shut down.

25. T F One result of permitting an airplane engine to idle for a long period of time while on the ground is that the spark plugs may become fouled.

26. T F If an airplane engine continues to run after the ignition switch is turned to the OFF position, the probable cause is a broken magneto ground wire.

27. T F Applying carburetor heat results in more air going through the throttle valve.

28. T F In case of electrical fire, one first turns off the master switch.

29. T F The main reason to avoid long engine runups on the ground is to help avoid excessive vibration caused by air moving over the control surfaces.

30. T F During long descents, one should gradually increase the air/fuel mixture.

REVIEW QUESTIONS

1.   (FAA 1055) In comparison to fuel injection systems, float-type carburetor systems are generally considered to be

1--equally susceptible to icing as a fuel injection unit.
2--susceptible to icing only when visible moisture is present.
3--more susceptible to icing than a fuel injection unit.
4--less susceptible to icing than a fuel injection unit.

2.   (FAA 1056) The operating principle of float-type carburetors is based on

1--automatic metering of air at the venturi as the aircraft gains
    altitude.
2--difference in air pressure at the venturi throat and the air
    inlet.
3--increase in air velocity in the throat of a venturi causing an
    increase in air pressure.
4--measurement of the fuel flow into the induction system.

3.   (FAA 1057) The presence of carburetor ice, in an airplane equipped with a fixed-pitch propeller, can be verified by applying carburetor heat and noting

1--an increase in RPM and then a gradual decrease in RPM.
2--a decrease in RPM and then a constant RPM indication.
3--an immediate increase in RPM with no further change in RPM.
4--a decrease in RPM and then a gradual increase in RPM.

4.   (FAA 1058) Applying carburetor heat will

1--result in more air going through the carburetor.
2--not affect the mixture.
3--enrich the fuel/air mixture.
4--lean the fuel/air mixture.

5.   (FAA 1059) What change occurs in the fuel/air mixture when carburetor heat is applied?

1--A decrease in RPM results from the lean mixture.
2--No change occurs in the fuel/air mixture.
3--The fuel/air mixture becomes leaner.
4--The fuel/air mixture becomes richer.

6.   (FAA 1060) If an airplane is equipped with a fixed-pitch propeller and a float-type carburetor, the first indication of carburetor ice would most likely be

1--a drop in oil temperature and cylinder head temperature.
2--engine roughness.
3--a drop in manifold pressure.
4--loss of RPM.

7.   (FAA 1061) The use of carburetor heat tends to

1--decrease engine output and increase operating temperature.
2--decrease engine output and decrease operating temperature.
3--increase engine output and increase operating temperature.
4--increase engine output and decrease operating temperature.

8.   (FAA 1062) If an engine continues to run after the ignition switch is turned to the OFF position, the probable cause may be

1--the mixture is too lean and this causes the engine to diesel.
2--the voltage regulator points are sticking closed.
3--a broken magneto ground wire.
4--fouled spark plugs.

9.   (FAA 1063) One purpose of the dual ignition system on an aircraft engine is to provide for

1--improved engine performance.
2--uniform heat distribution.
3--balanced cylinder-head pressure.
4--easier starting.

10. (FAA 1064) An abnormally high engine oil temperature indication may be caused by

1--the oil level being too low.
2--a defective bearing.
3--operating with an excessively rich mixture.
4--the oil level being too high.

11. (FAA 1065) Excessively high engine temperatures, either in the air or on the ground, will

1--cause damage to heat-conducting hoses and warping of the
   cylinder cooling fins.
2--cause loss of power, excessive oil consumption, and possible
   permanent internal engine damage.
3--not appreciably affect an aircraft engine in either
   environment.
4--increase fuel consumption and may increase power due to the
   increased heat.

12. (FAA 1066) For internal cooling, reciprocating aircraft engines are especially dependent on

1--a properly functioning thermostat.
2--air flowing over the exhaust manifold.
3--the circulation of lubricating oil.
4--a lean fuel/air mixture.

13. (FAA 1067) If the engine oil temperature and cylinder head temperature gauges have exceeded their normal operating range, the pilot may have been

1--operating with the mixture set too rich.
2--operating with higher-than-normal oil pressure.
3--using fuel that has a higher-than-specified fuel rating.
4--operating with too much power and with the mixture set too lean.

14. (FAA 1070) Filling the fuel tanks after the last flight of the day is considered a good operating procedure because this will

1--force any existing water to the top of the tank away from the fuel lines to the engine.
2--prevent expansion of the fuel by eliminating airspace in the tanks.
3--prevent moisture condensation by eliminating airspace in the tanks.
4--eliminate vaporization of the fuel.

15. (FAA 1071) Detonation occurs in a reciprocating aircraft engine when

1--the spark plugs are fouled or shorted out or the wiring is defective.
2--hot spots in the combustion chamber ignite the fuel/air mixture in advance of normal ignition.
3--there is too rich a fuel/air mixture.
4--the unburned charge in the cylinder explodes instead of burning normally.

16. (FAA 1072) If a pilot suspects that the engine (with a fixed-pitch propeller) is detonating during climb-out after takeoff, normally the corrective action to take would be to

1--increase the rate of climb.
2--retard the throttle.
3--lean the mixture.
4--apply carburetor heat.

17. (FAA 1073) If the grade of fuel used in an aircraft engine is lower than specified for the engine, it will most likely cause

1--a mixture of fuel and air that is not uniform in all cylinders.
2--lower cylinder head temperatures.
3--an increase in power which could overstress internal engine components.
4--detonation.

18. (FAA 1074) The uncontrolled firing of the fuel/air charge in advance of normal spark ignition is known as

1--combustion
2--preignition
3--atomizing
4--detonation

19. (FAA 1075) The practice of running a fuel tank dry before switching tanks is considered unwise because

1--the engine-driven fuel pump or electric fuel boost pump may draw air into the fuel system and cause vapor lock.
2--the engine-driven fuel pump is lubricated by fuel and operating on a dry tank may cause pump failure.
3--any foreign matter in the tank will be pumped into the fuel system.
4--the fuel pump is located above the bottom portion of the fuel tank.

20. (FAA 1076) Which of the following would most likely cause the cylinder head temperature and engine oil temperature gauges to exceed their normal operating range?

1--Using fuel that has a lower-than-specified fuel rating.
2--Using fuel that has a higher-than-specified fuel rating.
3--Operating with higher-than-normal oil pressure.
4--Operating with the mixture control set too rich.

21. (FAA 1077) What type fuel can be substituted for an aircraft if the recommended octane is not available?

1--The next higher octane aviation gas.
2--The next lower octane aviation gas.
3--Unleaded automotive gas of the same octane rating.
4--Unleaded automotive gas of the next higher rating.

22. (FAA 1081) The basic purpose of adjusting the fuel/air mixture control at altitude is to

1--decrease the amount of fuel in the mixture in order to compensate for increased air density.
2--decrease the fuel flow in order to compensate for decreased air density.
3--increase the amount of fuel in the mixture to compensate for the decrease in pressure and density of the air.
4--increase the fuel/air ratio for flying at altitude.

23. (FAA 1082) While cruising at 9,500 ft. MSL the fuel/air mixture is properly adjusted. If a descent to 4,500 ft. MSL is made without readjusting the mixture control

1--the fuel/air mixture may become excessively lean.
2--there will be more fuel in the cylinders than is needed for normal combustion, and the excess fuel will absorb heat and cool the engine.
3--the excessively rich mixture will create higher cylinder head temperatures and may cause detonation.
4--the fuel/air mixture may become excessively rich.

24. (FAA 1083) During the runup at a high-elevation airport a pilot notes a slight engine roughness that is not affected by the magneto check, but grows worse during the carburetor heat check. Under these circumstances, which of the following would be the most logical initial action?

1--Check the results obtained with a leaner setting of the mixture control.
2--Taxi back to the flight line for a maintenance check.
3--Reduce the manifold pressure to control detonation.
4--Check to see that the mixture control is in the full-rich position.

25. (FAA 1348) The first action after starting an aircraft engine should be to

1--adjust for proper RPM and check for desired indications on the engine gauges.
2--check the magneto or ignition switch momentarily in the OFF position for proper grounding.
3--test each brake and the parking brake.
4--visually clear the area for people and obstacles.

26. (FAA 1361) Is it necessary to preflight an aircraft that was hangared the night before in ready-to-fly condition?

1--No, if the aircraft has not been handled since hangaring.
2--Yes, because fuel contamination from condensation is possible.
3--Yes, because the oil level should always be checked.
4--No, if the same person who hangared the aircraft will act as pilot-in-command.

27. (FAA 1362) Prior to every flight, a pilot should at least

1--check the operation of the ELT.
2--drain fuel from each quick drain.
3--perform a walk-around inspection of the aircraft.
4--check the required documents aboard the aircraft.

28. (FAA 1363) What special check should be made on an aircraft during preflight after it has been stored for an extended period of time?

1--ELT batteries and operation.
2--Condensation in the fuel tanks.
3--Damage or obstruction caused by animals, birds, or insects.
4--Lubrication of control systems and proper inflation of landing gear struts.

29. (FAA 1364) The use of a written checklist for preflight inspection and starting the engines is recommended

1--as an excellent crutch for those pilots with a faulty memory.
2--for memorizing the procedures in an orderly sequence.
3--as a procedure to instill confidence in the passengers.
4--to ensure that all necessary items are checked in a logical sequence.

30. (FAA 1369) How is engine operation controlled on an engine equipped with a constant-speed propeller?

1--The throttle controls power output as registered on the manifold pressure gauge and the propeller control regulates the engine RPM.
2--The throttle controls power output as registered on the manifold pressure gauge and the propeller control regulates a constant blade angle.
3--The throttle controls engine RPM as registered on the tachometer and the mixture control regulates the power output.
4--The throttle controls engine RPM as registered on the tachometer and the propeller control regulates the power output.

31. (FAA 1370) A precaution for the operation of an engine equipped with a constant-speed propeller is to

1--avoid high RPM settings with high manifold pressure.
2--avoid high RPM settings with low manifold pressure.
3--always use a rich mixture with high RPM settings.
4--avoid high manifold pressure settings with low RPM.

32. (FAA 1371) What is an advantage of a constant-speed propeller?

1--Permits the pilot to select and maintain a desired cruising speed.
2--Allows a higher cruising speed than possible with a fixed-pitch propeller.
3--Provides a smoother operation with stable RPM and eliminates vibration.
4--Permits the pilot to select the blade angle for the most efficient performance.

33. (FAA 1423) When starting an airplane engine by hand, it is extremely important that a competent pilot

1--call "contact" before touching the propeller.
2--be at the controls in the cockpit.
3--in the cockpit be in charge and call out all commands.
4--turn the propeller and call out all commands.

34. (FAA 1427) What action can a pilot take to aid in cooling an engine that is overheating during a climb?

1--Lean the mixture to best power condition.
2--Increase RPM and reduce climb speed.
3--Reduce rate of climb and increase airspeed.
4--Increase RPM and climb speed.

35. (FAA 1428) What is one procedure to aid in cooling an engine that is overheating?

1--Enrichen the fuel mixture.
2--Increase the RPM.
3--Reduce the airspeed.
4--Use alternate air.

36. (FAA 1429) The most important rule to remember in the event of a power failure after becoming airborne is to

1--quickly check the fuel supply for possible fuel exhaustion.
2--determine the wind direction to plan for the forced landing.
3--turn back immediately to the takeoff runway.
4--maintain safe airspeed.

37. (FAA 1790) Which condition is favorable to the development of carburetor icing?

1--Any temperature below freezing and a relative humidity of less
     than 50 percent.
2--Temperature between 32° F and 50° F and low humidity.
3--Temperature between 0° F and 20° F and high humidity.
4--Temperature between 20° F and 70° and high humidity.

38. (FAA 1791) The possibility of carburetor icing should always be considered when operating in conditions where the

1--temperature is as high as 95° F and the relative humidity is
     30 percent or greater.
2--relative humidity range is from 25 percent to 100 percent,
     regardless of temperature.
3--relative humidity is between 30 percent and 100 percent and
     the temperature is between 0° F and 32° F.
4--temperature is as high as 70° F and the relative humidity is
     greater than 50 percent.

39. Piston movement in a reciprocating engine is translated into torque by a series of _____ rods that attach to the crankshaft.

1--connecting   3--propeller
2--intake   4--reciprocating

40. Preignition occurs in a reciprocating airplane engine when

1--hot spots in the combustion chamber ignite the fuel/air
   mixture prematurely.
2--the fuel/air mixture is too rich.
3--the spark plugs are disconnected.
4--the unburned fuel/air mixture in the cylinder explodes rather
   than burning evenly.

41. What precaution should you take to avoid fire danger when refueling an airplane?

1--Attach a grounding wire to the airplane.
2--Open the fuel vents.
3--Put the mixture control at full rich.
4--Turn off the carburetor heat.

42. What part of the carburetor opens when you increase the power setting?

1--Float mechanism.   3--Mixture control valve.
2--Fuel strainer.   4--Throttle valve.

43. Central to the airplane ignition system is the _____, which typically runs off the engine accessory drive pad and which produces a high-voltage electrical pulse.

1--alternator   3--generator
2--battery   4--magneto

44. Which of the following is not part of the airplane engine's lubrication system?

1--Connecting rods.   3--Sump.
2--Pressure sensor.   4--Temperature sensing device.

45. What part of the airplane's electrical system controls the variable output of the generator or alternator?

1--Ammeter.   3--Master switch.
2--Circuit breaker.   4--Voltage regulator.

46. With a fixed-pitch propeller, what instrument registers engine power?

1--Ammeter.   3--Suction gauge.
2--Manifold pressure gauge. 4--Tachometer.

47. With a constant-speed propeller airplane,

1--propeller speed should be increased before power is added.
2--propeller speed should be increased before power is reduced.
3--propeller speed should be reduced before power is added.
4--propeller speed should be reduced before power is reduced.

48. What part of the carburetor is most susceptible to ice?

1--Accelerating pump.   3--Mixture control.
2--Float chamber.   4--Throttle valve.

49. The carburetor heat system directs

1--cold air away from the carburetor.
2--moisture away from the carburetor.
3--warm air to the carburetor.
4--prewarmed fuel to the carburetor.

50. To properly shut down an airplane engine, first you

1--pull the mixture control to full lean.
2--switch the ignition switch to the OFF position.
3--turn off the master switch.
4--apply carburetor heat.

51. In what position should the ignition switch be during
takeoff?

1--Both.   2--L.   3--R.   4--Start.

52. Suppose that the electrical system (battery and alternator)
fails during flight. In this situation,

1--avionics equipment would also fail.
2--cylinder head temperature would increase and oil pressure
   would decrease.
3--the engine-driven fuel boost pump would fail, leading to
   engine failure, as well as loss of all avionics equipment,
   lights, and AC instruments.
4--the engine ignition system, fuel gauges, lighting system, and
   all avionics would fail.

53. Which of the following statements is correct?

1--Carburetor icing most likely would form when the air
   temperature is between 20° F and 70° F, with visible moisture
   or high humidity.
2--The carburetor heater is a deicing device that heats the air
   after it enters the carburetor.
3--The first indication of carburetor icing in a fixed-pitch
   propeller airplane is an increase in RPM, followed by a rapid
   decrease in RPM.
4--Carburetor icing occurs whenever the temperature falls below
   freezing (32° F).

54. Which of the following statements regarding fouling of the
spark plugs is correct?

1--Carbon fouling of the plugs is caused primarily by operating
   the engine at excessively high cylinder head temperatures.
2--Excessive heat in the combustion chamber of a cylinder causes
   oil to form on the center electrode of the plug, causing it to
   preignite.
3--Permitting the engine to idle for a long period of time on the
   ground is the best way to clean fouled spark plugs.
4--Spark plug fouling results from operating with an excessively
   rich mixture.

ANSWERS

Key Terms and Concepts, Part 1

1.	A	7.	E	13.	K	19.	F
2.	T	8.	Q	14.	D	20.	L
3.	H	9.	M	15.	R	21.	X
4.	J	10.	B	16.	G	22.	W
5.	C	11.	S	17.	N	23.	V
6.	O	12.	P	18.	I	24.	U

Key Terms and Concepts, Part 2

1.	I	7.	A	13.	H	19.	J
2.	G	8.	N	14.	Q	20.	M
3.	P	9.	T	15.	D	21.	U
4.	C	10.	B	16.	S	22.	V
5.	R	11.	O	17.	K	23.	W
6.	L	12.	E	18.	F	24.	X

Discussion Questions and Exercises

14. F--You can use a higher octane, but not a lower one.
15. T--Self-explanatory.
16. F--It should be checked before **every** flight.
17. F--Fuel injection systems distribute fuel more evenly and
         are faster in response to throttle changes.
18. T--Self-explanatory.
19. F--It is considered dual because it has two separate and
         independent magneto systems.
20. F--This defines the magneto, not the alternator.
21. T--Self-explanatory.
22. F--This is a value that should **never** be exceeded.
23. F--It almost always occurs near the throttle valve.
24. T--Self-explanatory.
25. T--Spark plug fouling may occur due to the rich mixture used
         for most ground operations.

26. T--One reason for the runup is to make certain the magnetos are properly grounded; if they do not decrease in RPM when checked, they may be improperly grounded (that is, any movement of the propeller may cause the engine to start).
27. F--It causes the mixture to become richer. Since warm air is less dense than cold air, the mixture runs richer.
28. T--This causes all electrical systems to stop (except the ignition system), after which you can test them individually using the various circuit breakers.
29. F--One reason is to keep the engine from overheating; it also helps to prevent spark plug fouling.
30. T--As altitude decreases, air density increases, so one has to increase the amount of fuel in the mixture to compensate.

Review Questions

1. 3--Fuel injection systems are less susceptible to icing since fuel is injected directly into the cylinders. With float-type carburetors, the fuel is mixed with air in the carburetor.
2. 2--Most carburetors operate on the principle of differential pressure between the inlet and the venturi throat.
3. 4--Initially, RPMs decrease as the ice melts and the resulting water goes through the carburetion system. After all of the water is gone, RPMs increase.
4. 3--Carburetor heat leads to an increase in the temperature of the air/fuel mixture. Heated air is less dense, thus the air/fuel mixture is richer.
5. 4--See answer 4.
6. 4--See answer 3.
7. 1--Carburetor heat reduces the engine's power output; the engine also operates at a higher temperature.
8. 3--An improperly grounded magneto allows one spark plug in each cylinder to provide ignition. This is important to remember since anything that causes the propeller to turn will start the engine.
9. 1--The dual ignition system provides redundancy and leads to more efficient airplane engine operation.
10. 1--Oil serves not only to lubricate the moving components of the engine, but also to remove heat from the cylinders. Thus, a low oil level reduces the ability of the oil to do its job.
11. 2--Excessive temperatures should be avoided as they may do permanent damage to the engine as well as increase oil consumption and lead to a loss of power.
12. 3--See answer 10.
13. 4--Too much power and having the mixture set too lean will lead to increases in engine temperature, as will low levels of oil and using aviation fuel with a lower-than-recommended octane value.

14. 3--Condensation inside the fuel tanks is the most common way
        water enters the fuel system. This can be avoided by
        filling the tanks after the last flight of the day.
15. 4--Detonation, which literally means to explode, occurs as a
        result of using low-grade fuel or having the mixture set
        too lean.
16. 2--See answer 15. In addition, keeping the engine cool
        helps reduce the rate of detonation, so reducing power by
        retarding the throttle would help.
17. 4--See answers 15 and 16.
18. 2--This is the definition of preignition.
19. 1--Pumping air into the fuel lines may lead to vapor lock.
20. 1--Several factors may lead to increased engine temperature
        such as using a low grade of aviation fuel, having a low
        level of oil, and running with the mixture too lean.
21. 1--You can use the next higher grade, but even this may not
        be desirable, particularly for long periods of time. If
        you must use an alternate, however, remember that you can
        use a higher grade of aviation fuel, but not a lower
        grade.
22. 2--As altitude increases, the air becomes less dense. Thus,
        the mixture becomes richer. To compensate, one must
        decrease the fuel flow through a process called leaning.
23. 1--This procedure is the opposite of that described in
        answer 22. As altitude decreases, the air becomes more
        dense. Thus, the mixture becomes leaner; and, to
        compensate, one must increase the fuel flow or make the
        mixture richer.
24. 1--See answers 4 and 22. This is a combination of two
        concepts, altitude and carb heat. Carb heat leads to a
        richer mixture, as does increased altitude. Thus, you
        may have to lean the mixture to get the engine to run
        smoothly.
25. 1--Once the engine starts, set the proper idle setting and
        check for oil pressure, oil temperature, and other engine
        gauges.
26. 2--Fuel contamination from condensation is always possible,
        especially after the airplane sits overnight. So, it
        should be preflighted.
27. 3--Always preflight the airplane.
28. 3--The key word to this question is **special**. The question
        is a little tricky since you should obviously check the
        airplane carefully, but you should be particularly
        attuned to anything out of the ordinary such as bird or
        squirrel nests.
29. 4--Using a written checklist ensures that everything is
        covered in a logical sequence.
30. 1--With a constant-speed propeller, the manifold pressure
        gauge indicates engine power and the propeller control
        governs RPM. Avoid high RPM settings with low manifold
        pressure. The propeller control allows the pilot to
        change blade angle to obtain the most efficient
        performance.
31. 2--See answer 30.
32. 4--See answer 30.

33. 2--A competent pilot should be seated at the controls in the cockpit when starting an airplane by hand.
34. 3--Reducing the rate of climb will tax the engine less. Also, increasing airspeed will cause more air to flow over the engine to cool it.
35. 1--Anything that increases airspeed will help. Enriching the mixture will lead to cooler combustion.
36. 4--This may be a little early for this question, but it is worth testing you here. You **must** maintain your airspeed in the event of a power failure or in any other emergency situation. Without positive attitude control, nothing else will help.
37. 4--Carb ice is most likely to form under conditions of high relative humidity (50 percent or greater) when the temperature is between 20° and 70° F.
38. 4--See answer 37.
39. 1--Definition.
40. 1--Definition.
41. 1--Always ground the airplane before refueling it.
42. 4--Definition.
43. 4--Definition.
44. 1--Connecting rods are part of the crankshaft system.
45. 4--Definition.
46. 4--Definition.
47. 1--Increase RPM before adding power. Decrease power before decreasing RPM.
48. 4--The throttle valve (butterfly valve) is the most susceptible place for carburetor ice to form.
49. 3--This is the main function of the carb heat system. Remember that the air is unfiltered, so avoid using carb heat on the ground, particularly in dusty conditions.
50. 1--The first thing to do is to starve the cylinders of fuel by pulling the mixture control to full lean.
51. 1--Definition.
52. 1--Remember that the electrical system is independent of the ignition system. The ignition system would continue to function properly, but things such as the avionics, which are part of the electrical system, would fail.
53. 1--Carburetor ice is most likely to form in conditions of high humidity and when the temperature is between 20° and 70° Fahrenheit.
54. 4--Excessively rich mixtures lead to spark plug fouling. Permitting the engine to idle for long periods of time also contributes.

4/FLIGHT INSTRUMENTS

MAIN POINTS

1.   Basic flight instruments are classified by what they do
and by how they work. They provide information about the
airplane's attitude, flight path, and performance.

2.   The **pitot-static system** measures the pressure of air
that hits the airplane head-on (ram air) and the pressure of the
still air (static air). The **pitot tube** captures ram air and
transmits that pressure to the **airspeed indicator**. The **static
ports**, mounted at a 90° angle from the direction of flight,
receive air at outside atmospheric pressure. Static pressure is
used as a reference for determining the airplane's speed through
the air. Static air is also routed to the **altimeter** and the
vertical-speed indicator.

3.   The difference between pitot pressure and static
pressure is called **dynamic pressure** and is represented on the
airspeed indicator as **indicated airspeed**. This represents speed
through the air, not over the ground, and is the same for takeoffs and landings, regardless of the airplane's altitude.
Calibrated airspeed is indicated airspeed corrected for erroneous
pressures read by the static port. **True airspeed** is airspeed
corrected for altitude and temperature. True airspeed can be
approximated by increasing indicated airspeed by about 2 percent
per 1,000 ft. of altitude above mean sea level (MSL).

4.   There are several airspeeds, or V (velocity) speeds, on
the airspeed indicator with which you should be familiar. The
white arc is the flap operating range. The low end of the scale
represents **Vso**, the stall speed in the landing configuration.
The high end of the white arc represents **Vfe**, the maximum flap
extension speed. The **green arc** indicates the normal operating

range. **Vs1** is the stall speed with flaps and gear up and is at the low end of the green arc. At the upper end is **Vno**, the maximum structural cruising speed. The **yellow arc** represents caution. You may operate in this range as long as there is no turbulence. The yellow arc ends at the **red line** speed. **Vne** indicates that this is the never-exceed speed.

5.   The altimeter measures the height of the airplane above some constant reference point (for example, sea level). It does <u>not</u> measure the height above the ground; but, by knowing the height of the terrain and the obstacles upon it, you can calculate the distance between you and the terrain. The altimeter is a barometer that measures changes in atmospheric pressure. As altitude increases, pressure decreases. Pressure also changes as the weather changes, and so it is important to keep adjusting the altimeter to reflect these changing conditions.

6.   There are several types of altitude. **Pressure altitude** is the reading on the altimeter when the Kollsman window is set to standard sea level pressure of 29.92. **True altitude** is the height above **mean sea level (MSL)**. **Indicated altitude** is what the altimeter reads at any point in time. **Absolute altitude** is the airplane's altitude **above ground level**. The altimeter translates changes in pressure to changes in altitude: The conversion formula is approximately 1 in. of mercury for each 1,000 ft. in elevation. Altimeters are not always perfect, however, due to inherent limitation such as scale error, friction error, and hysteresis. Finally, remember that as atmospheric conditions change, so does the barometric pressure, which in turn affects the altitude reading. The saying "From a high to a low, look out below" refers to the condition in which you fly from a high-pressure area to a low-pressure area. Unless you reset the value in the Kollsman window, the altimeter will indicate that you are higher than you in fact are, a condition that can have disastrous consequences when you get closer to an obstruction or the ground. The same saying also applies to changes that occur in the temperature along your route of flight.

7.   Closely allied to the altimeter is the **vertical-speed indicator (VSI)**, which measures the rate of descent or climb, typically in feet per minute. Due to their construction, most VSIs lag behind the airplane's performance and should not be relied upon during the entry into a climb or descent.

8.   The **magnetic compass** indicates direction relative to magnetic north. Its primary practical value is a reference force for the heading indicator. There are several limitations to its use. First, it points to magnetic north rather than true north. This difference is called **variation** (to be discussed in a later chapter in detail). Second, it can be affected by other instruments and radios, referred to as **deviation** errors. Third,

the compass will dip when the airplane accelerates or decelerates on easterly or westerly headings. A handy way to remember these **acceleration-deceleration** errors is ANDS--Accelerate-North; Decelerate-South. Fourth, there are **northerly-southerly turning errors**. When you turn from a **northerly** heading,the compass will initially indicate a turn in the **opposite** direction and continue to lag the turn. When you turn from a **southerly** heading, it will register a turn in the **same** direction but at an accelerated rate. Finally, in turbulent air, the magnetic compass may be subject to **oscillation** error.

    9. **Gyroscopic instruments** include the attitude indicator, turn coordinator, and heading indicator. A gyroscope operates according to two principles: rigidity in space and precession. The **attitude indicator** or **artificial horizon** is a graphic display of the airplane, sky, and ground. It represents the airplane with reference to the horizon and indicates changes exactly when they occur, as when the airplane banks or pitches up or down. The **turn coordinator** indicates the rate of turn and provides information about yaw (inclinometer). The ball in the inclinometer indicates whether the airplane is slipping or skidding in a turn. A **slip** results when the rate of turn is too slow for the angle of bank. A **skid** results when the rate of turn is too fast for the angle of bank. Corrections can be effected by "stepping on the ball" with the appropriate rudder pedal. The **heading indicator, or directional gyro,** displays the airplane's heading. It must be set by reference to an established heading such as the magnetic compass either on the ground or in straight-and-level flight.

    10. Airplane gyro instruments are typically powered by either electrical power or vacuum (air pressure) systems. In most airplanes, the turn coordinator and attitude indicator are powered independently so that, if one system fails, the pilot still has a backup source of information, referred to as the concept of **redundancy**.

    11. Most instrument panels have a **T arrangement** with the attitude indicator in the middle, flanked on either side by the airspeed indicator and altimeter. The heading indicator completes the leg of the **T**. It is typically flanked by the turn coordinator and vertical-speed indicator. The process of checking and rechecking instruments against one another is called **cross-check**. It is important to safe flight, but it is not a substitute for constantly referring to what is going on outside the airplane. Checking back and forth between the outside world and the flight instruments is called **composite flying**.

KEY TERMS AND CONCEPTS, PART 1

Match each term or concept (1-24) with the appropriate description (A-X) below. Each item has only one match.

___ 1. deviation
___ 2. Vfe
___ 3. magnetic compass
___ 4. indicated altitude
___ 5. altimeter
___ 6. precession
___ 7. north
___ 8. cross-check
___ 9. variation
___ 10. ram air
___ 11. composite flying
___ 12. absolute altitude

___ 13. 1,000
___ 14. knots
___ 15. gyroscope
___ 16. vertical-speed indicator
___ 17. pitot-static system
___ 18. true airspeed
___ 19. Vso
___ 20. Vne
___ 21. static ports
___ 22. Kollsman window
___ 23. south
___ 24. oscillation errors

A. maximum flap extension speed
B. altitude of an airplane above ground level (AGL)
C. movement of a gyro's spin axis due to some external force
D. turning the airplane from this heading causes the magnetic
    compass to initially indicate a turn in the opposite
    direction
E. what the altimeter indicates at any point in time
F. air that impacts the airplane head-on
G. modern aviation uses this to measure speed and distances
H. instrument that measures altitude
I. instrument that measures the rate of climb or descent
J. magnetic compass errors due to metallic or electrical cockpit
    components
K. process of checking and rechecking flight instruments against
    one another
L. system that measures and displays information about
    differential air pressure
M. 1 in. of mercury equals a change in altitude of _____ ft.
N. angle between magnetic north and true north
O. a mass spinning about an axis
P. cross-referencing the world outside the cockpit with the
    flight instruments
Q. instrument that measures the airplane's heading with respect
    to magnetic north
R. receptacles where air at outside atmospheric pressure enters
S. stall speed in the landing configuration
T. errors in magnetic compass readings during turbulent flight
U. speed through the air, corrected for altitude and temperature
V. never-exceed airspeed
W. direction the magnetic compass turns when decelerating
X. face of the altimeter

KEY TERMS AND CONCEPTS, PART 2

Match each term or concept (1-20) with the appropriate description (A-T) below. Each item has only one match.

___ 1. pitot tube
___ 2. heading indicator
___ 3. barometer
___ 4. airspeed indicator
___ 5. turn coordinator
___ 6. indicated airspeed
___ 7. gimbals
___ 8. calibrated airspeed
___ 9. redundancy
___ 10. attitude indicator

___ 11. inclinometer
___ 12. pressure altitude
___ 13. true altitude
___ 14. acceleration-deceleration
___ 15. 75
___ 16. skid
___ 17. vacuum system
___ 18. northerly turning error
___ 19. 3° per second
___ 20. slip

A. difference between ram air pressure and static air pressure
B. altitude read from the altimeter when it is set at standard conditions
C. another name for the artificial horizon
D. gyroscopic instrument that displays the airplane's heading
E. the magnetic compass indicates a turn in the <u>opposite</u> direction
F. airspeed corrected for pitot-static source vent position and other mechanical losses
G. results when the rate of turn is too fast for the bank angle
H. instrument at the heart of most altimeters
I. number of feet of tolerance suggested by the FAA in altitude difference between the altimeter and field elevation
J. instrument that tells how fast you are moving through the air
K. object's height above mean sea level (MSL)
L. power system that converts engine motion to air pressure
M. interconnected frames used to suspend a rotating mass
N. device that captures ram air
O. concept: two independent systems perform the same function
P. instrument that provides the pilot with yaw information
Q. ball in the turn coordinator that indicates skids and slips
R. magnetic compass errors that occur during changes in airspeed
S. standard rate turn
T. results when the rate of turn is too slow for the bank angle

DISCUSSION QUESTIONS AND EXERCISES

1. What flight instruments require pitot air pressure to operate? Which ones require static pressure? Which ones require both?

2.   T   F   Airspeed is not the same as groundspeed.

3.   What is the difference between pitot pressure and static
pressure? How is this related to calibrated airspeed?

4.   What happens to air pressure as measured by the static ports
as altitude increases? Why?

5.   Name and define the four types of altitude necessary for the
safe operation of an airplane.

6.   Suppose that you depart Denver (elevation 5,280 ft. MSL) and
fly to Vinland Valley, Kansas (elevation 890 ft. MSL). Suppose
further that the altimeter reads 29.96 when you depart from
Denver and 29.96 when you land at Vinland. How many feet have
you gained or lost in actual altitude during your trip due to
changes in atmospheric conditions?

7.   Suppose that in question 6 Vinland Valley had a barometric
reading of 28.74 when you landed. Had you not changed your
altimeter to reflect this change, something that is _required_ by
the FAA, what would your altimeter have read when you landed at
Vinland Valley?

8.   Suppose in question 6 the temperature in Denver was 95° F
when you left and 67° F at Vinland when you landed. Would your
altimeter have registered higher or lower when you landed? Why?

9.   What does the vertical-speed indicator measure? What is its
major limitation?

10.   What are the two fundamental properties of a gyroscope?
Name three major gyroscopic instruments and describe their
functions.

11.   What are two major functions of the magnetic compass?
Define each of the following magnetic compass errors:
acceleration-deceleration error, northerly-southerly turn error,
variation, deviation, and oscillation.

12.   Why do most general aviation airplanes have both a vacuum
power system and an electrical system for their instruments?

13.   What is the concept of **cross-check**? How is it related to
the basic T formation of airplane instruments? Be sure to state
what instruments are included in the basic T and why they are
aligned the way they are. Drawing a picture may help you
visualize the answer.

REVIEW QUESTIONS

1.   (FAA 1003) If a pilot plans to land at an airport where the
elevation is 7,500 ft., the indicated approach speed should be

1--higher than that used for a sea level airport, and some power
     should be used until touchdown.
2--the same as that used at a sea level airport.
3--lower than that used at a sea level airport.
4--higher than that used at a sea level airport.

2.   (FAA 1004) How can pressure altitude be determined?

1--Set the field elevation in the altimeter setting window and
     read the indicated altitude.
2--Set the altimeter to the field elevation and read the value in
     the altimeter setting window.
3--Set the altimeter to zero and read the value in the altimeter
     setting window.
4--Set 29.92 in the altimeter setting window and read the
     indicated altitude.

3.   (FAA 1006) Absolute altitude is the

1--altitude read directly from the altimeter.
2--altitude above the surface.
3--altitude reference to the standard datum plane.
4--indicated altitude corrected for instrument error.

4.   (FAA 1007) What is true altitude?

1--Actual height above sea level corrected for all errors.
2--Altitude above the surface.
3--Altitude reference to the standard datum plane.
4--Altitude shown on a radar altimeter.

5.   (FAA 1087 DI) Refer to the attitude indicator in Figure 4.1.
The proper adjustment to make on the attitude indicator in Figure
4.1 during level flight is to align the

1--horizon bar to the level-flight indication.
2--horizon bar to the miniature airplane.
3--miniature airplane to the horizon bar.
4--banking indicator to the zero-bank indication.

6.   (FAA 1088 DI) Refer to the attitude indicator in Figure 4.1.
How should a pilot determine the direction of bank from the
attitude indicator illustrated in Figure 4.1?

1--The direction of deflection of the banking scale (A).
2--The direction of deflection of the horizon bar (B).
3--The direction of deflection of the miniature airplane (C).
4--The relationship of the miniature airplane (C) to the
     deflecting horizon bar (B).

Figure 4.1

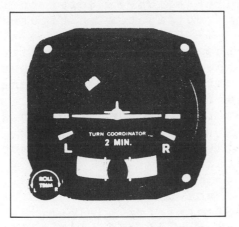

Figure 4.2

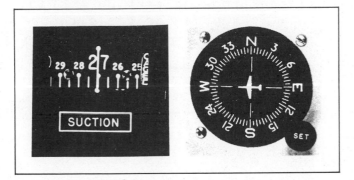

Figure 4.3

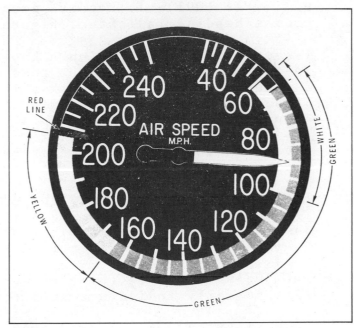

Figure 4.4

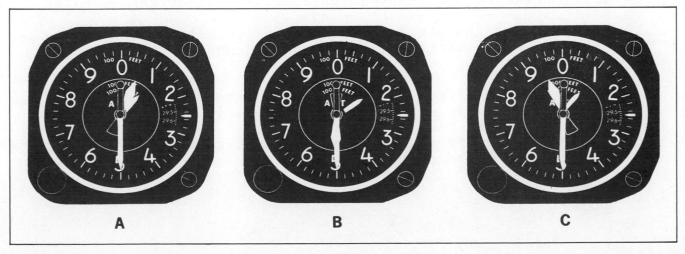

Figure 4.5

7.   (FAA 1089) Refer to the turn coordinator in Figure 4.2 The turn coordinator in Figure 4.2 provides an indication of

1--the movement of the airplane about the yaw and roll axes.
2--the angle of bank to but not exceed 30°.
3--attitude of the airplane with reference to the longitudinal axis.
4--motion of the airplane about the lateral and vertical axes.

8.   (FAA 1090 DI) For accurate operation, the attitude indicator depends upon

1--a vacuum
2--electricity.
3--proper coordination of controls.
4--rigidity in space.

9.   (FAA 1091) To receive accurate indications during flight from a heading indicator, Figure 4.3, the instrument must be

1--set prior to flight on a known heading.
2--calibrated on a compass rose at regular intervals.
3--adequately powered so that it seeks the proper direction.
4--periodically realigned with the magnetic compass as the gyro precesses.

10.   (FAA 1092 DI) The accuracy of the operation of either heading indicator illustrated in Figure 4.3 depends upon the

1--coordination of the flight controls.
2--smoothness of the air mass.
3--principle of rigidity in space.
4--precise amount of electrical suction power.

11.   (FAA 1095) In the Northern Hemisphere, a magnetic compass will normally indicate a turn toward the north if

1--a right turn is entered from an east heading.
2--a left turn is entered from a west heading.
3--an aircraft is decelerated while on an east or west heading.
4--an aircraft is accelerated while on an east or west heading.

12.   (FAA 1096) In the Northern Hemisphere, if an airplane is accelerated or decelerated, the magnetic compass will normally indicate

1--a turn momentarily, with changes in airspeed on any heading.
2--correctly when on a north or south heading while either accelerating or decelerating.
3--a turn toward the south while accelerating on a west heading.
4--a turn toward the north while decelerating on an east heading.

13. (FAA 1098) Deviation in a magnetic compass is caused by

1--presence of flaws in the permanent magnets of the compass.
2--the difference in the location between true north and magnetic
   north.
3--magnetic ore deposits in the Earth distorting the lines of
   magnetic force.
4--magnetic fields within the airplane distorting the lines of
   magnetic force.

14. (FAA 1100) In the Northern Hemisphere, a magnetic compass
will normally indicate initially a turn toward the west if

1--an aircraft is decelerated while on a south heading.
2--an aircraft is accelerated while on a north heading.
3--a left turn is entered from a north heading.
4--a right turn is entered from a north heading.

15. (FAA 1101) In the Northern Hemisphere, a magnetic compass
will normally indicate initially a turn toward the east if

1--an aircraft is decelerated while on a south heading.
2--an aircraft is accelerated while on a north heading.
3--a right turn is entered from a north heading.
4--a left turn is entered from a north heading.

16. (FAA 1102) During flight, when are the indications of a
magnetic compass accurate?

1--Only in straight-and-level unaccelerated flight.
2--As long as the airspeed is constant.
3--During turns if the bank does not exceed 18°.
4--In all conditions of flight.

17. (FAA 1104) Under what conditions is the indicated altitude
the same as true altitude?

1--If the altimeter has no mechanical error.
2--When at sea level under standard conditions.
3--When at 18,000 ft. with the altimeter set at 29.92.
4--At any altitude if the indicated altitude is corrected for
   non-standard sea level temperature and pressure.

18. (FAA 1105) How do variations in temperature affect the
altimeter?

1--Pressure levels are raised on warm days and the indicated
   altitude is lower than true altitude.
2--Higher temperatures expand the pressure levels and the
   indicated altitude is higher than true altitude.
3--Lower temperatures lower the pressure levels and the indicated
   altitude is lower than true altitude.
4--Indicated altitude varies directly with the temperature.

19. (FAA 1106) If it is necessary to set the altimeter from
29.15 to 29.85, what change is made on the indicated altitude?

1--70-ft. increase.   3--700-ft. decrease.
2--700-ft. increase.   4--70-ft. decrease.

20. (FAA 1107) Of what practical value is pressure altitude?

1--A pilot should use it to check the accuracy of the altimeter
   during standard conditions.
2--To use on all aircraft performance charts since the charts are
   based on pressure altitude.
3--To use for obstacle clearance at higher altitudes where
   accurate altimeter settings are not available.
4--To use for computer solutions to determine density altitude,
   true altitude, true airspeed, etc.

21. (FAA 1108) Refer to the color-coded markings on the
airspeed indicator in Figure 4.4. What is the normal flap
operating range for the airplane?

1--60 to 100 MPH.   3--60 to 208 MPH.
2--65 to 165 MPH.   4--165 to 208 MPH.

22. (FAA 1109) Refer to the airspeed indicator in Figure 4.4.
The maximum speed at which the airplane can be operated in smooth
air is

1--100 MPH.   3--65 MPH.
2--165 MPH.   4--208 MPH.

23. (FAA 1110) Refer to the airspeed indicator in Figure 4.4.
Which of the color-coded markings on the airspeed indicator
identifies the never-exceed speed?

1--Lower A/S limit of the yellow arc.
2--Upper A/S limit of the white arc.
3--Upper A/S limit of the green arc.
4--The red radial line.

24. (FAA 1111) Refer to the airspeed indicator in Figure 4.4.
Which color-coded marking identifies the power-off stalling speed
with flaps and landing gear in the retracted position?

1--Upper A/S limit of the green arc.
2--Upper A/S limit of the white arc.
3--Lower A/S limit of the green arc.
4--Lower A/S limit of the white arc.

25. (FAA 1112) Refer to the airspeed indicator in Figure 4.4.
What is the maximum flaps-extended speed?

1--165 MPH.   3--65 MPH.
2--100 MPH.   4--60 MPH.

26. (FAA 1113) Refer to the airspeed indicator in Figure 4.4.
Which of the color-coded markings identifies the normal flap
operating range?

1--The lower limit of the white arc to the upper limit of the
   green arc.
2--The green arc.
3--The white arc.
4--The yellow arc.

27. (FAA 1114) Refer to the airspeed indicator in Figure 4.4.
Which of the color-coded markings identifies the power-off
stalling speed with wing flaps and leading gear in the landing
position?

1--Upper A/S limit of the green arc.
2--Upper A/S limit of the white arc.
3--Lower A/S limit of the green arc.
4--Lower A/S limit of the white arc.

28. (FAA 1115) Refer to the airspeed indicator in Figure 4.4.
What is the maximum structural cruising speed?

1--100 MPH.   3--208 MPH.
2--165 MPH.   4--65 MPH.

29. (FAA 1116) Refer to the color-coded markings on the
airspeed indicator in Figure 4.4 What is the caution range of
the airplane?

1--0 to 60 MPH.   3--165 to 208 MPH.
2--100 to 165 MPH.   4--60 to 100 MPH.

30. (FAA 1117) The red line on an airspeed indicator means a
maximum airspeed that

1--may be exceeded only if gear and flaps are retracted.
2--may be exceeded if abrupt maneuvers are not attempted.
3--may be exceeded only in smooth air.
4--should not be exceeded.

31. (FAA 1118) The pitot system provides impact pressure for
only the

1--airspeed indicator, vertical-speed indicator, and altimeter.
2--altimeter and vertical-speed indicator.
3--vertical-speed indicator.
4--airspeed indicator.

32. (FAA 1119) What altitude does altimeter A in Figure 4.5
indicate?

1--500 ft.   3--10,500 ft.
2--1,500 ft.   4--15,000 ft.

33. (FAA 1120 DI) Altimeter B in Figure 4.5 indicates

1--1,500 ft.   3--14,500 ft.
2--4,500 ft.   4--15,500 ft.

34. (FAA 1121) Altimeter C in Figure 4.5 indicates

1--9,500 ft.   3--15,940 ft.
2--10,950 ft.   4--19,500 ft.

35. (FAA 1122) Which altimeter(s) in Figure 4.5 indicate(s)
more than 10,000?

1--A, B, and C.   3--A only.
2--A and B only.   4--B only.

36. (FAA 1123) What is an important airspeed limitation that is
not color-coded on airspeed indicators?

1--Never-exceed speed.
2--Maximum structural cruising speed.
3--Maneuvering speed.
4--Maximum flaps-extended speed.

37. (FAA 1124) If the pitot tube and outside static vents or
ports were clogged, which instrument or instruments would be
affected?

1--The airspeed indicator, altimeter, and turn-and-slip
   indicator.
2--The altimeter, vertical-speed indicator, and airspeed
   indicator would provide inaccurate instrument readings.
3--The only instruments that would provide erroneous indications
   would be the airspeed indicator and altimeter.
4--The airspeed indicator would indicate excessively high
   airspeeds.

38. (FAA 1125) Which instrument(s) will become inoperative if
the pitot tube becomes clogged?

1--Altimeter.   3--Airspeed.
2--Vertical speed.   4--Altimeter and airspeed.

39. (FAA 1126) Which instrument(s) will become inoperative if
the static vents become clogged?

1--Airspeed only.
2--Altimeter only.
3--Airspeed and altimeter only.
4--Airspeed, altimeter, and vertical-speed.

40. (FAA 1373) If severe turbulence is encountered, the
airplane's airspeed should be reduced to

1--maneuvering speed.
2--the minimum steady flight speed in the landing configuration.
3--normal operation speed.
4--maximum structural cruising speed.

41. Static ports are aligned _____ the line of flight.

1--at a 45° angle away from
2--directly away from
3--directly into
4--perpendicular to

42. Total pressure (from the pitot head) less static pressure
(from the static ports), when corrected for minor installation
and mechanical errors, yields a measure of the airplane's

1--calibrated airspeed.   3--indicated airspeed.
2--dynamic airspeed.   4--true airspeed.

43. Which of the following instruments rely on the pitot-static
system as a source of pressure?

     A--airspeed indicator   D--magnetic compass
     B--altimeter   E--turn coordinator
     C--heading indicator   F--vertical-speed indicator

1--A only.   3--B, D, and E.
2--A, B, and F.   4--C and E only.

44. Atmospheric pressure _____ as altitude _____.

1--decreases; decreases.   3--increases; increases.
2--increases; decreases.   4--remains constant; increases.

45. Which of the following conditions, if any, will lead to an
indicated altitude greater than the airplane's true altitude?

1--Flying from an area of high ground elevation to an area of low
   ground elevation.
2--Decreasing air temperature.
3--Both (1) and (2).
4--Neither (1) nor (2).

46. When flying from air that is warm to air that is cold, the
airplane will be _____ the altitude indicated on the altimeter.

1--higher than
2--lower than
3--equal to
4--impossible to predict since air pressure does not change
   consistently with changes in temperature.

47. Suppose that you are flying over Cleveland, Ohio (field
elevation 1,850 ft.), and that you have just set your altimeter
to 29.04 based on radio contact with nearby Cleveland Flight
Service Station. Your altimeter reads 4,500 ft. This would be
defined as

1--absolute altitude.   3--pressure altitude.
2--indicated altitude.   4--true altitude.

48. Suppose that you are flying over Cleveland, Ohio (field
elevation 1,850 ft.), and that you have just set your altimeter
to 29.04 based on radio contact with nearby Cleveland Flight
Service Station. Your altimeter reads 4,500 ft. Next, suppose
that you had a second altimeter and set it to standard
conditions. What would the second altimeter show?

1--3,620 ft.
2--4,500 ft.
3--5,380 ft.
4--Impossible to determine since we do not know Cleveland's true
   altitude.

49. Suppose that you are flying over Cleveland, Ohio (field
elevation 1,850 ft.), and that you have just set your altimeter
to 29.04 based on radio contact with nearby Cleveland FSS. Your
altimeter reads 4,500 ft. Upon landing at Cleveland, you notice
that your altimeter reads 1,975 ft. You check with Cleveland FSS
again and find that the current altimeter is still 29.04. What
should you do?

1--Change your altimeter to 1,850 ft. since that is the true
   altitude and does not change with changes in atmospheric
   conditions.
2--Have your altimeter checked since the discrepancy is outside
   the tolerance limits suggested by the FAA.
3--Ignore the discrepancy since it is well within the tolerance
   limits set by the FAA.
4--Leave your altimeter at 1,975 ft. and report the discrepancy
   to the nearest FSS Station so they can issue a Notice to
   Airmen (NOTAM).

50. Suppose that you depart from Superior, Wisconsin (barometer
29.34), and fly direct to Minot, North Dakota (field elevation
1,800 ft.). As you approach Minot at an indicated altitude of
6,500 ft. you receive a new altimeter setting of 29.62. When you
change your altimeter to the new setting, approximately what
indicated altitude will your altimeter read?

1--4,420 ft.   4--6,220 ft.
2--4,700 ft.   5--6,500 ft.
3--4,980 ft.   6--6,780 ft.

51. Vertical speed indicators found in most general aviation airplanes tend to

1--accelerate during an altitude change.
2--accelerate during climbs and lag behind during descents.
3--lag behind an altitude change.
4--lag behind during climbs and accelerate during turns.

52. The movement of a gyroscope's spin axis from its original position due to some external force is referred to as

1--gimballing.   3--precession.
2--inclination.   4--rigidity in space.

53. Which of the following instruments is used to indicate changes in an airplane's pitch?

    A--Attitude indicator   C--Heading indicator
    B--Turn coordinator   D--Magnetic compass

1--A only.   3--C and D.
2--A and B.   4--B, C, and D.

54. The difference between magnetic north and true north is called

1--acceleration-deceleration error.
2--deviation.
3--oscillation error.
4--variation.

55. Which of the following instruments constitute the basic T arrangement?

    A--attitude indicator   E--magnetic compass
    B--vertical speed indicator   F--airspeed indicator
    C--altimeter   G--turn coordinator
    D--heading indicator

1--A, C, D, and F.
2--A, D, F, and G.
3--A, B, C, D, and F.
4--C, D, F, and G.
5--A, B, C, D, E, F, and G.

56. The process of checking and rechecking flight instruments against one another is called

1--composite flying.   3--oscillation.
2--cross-check.   4--redundancy.

ANSWERS

Key Terms and Concepts, Part 1

1.	J	7.	D	13.	M	19.	S
2.	A	8.	K	14.	G	20.	V
3.	Q	9.	N	15.	O	21.	R
4.	E	10.	F	16.	I	22.	X
5.	H	11.	P	17.	L	23.	W
6.	C	12.	B	18.	U	24.	T

Key Terms and Concepts, Part 2

1.	N	6.	A	11.	Q	16.	G
2.	D	7.	M	12.	B	17.	L
3.	H	8.	F	13.	K	18.	E
4.	J	9.	O	14.	R	19.	S
5.	P	10.	C	15.	I	20.	T

Discussion Questions and Exercises

2. T--Airspeed is speed relative to the air.
6. None, since the barometric conditions are the same in both places.
7. 2,100 ft. Since you are flying from high pressure to low pressure without changing the altimeter setting, actual MSL altitude would be lower than the altimeter reading. 29.96 - 28.74 = 1.22 in. of mercury. Since 1 in. = 1,000 ft., the altimeter would be off by 1,220 ft., and it would have registered 2,100 ft. when you landed! Further, had you used your altimeter to enter the pattern (1,880 ft. MSL), you would have come in contact with the ground before you ever reached the pattern. "When flying from a high to a low, look out below!"
8. It would have registered higher, since moving from high to low temperature results in your being lower than your altimeter indicates.

Review Questions

1. 2--Indicated airspeed for landings and takeoff is the same regardless of altitude. At higher altitudes, however, true airspeed will be faster as the air is less dense and the plane must fly faster to develop an equivalent amount of lift.
2. 4--This is the definition of pressure altitude, which is most useful for determining answers to computational problems involving true altitude and true airspeed.
3. 2--Absolute altitude is the height above ground level (AGL). It is the difference between the true elevation of the surface and your (true) altitude.

4. 1--This is the definition of true altitude--true height
    above sea level.
5. ?--This question has been designated as unusable by the FAA
    and removed from the FAA Question Selection Sheets.
6. ?--This question has been designated as unusable by the FAA
    and removed from the FAA Question Selection Sheets.
7. 1--The turn coordinator shows both rate of turn and roll.
    It also indicates yaw, as when the inclinometer moves in
    response to P-factor and torque on takeoff.
8. ?--This question has been designated as unusable by the FAA
    and removed from the FAA Question Selection Sheets.
9. 4--The heading indicator tends to drift off its setting due
    to the gyroscopic principle of precession.
10. ?--This question has been designated as unusable by the FAA
    and removed from the FAA Question Selection Sheets.
11. 4--Acceleration will cause the magnetic compass to indicate
    a turn to the north when the airplane is flying on an
    east or west heading.
12. 2--See answer 11.
13. 4--Metallic and electrical components in the airplane may
    distort the lines of magnetic force. This type of error
    is referred to as deviation.
14. 4--The initial indication of the magnetic compass indicates
    a turn in the opposite direction when a turn toward east
    or west is made from a northerly heading.
15. 4--See answer 14.
16. 1--The magnetic compass should be used for accurate readings
    only when the aircraft is in unaccelerated, straight-and-
    level flight.
17. 2--True altitude is the measured height above MSL and
    indicated altitude is what your altimeter reads at any
    point in time. If the altimeter is working correctly and
    you have the current altimeter setting in the Kollsman
    window, indicated altitude is the same as true altitude.
    For the readings to be identical, however, you must be at
    sea level under standard conditions.
18. 1--A warm day increases the true altitude for any given
    indicated altitude, so the aircraft will be slightly
    higher than indicated.
19. 2--1 in. of mercury = 1,000 ft. 29.15 - 29.85 = -0.7. So,
    -0.7 x 1,000 = -700 ft.
20. 4--Pressure altitude is obtained by setting 29.92 in the
    Kollsman window. It is used to compute such things as
    true airspeed, density altitude, and true altitude.
21. 1--The white arc indicates the flap operating range.
22. 4--The yellow arc is the caution range; it includes speeds
    that should only be flown in smooth air.
23. 4--The red line indicates the never-exceed speed.
24. 3--The green arc indicates the normal operating range with
    flaps and gear retracted.
25. 2--The white arc indicates the flap operating range.
26. 3--The white arc indicates the flap operating range.
27. 4--The white arc indicates the flap operating range.

28. 2--The upper limit of the green arc (normal operating
       speeds) is defined as the "maximum structural cruising
       speed."
29. 3--The yellow arc is the caution range; it includes speeds
       that should only be flown in smooth air.
30. 4--The red line defines the speed that should never be
       exceeded.
31. 4--The pitot tube measures the pressure produced by ram air.
       Only the airspeed indicator utilizes this input.
32. 3--Always check to see where each hand is. The small hand
       is slightly above 10,000 ft. The middle-sized hand is
       between 0 and 1, and the large hand is on 500. If you
       begin with the small hand and move up in size, **always
       round down**. The small hand (10,000 ft. increments) is
       greater than 1, so round down to 1 (+10,000). The middle
       hand (1,000 ft. increments) is greater than 0 but less
       than 1, so round down (+0000). Finally, the large hand
       (100 ft. increments) is exactly on 5, so round down to 5
       (+500). The sum equals 10,500.
33. ?--This question has been designated as unusable by the FAA
       and removed from the FAA Question Selection Sheets.
34. 1--Let's try the logic one more time. It looks as though
       the small hand is between 1 and 2, so this could be
       tough. If it were closer to 2, you would suspect that
       the reading would approximate 19,000 ft. However, it is
       very close to 1 (10,000). Using **logic**, the marker must
       be between 0 and 1, so rounding down yields +00,000. The
       middle hand is between 9 and 10, so round down to 9
       (+9,000). The large hand is on 5, so add another +500
       ft.
35. 2--The small (10,000 ft.) hands tell it all on this one.
       See answers 32 and 34.
36. 3--Only the maneuvering speed is not color-coded. The never
       exceed speed is indicated by the red line at the end of
       the yellow arc, the maximum structural cruising speed is
       indicated at the upper end of the green arc, and the
       maximum flaps-extended speed is shown at the upper end of
       the white arc.
37. 2--The altimeter, vertical-speed indicator, and airspeed
       indicator all rely on static air pressure to operate
       accurately. The airspeed indicator relies on ram air
       from the pitot tube. If both sources are clogged, none
       of these instruments will provide accurate information.
38. 3--See answer 37.
39. 4--See answer 37.
40. 1--The best speed at which to penetrate turbulent air is the
       airplane's maneuvering speed.
41. 4--So as not to be affected by ram air, the static ports are
       located at a 90° angle to the line of flight.
42. 1--This defines calibrated airspeed.
43. 2--See answer 37.
44. 2--The lower you go, the greater the pressure exerted by the
       air.

45.   2--Ground elevation has nothing to do with indicated
       altitude. However, as air temperature decreases, all
       things remaining equal, indicated altitude increases.
46.   2--The altimeter reads erroneously high when you fly from
       warm to cold air, or from high pressure to low pressure.
47.   2--This is indicated altitude. If you landed at Cleveland,
       your altimeter would read Cleveland's actual height above
       sea level, assuming there were no mechanical problems
       with the altimeter.
48.   3--It would indicate higher than you are by about 880 ft.
       (29.92 - 29.04 = .88 in. of mercury). Since 1 in. of
       mercury equals 1,000 ft., .88 in. equals 880 ft.
49.   2--The tolerance suggested by the FAA is plus or minus 75
       ft. between the surveyed altitude and the height
       indicated on your altimeter when corrected for current
       ground level atmospheric conditions. Since the altimeter
       is off by 125, it is outside of the tolerance limits set
       by the FAA.
50.   6--Your altimeter would indicate that you are lower than you
       actually are by (29.62 - 29.34 = .28) x 100 = 280 ft.
       So, while the altimeter indicates 6,500 ft., your
       airplane is actually 6,780 ft. above sea level.
51.   3--VSIs have a tendency to lag behind altitude changes,
       which means that they are most helpful once a flight
       attitude has been established.
52.   3--This defines precession, one of the two fundamental
       properties of a gyroscope. It also explains why the
       heading indicator will slowly change its indication
       during flight, something which must be corrected by
       reference to the magnetic compass in unaccelerated,
       straight-and-level flight.
53.   1--Pitch, movement about the airplane's lateral axis, is
       indicated on the attitude indicator, as is the degree of
       bank (roll, or movement about the longitudinal axis).
54.   4--This defines variation.
55.   1--The attitude indicator, altimeter, heading indicator, and
       airspeed indicator define the basic T. The attitude
       indicator is at the center.
56.   2--Cross-check involves checking and rechecking the
       instruments in the basic T against one another.

5/AIRPLANE WEIGHT AND BALANCE

MAIN POINTS

1.   A teeter-totter is a lever on either side of a **fulcrum**. The length of the lever is called the **arm**. The force of a weight acting at the end of the arm is called a **moment**. Arm, moment, and weight are related such that Weight x Arm = Moment, Arm = Moment/Weight, or Weight = Moment/Arm.

2.   Airplanes balance at a single point called the **center of gravity (CG)**, around which all moments are equal. The CG must be kept within certain limits to maintain control stability.

3.   The reference point from which moment arms are measured is called the **datum** or **datum line**. The datum in most planes is located near the firewall. The CG and location of all components in the plane are measured in inches from the datum. Weight times distance from the datum yields a moment. The sum of the airplane's empty weight plus everything else you put in it is called **gross weight**. If you sum the moments and divide the gross weight, you have the new CG location relative to the datum.

4.   **Empty weight** is the airplane's weight with all listed equipment, hydraulic fluid, undrainable engine oil, and unusable fuel (fuel not available to the engine due to aircraft design). **Basic empty weight** is empty weight plus full oil. Gas weighs 6 lb. per gal., and oil weigh 7.5 lb. per gal.   **Useful load** includes usable fuel and the payload (occupants, cargo, and baggage). **Maximum ramp weight** is the manufacturer's maximum ground maneuvering weight. **Maximum takeoff weight** is the maximum for the start of the takeoff run. **Maximum landing weight** is the maximum weight approved for touchdown.

5.   The two basic questions you must ask as a pilot are: Is gross weight within allowable limits? Is the CG within allowable limits? After determining how much useful load you have and where it will go, you use the Pilot's Operating Handbook to

determine the fuselage station number (arm) for each item, which
is listed conveniently in inches from the datum. Multiply weight
by its arm and you have the moment; add all moments and divide by
the gross weight to obtain the new CG. The Flight Manual for
your airplane provides specific total moment limits for each
airplane load condition.

6.   Changing the weight and/or CG changes the airplane's
performance characteristics. For example, high gross weights
require a greater angle of attack to maintain level flight.
Thus, the airplane will stall at higher airspeeds and have poorer
climb and cruise performance. Forward CGs make the plane more
stable, but extra elevator pressure loads are required to hold a
desired attitude. Also, the airplane will stall at a slightly
higher airspeed since the added tail-down load adds to the lift
required from the wings. An aft CG makes the plane less stable
and increases the chances of serious stalls or spins. Since the
tail provides less download, the wind operates at a lower angle
of attack, which in turn reduces stalling speed. An aft CG also
increases fuel efficiency. All of these changes are minor and
acceptable when the airplane is loaded **within** the limits defined
in the Flight Manual, but they become extremely dangerous when
those limits are exceeded.

7.   There are two common formats for solving weight and
balance problems: tabular and graphical. Both methods require
that you first figure the **zero fuel condition**, then the **fuel
loading**, which together yield the **ramp condition**. The **takeoff
condition** compensates for fuel burned during taxi and runup. The
landing condition takes into account fuel burned during flight.
The tabular format refers you to a table of moment limits versus
weight to determine if you are within an acceptable range of CG
values. With the tabular format, you may have to **interpolate**
between the numbers listed in the table. The graphical method
presents moment values as index units for each compartment of the
plane. The sum of the moments and weights is used to determine
if you fall within the **CG envelope**.

8.   Good pilots make weight and balance planning a **habit**!
Although it may be a chore to rearrange the payload or embarras-
sing to ask someone for their actual weight, remember that your
safety and the safety of your passengers is at stake. Finally,
remember that most people, particularly those who are overweight,
tend to **underestimate** their weight. Adding 10 percent to the
reported weight typically compensates for this discrepancy.

KEY TERMS AND CONCEPTS

Match each term or concept (1-16) with the appropriate description (A-P) below. Each item has only one match.

___ 1. ramp weight   ___ 9. useful load
___ 2. arm   ___ 10. fuselage station number
___ 3. empty weight   ___ 11. maximum gross weight
___ 4. moment   ___ 12. usable fuel
___ 5. payload   ___ 13. datum
___ 6. takeoff condition   ___ 14. fulcrum
___ 7. CG envelope   ___ 15. zero fuel condition
___ 8. landing condition   ___ 16. center of gravity (CG)

A. reference point from which all moments are measured
B. ramp condition, less fuel used for taxi and runup
C. number of inches an item is from its datum
D. weight of fuel aboard that is available to the engine
E. airplane's weight with all equipment and unusable fuel
F. weight of cargo, occupants, and baggage
G. weight of an item times its distance from the datum (arm)
H. takeoff condition, less fuel burned in flight
I. weight and moment of an airplane with payload, but no fuel load
J. point in an aircraft about which all moments are equal
K. distance of a station or item from the datum
L. occupants, cargo, baggage, and usable fuel
M. maximum weight to which an aircraft is certified by the FAA
N. support point of a level
O. weight and moment of an airplane with payload and fuel included
P. range of weight and balance operating limits

DISCUSSION QUESTIONS AND EXERCISES

1. What is the relationship among arm, moment, and weight? How does this apply to general aviation aircraft?

2. What are the two major factors about the airplane's weight and balance that a pilot must attend to before the airplane flies?

3.   What is **basic empty weight**? How is it different from **empty weight**?

4.   Name three effects of each of the following on the airplane's flight characteristics:

   a. high gross weight

   b. operation with a forward CG

   c. operation with an aft CG

Table 5.1 Excerpt from a Pilot's Operating Handbook

Aircraft Designation: Four-place, single-engine, land monoplane
Engine operating limits: 150 horsepower at 2,700 RPM
Fuel system: Float-type carburetor
Fuel capacity: Standard tanks--42 gal., 38 usable
               Long-range tanks--52 gal., 48 usable
Oil capacity: 8 qts. (not included in licensed empty weight)
Propeller: Fixed-pitch
Landing gear: Fixed tricycle gear
Wing flaps: Electrically operated, 0-40°
Licensed empty weight: 1,364 lb.
Maximum gross weight: 2,300 lb.
Maximum weight in baggage compartment: 120 lb.

---

6.   To answer questions a-e, refer to Table 5.1 and assume the airplane is loaded as follows:

Pilot	160 lb.
Front seat passenger	148 lb.
Rear seat passenger	122 lb.
Rear seat passenger	176 lb.
Baggage	80 lb.
Oil	Full
Fuel (standard tanks)	Full

a.   How is the airplane loaded with respect to maximum gross weight limits?

b.   Assume that you plan to load the airplane with 120 lb. of baggage, 8 qts. of oil, and four persons whose total weight is 698 lb. What is the total amount of usable fuel (standard tanks) that can be aboard without exceeding the maximum certified gross weight?

c.   What is the combined maximum weight of four persons and baggage that can be loaded without exceeding the maximum certified gross weight if the airplane is serviced to capacity with oil and fuel (including long-range tanks)?

d.   Suppose you have filled the airplane's long-range fuel tanks to capacity and there are 8 qts. of oil in the engine. You wish to carry four persons whose total weight is 680 lb. There will be no baggage aboard. How close would the airplane be to maximum certified gross weight limits?

e.   During the preflight inspection, you notice that there are 8 qts. of oil in the engine and the standard tanks are filled to capacity. The total weight of the pilot and passengers is 670 lb. What is the total weight of the baggage, if any, that can be loaded aboard without exceeding the maximum certified gross weight of the airplane?

REVIEW QUESTIONS

1.   (FAA 1171) An airplane has been loaded in such a manner
that the CG is located aft of the CG limit. One undesirable
flight characteristic a pilot might experience with this airplane
would be

1--a longer takeoff run.
2--the inability to recover from a stalled condition.
3--stalling at higher-than-normal airspeeds.
4--the inability to flare during landings.

2.   (FAA 1172) What effect does loading an airplane to the most
aft CG have on the airplane stability?

1--The airplane will be less stable at slow speeds, but more
    stable at high speeds.
2--The airplane will be less stable at high speeds, but more
    stable at low speeds.
3--The airplane will be more stable at all speeds.
4--The airplane will be less stable at all speeds.

3.   (FAA 1173) Frost which has not been removed from the wings
of an airplane before flight

1--may cause the airplane to become airborne with a lower angle
    of attack and at a lower indicated airspeed.
2--may make it difficult or impossible to become airborne.
3--would present no problem since frost will blow off when the
    airplane starts moving during takeoff.
4--will change the camber (curvature of the wing) thereby
    increasing lift during takeoff.

4.   (FAA 1198) Which of the following items are included in the
licensed empty weight of an airplane?

1--Hydraulic fluid and usable fuel.
2--Only the airframe, powerplant, and equipment installed by the
    manufacturer.
3--Full fuel tanks and engine oil to capacity, but excluding crew
    and baggage.
4--Unusable fuel and optional equipment.

5.   (FAA 1199) An airplane is loaded 110 lb. over maximum
certified gross weight. If fuel (gasoline) is drained to bring
the aircraft weight within limits, how much fuel should be
drained?

1--16.2 gal.   3--15.7 gal.
2--18.4 gal.   4--17.1 gal.

6.   (FAA 1200) If an airplane is loaded 90 lb. over maximum certified gross weight and fuel (gasoline) is drained to bring the aircraft weight within limits, how much fuel should be drained?

1--9 gal.   3--6 gal.
2--12 gal.   4--15 gal.

7.   (FAA 1207) What is the maximum amount of baggage that may be loaded aboard the airplane for the CG to remain within the loading envelope? (See Figure 5.1).

	WEIGHT (lb.)	MOM/1000
Empty weight	1,350	51.5
Pilot and front passenger	250	
Rear passengers	400	——
Baggage	——	——
Fuel 30 gal.		
Oil 8 qts.	15	-0.2

1--120 lb.   3--90 lb.
2--105 lb.   4--75 lb.

8.   (FAA 1208 DI) Calculate the loaded aircraft moment/1000 of the airplane loaded as follows and determine which category when plotted on the CG moment envelope. (See Figure 5.1.)

	WEIGHT (lb.)	MOM/1000
Empty Weight	1,350	51.5
Pilot and front passenger	310	——
Rear passengers	196	——
Fuel 38 gal.		
Oil 8 qts.	15	-0.2

1--79.2, utility category.   3--81.2, normal category.
2--80.8, utility category.   4--82.0, normal category.

9.   (FAA 1209) What is the maximum amount of fuel that may be aboard the airplane on takeoff if it is loaded as follows? (See Figure 5.1.)

	WEIGHT (lb.)	MOM/1000
Empty weight	1,350	51.5
Pilot and front passenger	340	——
Rear passengers	310	——
Baggage	45	——
Oil 8 qts.	15	-0.2

1--24 gal.   3--40 gal.
2--34 gal.   4--46 gal.

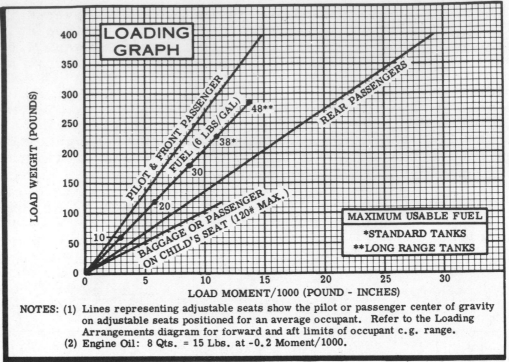

NOTES: (1) Lines representing adjustable seats show the pilot or passenger center of gravity
           on adjustable seats positioned for an average occupant. Refer to the Loading
           Arrangements diagram for forward and aft limits of occupant c.g. range.
       (2) Engine Oil: 8 Qts. = 15 Lbs. at -0.2 Moment/1000.

NOTE: The empty weight of this airplane does not include
        the weight of the oil.

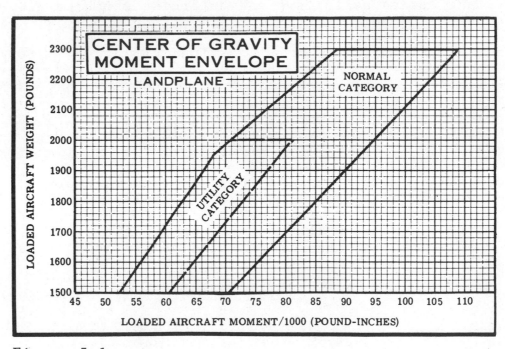

Figure 5.1

10. (FAA 1210) Determine the CG moment/1000 with the following data using the graph in Figure 5.1.

	WEIGHT (lb.)	MOM/1000
Empty weight	1,350	51.5
Pilot and front passenger	340	———
Fuel (std. tanks)	Capacity	———
Oil 8 qts.	15	-0.2

1--38.7 lb.-in.   3--74.9 lb.-in.
2--69.9 lb.-in.   4--77.0 lb.-in.

11. (FAA 1211) Calculate the CG and determine the plotted position on the CG moment envelope chart, Figure 5.1.

	WEIGHT (lb.)	MOM/1000
Empty weight	1,350	51.5
Pilot and front passenger	380	———
Fuel 48 gal.	288	———
Oil 8 qt.	15	———

1--CG 38.9, out of limits forward.
2--CG 39.9, utility category.
3--CG 38.9, normal category.
4--CG 39.9, normal category.

12. (FAA 1222) Determine if the airplane weight and balance is within limits. (See Figure 5.2 and Figure 5.3.)

```
Front seat occupants..........340 lb.
Rear seat occupants...........295 lb.
Fuel......................... 44 gal.
Baggage...................... 56 lb.
```

1--Within limits.
2--20 lb. overweight, CG within limits.
3--Weight within limits, CG out of limits forward.
4--39 lb. overweight, CG out of limits forward.

13. (FAA 1223) Which is the maximum amount of baggage that can be carried when the airplane is loaded as follows? See Figures 5.2 and 5.3.

```
Front seat occupants.......... 387 lb.
Rear seat occupants........... 293 lb.
Fuel......................... 35 gal.
```

1--45 lb.   3--220 lb.
2--63 lb.   4--255 lb.

Weight	Minimum Moment/100	Maximum Moment/100
2600	2037	2224
2610	2048	2232
2620	2058	2239
2630	2069	2247
2640	2080	2255
2650	2090	2263
2660	2101	2271
2670	2112	2279
2680	2123	2287
2690	2133	2295
2700	2144	2303
2710	2155	2311
2720	2166	2319
2730	2177	2326
2740	2188	2334
2750	2199	2342
2760	2210	2350
2770	2221	2358
2780	2232	2366
2790	2243	2374
2800	2254	2381
2810	2265	2389
2820	2276	2397
2830	2287	2405
2840	2298	2413
2850	2309	2421
2860	2320	2428
2870	2332	2436
2880	2343	2444
2890	2354	2452
2900	2365	2460
2910	2377	2468
2920	2388	2475
2930	2399	2483
2940	2411	2491
2950	2422	2499

Weight	Minimum Moment/100	Maximum Moment/100
2100	1617	1800
2110	1625	1808
2120	1632	1817
2130	1640	1825
2140	1648	1834
2150	1656	1843
2160	1663	1851
2170	1671	1860
2180	1679	1868
2190	1686	1877
2200	1694	1885
2210	1702	1894
2220	1709	1903
2230	1717	1911
2240	1725	1920
2250	1733	1928
2260	1740	1937
2270	1748	1945
2280	1756	1954
2290	1763	1963
2300	1771	1971
2310	1779	1980
2320	1786	1988
2330	1794	1997
2340	1802	2005
2350	1810	2014
2360	1817	2023
2370	1825	2031
2380	1833	2040
2390	1840	2048
2400	1848	2057
2410	1856	2065
2420	1863	2074
2430	1871	2083
2440	1879	2091
2450	1887	2100
2460	1894	2108
2470	1902	2117
2480	1911	2125
2490	1921	2134
2500	1932	2143
2510	1942	2151
2520	1953	2160
2530	1963	2168
2540	1974	2176
2550	1984	2184
2560	1995	2192
2570	2005	2200
2580	2016	2208
2590	2026	2216

Figure 5.3

USEFUL LOAD WEIGHTS AND MOMENTS

OCCUPANTS

FRONT SEATS ARM 85

Weight	Moment/100
120	102
130	110
140	119
150	128
160	136
170	144
180	153
190	162
200	170

REAR SEATS ARM 121

Weight	Moment/100
120	145
130	157
140	169
150	182
160	194
170	206
180	218
190	230
200	242

USABLE FUEL

MAIN WING TANKS ARM 75

Gallons	Weight	Moment/100
5	30	22
10	60	45
15	90	68
20	120	90
25	150	112
30	180	135
35	210	158
40	240	180
44	264	198

AUXILIARY WING TANKS ARM 94

Gallons	Weight	Moment/100
5	30	28
10	60	56
15	90	85
19	114	107

*OIL

Quarts	Weight	Moment/100
10	19	5

BAGGAGE OR 5TH SEAT OCCUPANT ARM 140

Weight	Moment/100
10	14
20	28
30	42
40	56
50	70
60	84
70	98
80	112
90	126
100	140
110	154
120	168
130	182
140	196
150	210
160	224
170	238
180	252
190	266
200	280
210	294
220	308
230	322
240	336
250	350
260	364
270	378

*Included in Basic Empty Weight

Empty Weight – 2015

MOM/100 – 1554

MOMENT LIMITS vs WEIGHT

Moment limits are based on the following weight and center of gravity limit data (landing gear down).

WEIGHT CONDITION	FORWARD CG LIMIT	AFT CG LIMIT
2950 lb. (take-off or landing)	82.1	84.7
2525 lb.	77.5	85.7
2475 lb. or less	77.0	85.7

Figure 5.2

14. (FAA 1224) Calculate the weight and balance and determine
if the CG and the weight of the airplane are within limits. See
Figures 5.2 and 5.3.

          Front seat occupants.......... 350 lb.
          Rear seat occupants........... 325 lb.
          Baggage....................... 27 lb.
          Fuel.......................... 35 gal.

1--81.7, out of limits forward.
2--83.4, within limits.
3--84.1, within limits.
4--84.8, out of limits aft.

15. (FAA 1225) Determine if the airplane weight and balance is
within limits. See Figures 5.2 and 5.3.

          Front seat occupants.......... 415 lb.
          Rear seat occupants........... 110 lb.
          Fuel, main aux. tanks......... 63 gal.
          Baggage....................... 32 lb.

1--19 lb. overweight, CG within limits.
2--19 lb. overweight, CG out of limits forward.
3--Weight within limits, CG out of limits.
4--Weight and balance within limits.

16. (FAA 1226) Which action can adjust the airplane's weight to
maximum gross weight and the CG within limits for takeoff? See
Figures 5.2 and 5.3.

          Front seat occupants.......... 425 lb.
          Rear seat occupants........... 300 lb.
          Fuel, main tanks.............. 44 gal.

1--Drain 12 gal. of fuel.
2--Drain 9 gal. of fuel.
3--Transfer 12 gal. of fuel from the main tanks to the auxiliary
   tanks.
4--Transfer 19 gal. of fuel from the main tanks to the auxiliary
   tanks.

17. (FAA 1227 DI) With an airplane loaded as follows, what
action can be taken to balance it? See Figures 5.2 and 5.3.

          Front seat occupants.......... 400 lb.
          Main wing tanks............... 44 gal.

1--Drain 30 gal. of fuel.
2--Fill the auxiliary wing tanks.
3--Transfer 19 gal. of fuel from the main tanks to the auxiliary
   tanks.
4--Add a 100-lb. weight to the baggage compartment.

18. (FAA 1228) Upon landing, the front passenger (180 1b.) departs the airplane. A rear passenger (204 1b.) moves to the front passenger position. What effect does this have on the CG if the airplane weighed 2,690 1b. and the MOM/100 was 2260 just prior to the passenger transfer? See Figures 5.2 and 5.3.

1--The weight changes but the CG is not affected.
2--The CG moves forward approximately 0.1 in.
3--The CG moves forward approximately 2.4 in.
4--The CG moves forward approximately 3.0 in.

19. (FAA 1229) What effect does a 35-gal. fuel burn have on the weight and balance if the airplane weighed 2,890 1b. and the MOM/100 was 2452 at takeoff? See Figures 5.2 and 5.3.

1--Weight is reduced by 210 1b. and the CG is unaffected.
2--Weight is reduced to 2,680 1b. and the CG moves forward.
3--Weight is reduced to 2,855 1b. and the CG moves aft.
4--Weight is reduced by 210 1b. and the CG is aft of limits.

20. (FAA 1794) Frost which has not been removed from the lifting surfaces on an airplane before flight

1--may prevent the airplane from becoming airborne.
2--will change the camber (curvature of the wing) thereby increasing lift during the takeoff.
3--may cause the airplane to become airborne with a lower angle of attack and a lower indicated airspeed.
4--would present no problems since frost will blow off when the airplane starts moving during takeoff.

21. (FAA 1792) Why is frost considered hazardous to flight operation?

1--The increased weight requires a greater takeoff distance.
2--Frost changes the basic aerodynamic shape of the airfoil.
3--Frost decreases control effectiveness.
4--Frost causes early airflow separation resulting in a loss of lift.

22. If a camera bag originally in the front of the plane is moved to the baggage compartment aft of the cabin, how would this affect the airplane's center of gravity?

1--The CG would change unpredictably as flight attitude changed.
2--The CG would move aft.
3--The CG would move forward.
4--The CG would remain the same.

23. If you have a 40-1b. weight and an 80-in. arm, what is the moment?

1--2 1b.-in.   3--120 1b.-in.
2--32 1b.-in.   4--3200 1b.-in.

24. The point in an airplane around which all moment-arms are equal is called the

1--ballast.   3--datum.
2--center of gravity.   4--point of equilibrium.

25. The set of weight and balance operational restrictions assigned to your airplane is called the

1--CG envelope.
2--fuselage station number.
3--load limit.
4--moment limit.

26. Which of the following conditions would probably lead to an increase in stalling speed?

1--An increased angle of bank.
2--Operation at a high gross weight.
3--Operation with a forward CG.
4--All of the above would tend to increase stalling speed.

27. Of the following preflight planning procedures, which should be considered **first** in completing a weight and balance form?

1--Payload.   3--Fuel to destination.
2--Fuel loading.   4--Ramp condition.

ANSWERS

Key Terms and Concepts

1.	O	5.	F	9.	L	13.	A
2.	K	6.	B	10.	C	14.	N
3.	E	7.	P	11.	M	15.	I
4.	G	8.	H	12.	D	16.	J

Discussion Questions and Exercises

6.   a. 2,293, or 7 lb. under gross weight. Empty weight
        (1,364), pilot and front seat passenger (298), baggage
        (80), oil (8 qts. = 2 gal. x 7.5 lb. per gal. = 15), and
        fuel (38 x 6 = 228).
     b. 17.1 gal. Use the data in question 6a. The total left
        for fuel is 103 lb., or 17.1 gal. at 6 lb. per gal.
     c. 633 lb. Use the total in question 6a. Fuel: 48 x 6 =
        288; 2,300 less fuel (288), less oil (15), less empty
        weight (1,364) equals 633 lb.
     d. 47 lb. over maximum certified gross weight. Use the data
        in question 6a and make the appropriate substitutions.
     e. 23 lb. are left for baggage. Again, use the data in
        question 6a and make the appropriate substitutions.

Review Questions

1. 2--An aft CG outside the limits will make the airplane less
        stable, which in turn will make it more difficult to
        recover from a stalled condition.
2. 4--See answer 1.
3. 2--Frost increases the weight of the airplane and may
        disrupt the flow of air over the wings, making it
        impossible to become airborne. Frost may cause early
        airflow separation resulting in a loss of lift. The
        formation of frost will be discussed in the chapter on
        aviation weather.
4. 4--All permanently installed equipment, unusable fuel and
        oil, and hydraulic fluids are included in the licensed
        empty weight.
5. 2--Since aviation gas weighs 6 lb. per gal., you divide 110
        lb. by 6 to arrive at 18.33 gal.
6. 4--Since aviation gas weighs 6 lb. per gal., you divide 90
        lb. by 6 to arrive at 15 gal.
7. 2--To complete this problem, you must use Figure 5.1. Use a
        straightedge to read down to the load moment axis. The
        amounts you read from the chart may not be exactly the
        same as those I've figured, but they should be close.
        Let's begin by calculating the airplane's weight. As you
        can see from the table, total weight without baggage is
        2,195 lb. The maximum amount of baggage allowable,
        therefore, will be 105 lb., if the airplane is still
        within CG limits. As you can see from the table, it is
        still within the CG envelope with 105 lb. of baggage.

	WEIGHT (lb.)	MOM/1000
Empty weight	1,350	51.5
Pilot and front passenger	250	9.4
Rear passenger	400	29.2
Baggage	105	10.0
Fuel 30 gal.	180	8.8
Oil 8 qts.	15	-0.2
Totals	2,300	108.7

8. ?--This question has been designated as unusable by the FAA
        and removed from the FAA Question Selection Sheets.

	WEIGHT (lb.)	MOM/1000
Empty weight	1,350	51.5
Pilot and front passenger	310	11.5
Rear passenger	196	14.4
Fuel 38 gal.	228	11.0
Oil 8 qts.	15	-0.2
Totals	2,099	88.2

9.   3--Use the data in the table that follows.

	WEIGHT (lb.)	MOM/1000
Empty weight	1,350	51.5
Pilot and front passenger	340	12.6
Rear passengers	310	22.5
Baggage	45	4.3
Oil 8 qts.	15	-0.2
Fuel 40 gal.	240	12.0
	-----	-----
Totals	2,300	102.7

10.   3--Use the analysis in the table that follows.

	WEIGHT (lb.)	MOM/1000
Empty weight	1,350	51.5
Pilot and front passenger	340	12.6
Fuel	228	11.0
Oil 8 qts.	15	-0.2
	-----	-----
Totals	1,933	74.9

11.   3--Use the data in the table that follows. Once you have
      computed weight (2,033 lb.) and MOM/1000 (79.0), there is
      a final computation that must be made. CG is equal to
      moment divided by weight, or 79,000/2,033 = 38.86.

	WEIGHT (lb.)	MOM/1000
Empty weight	1,350	51.5
Pilot and front passenger	380	14.0
Fuel 48 gal.	288	13.7
Oil 8 qts.	15	-0.2
	-----	-----
Totals	2,033	79.0

12.   2--The computations that follow reveal that the airplane is
      20 lb. overweight, but within CG limits. To determine
      the MOM/100 for the front seat occupants, use 340 lb. x
      85 = 28,900/100 = 289. To compute MOM/100 for the rear
      seat use 295 lb. x 121 = 35,695/100 = 357. To compute
      the MOM/100 for baggage, 56 x 140 = 7,840/100 = 78.

	WEIGHT (lb.)	MOM/100
Empty weight	2,015	1,554
Front seat occupants	340	289
Rear seat occupants	295	357
Fuel	264	198
Baggage	56	78
	-----	-----
Totals	2,970	2,476

13. 1--Total weight determined from the table that follows is
     2,905 lb., which leaves room for 45 lb. of baggage. When
     loaded with the baggage, the airplane is within both
     weight and CG limits.

	WEIGHT (lb.)	MOM/100
Empty weight	2,015	1,554
Front seat occupants	387	329
Rear seat occupants	293	355
Fuel 35 gal.	210	158
Subtotal	2,905	2,396
Baggage	45	63
Totals	2,950	2,459

14. 2--See the computations that follow. Remember that you must
     divide moment by weight to determine the CG.

	WEIGHT (lb.)	MOM/100
Empty weight	2,015	1,554
Front seat occupants	350	298
Rear seat occupants	325	393
Fuel 35 gal.	210	158
Totals	2,927	2,441

15. 3--For this problem, see the computations that follow.

	WEIGHT (lb.)	MOM/100
Empty weight	2,015	1,554
Front seat occupants	415	353
Rear seat occupants	110	133
Fuel - main tanks	264	198
Fuel - auxiliary tanks	114	107
Baggage	32	45
Totals	2,950	2,390

16. 2--First, calculate weight. You are 54 lb. over gross, so
     that eliminates options (3) and (4). The only action
     that will bring the airplane to maximum gross weight is
     to drain 54 lb. of fuel (9 gal.). See the computations
     that follow.

	WEIGHT (lb.)	MOM/100
Empty weight	2,015	1,554
Front seat occupants	425	361
Rear seat occupants	300	363
Fuel, main tanks	264	198
Subtotals	3,004	2,476
Less 9 gal. fuel	- 54	- 40
Totals	2,950	2,436

17. ?--This question has two correct answers, (1) and (4). In
      my initial calculation, I made a computational error on
      (4) which led me to compute each of the alternatives.
      Alternative (4) makes the most practical sense, so I
      would expect that is the one the FAA is looking for.
      First, let's compute the weight and balance for the
      airplane as described in the question.

	WEIGHT (lb.)	MOM/100
Empty weight	2,015	1,554
Front seat occupants	400	340
Main wing tanks	264	198
Totals	2,679	2,092

As you can see from the charts, the CG is forward of the
limit, so you must do something to move it aft. Looking
at the auxiliary wing tanks, adding weight to them does
not produce a great increase in aft CG, which is going to
be necessary. Adding baggage does produce a relatively
greater aft CG. In fact, adding 100 lb. produces a
MOM/100 of 140, which--added to the above figures--brings
the plane into the CG envelope. But, so does draining 30
gal. of fuel. Weight becomes (2,679 - 264 + 84) = 2,499
lb. MOM/100 becomes (2,092 - 198 + 63) = 1,957. As you
can see from the chart, these values are also within the
limits.

If you actually encountered this situation, the most
logical action would be to add the weight to the baggage
compartment, since you're not going to get very far on 14
gal. of fuel. At any rate, this question has been
designated as unusable by the FAA and removed from the
FAA Question Selection Sheets.

18. 4--First, you know that the CG will move forward by some
      amount since weight is being transferred to the front.
      But, how much will it move forward? The CG before the
      exchange is 226,000/2,690 = 84.01. See the computations
      that follow.

	WEIGHT (lb.)	MOM/100
Previous condition	2,690	2,260
Front passenger leaves	− 180	− 153
Rear passenger leaves	− 204	− 247
Old rear in front	+ 204	+ 173
	-----	-----
Totals	2,510	2,033

The new CG is 203,000/2,510 = 80.99. Subtracting the new CG from the old yields approximately 3 in.

19. 4--Weight is reduced by 210 lb. and MOM/100 is reduced by 158. The new weight is 2,680 lb. and the new MOM/100 is 2,294. Looking at the chart, the new CG is aft of the limits shown.

20. 1--See answer 3.

21. 4--See answer 3.

22. 2--CG changes anytime weight is shifted. If weight moves back, the CG moves back.

23. 4--Moment = arm x weight. In this case, moment = 80 x 40, or 3,200 lb.-in.

24. 2--This defines the center of gravity (CG).

25. 1--This defines the CG envelope, a range of values for relating weights and balances.

26. 4--As you may recall from Chapter 3, increasing the angle of bank increases stalling speed. Operation with a high gross weight or a forward CG destroys lift, which in turn increases stalling speed. A forward CG also makes it more difficult to recover from a stall.

27. 1--The first thing to figure is payload and empty condition.

6/PERFORMANCE: MEASURING AN AIRPLANE'S CAPABILITIES

MAIN POINTS

1.   The **demonstrated** capability of an airplane appears in the Performance Section of the Pilot's Operating Handbook (POH). These data were obtained with new airplanes flown under favorable conditions by experienced pilots.

2.   Atmospheric conditions influence both flight instruments and flight characteristics. **Atmospheric pressure** decreases as altitude increases due to the reduction of the density of the air. A 1000-ft. change in altitude corresponds to about a 1-in. change in mercury. Air temperature also decreases as altitude increases. The reference for measuring air properties is the **International Standard Atmosphere (ISA)**. Standard sea level conditions are 59° F (15° C) and 29.92 in. of mercury. When you set your altimeter to 29.92 and read the elevation from the altimeter, you have **pressure altitude**, which is the altitude above a pressure level of 29.92. The higher the barometric pressure, the lower the pressure altitude.

3.   **Density altitude** reflects three variables: pressure, temperature, and humidity. Warm temperatures make air expand and thus produce a higher density altitude, as do actual increases in altitude. Water vapor displacing air molecules also makes air less dense. A cold, dry day at sea level would have a low density altitude, while a warm, humid day in Mexico City (field elevation over 5,000 ft. MSL) would have a high density altitude. Pressure altitude is necessary to compute exact density altitude. Figure 6.2 shows a chart relating outside air temperature (OAT) and pressure altitude to density altitude. The solid diagonal line from upper left to lower right represents ISA values.

4.   Atmospheric factors affect takeoff performance. Less air density means longer takeoff runs and slower climb rates since there is less thrust and less lift. The pilot must calculate these effects of density altitude on the takeoff run.

Another atmospheric variable is the wind. Headwinds decrease
takeoff distances and tailwinds increase them. Airplane and
runway characteristics also affect takeoff performance. As
mentioned in an earlier chapter, more lift is required as weight
is increased and when the plane is loaded toward the forward CG
limit; thus, both increase takeoff distance. Flaps, when
partially extended, may increase lift and may be recommended for
some airplanes for departing from soft fields to decrease the
takeoff roll. The slope of the runway will also affect takeoff
distance. A downhill slope decreases it and an uphill slope
increases it. Rain, snow, and ice may also affect the takeoff
distance, as will your technique as a pilot.

     5.   The indicated airspeed for takeoffs and landings remains
the same regardless of changes in pressure and/or altitude. For
the purpose of calculating takeoff performance, we will use
indicated airspeed (uncorrected for errors in the instrument
itself).   First, determine pressure altitude, either by setting
the altimeter at 29.92 and reading it directly from the
altimeter, or by computing it from the pressure and altitude data
available. For example, in Lawrence, Kansas (field elevation 852
ft.), an atmospheric condition of 30.06 would have a pressure
altitude of 852 + ([-30.06 + 29.92] x 1,000) = 852 - 140, or 712
ft. Another way to determine pressure altitude is to look it up
on a pressure altitude correction chart. Next, you need to
calculate the effect of the wind. Runways are numbered according
to their magnetic direction; and, for the purposes of takeoff and
landing, surface wind directions are given as coming from a
magnetic heading. Note: Wind directions are always given as a
true direction, except for purposes of takeoffs and landings.
Using a wind components chart (see Figure 6.1), one finds the
angle between the wind direction and flight path and the wind
velocity to determine headwind and crosswind components.
Finally, using a takeoff distance chart such as that shown in
Figure 6.3, one integrates the effects of density altitude, gross
weight, wind components, and obstacle heights to determine the
takeoff distance.

     6.   There are two airspeeds you must know: Vx, the best
angle-of-climb, which yields the greatest altitude per distance
traveled; and Vy, the best rate-of-climb, which gives you the
greatest increase in altitude per unit of time. There are
performance charts to help calculate climb performance. Once
obstacles are cleared, use the recommended cruise climb speed to
help prevent engine overheating.

     7.   Cruise speed depends upon your objective: maximum
speed, maximum range, or maximum endurance (best ratio of fuel
burned per unit of time). Fuel mixture should be leaned as
recommended in the POH to enhance engine performance and to
realize the best fuel consumption. Once aloft, winds remain a
factor whether headwind, tailwind, or crosswind. Actual speed
through the air is called true airspeed (TAS); it is near
indicated airspeed (IAS) at low altitudes but higher than IAS at
high altitudes. Once again, there are graphs and charts to help

you determine true airspeed at various power settings and density altitudes. There are also graphs and charts that enable you to calculate range and endurance under different pressure altitudes and power settings.

   8.   Fuel efficiency can be improved by keeping your airplane clean, loading it lightly, loading it near (but not beyond) the aft CG limit, keeping ground operations to a minimum, climbing directly on course, leaning the engine as recommended, and using shallow glide paths.

   9.   One key to effective landings is to establish the proper final approach airspeed and attitude early. The use of flaps, for example, requires considerable planning because the airplane responds more sluggishly when flaps are extended. A typical landing chart is shown in Figure 6.5.

KEY TERMS AND CONCEPTS   *due monday*

   Match each term or concept (1-20) with the appropriate description (A-T) below. Each item has only one match.

___ 1. humidity
___ 2. short field
___ 3. best rate-of-climb
___ 4. atmospheric pressure
___ 5. tailwind
___ 6. true airspeed
___ 7. crabbing
___ 8. crosswind
___ 9. pressure altitude

___ 10. indicated airspeed
___ 11. abscissa
___ 12. maximum endurance
___ 13. ISA sea level
___ 14. ordinate
___ 15. magnetic north
___ 16. density altitude
___ 17. maximum range
___ 18. best angle-of-climb

A. horizontal axis on a graph
B. actual speed through the air
C. best ratio of fuel consumed per unit of time
D. reference for surface wind directions given by a control tower
E. wind condition that will increase the landing distance
F. amount of water vapor the air contains
G. pressure exerted by the air
H. vertical axis on a graph
I. airspeed that produces the greatest altitude gain per unit of time
J. wind from any other direction but the nose or tail
K. flying with the nose at an angle to the ground track
L. altitude corrected for non-standard air temperature and pressure
M. dry air, 59° F at a barometer setting of 29.92
N. airspeed that produces the greatest altitude gain for the distance traveled
O. airspeed uncorrected for density altitude or instrument error
P. best ratio of nautical miles per gallon of fuel consumed
Q. limited length runway with an obstacle at the departure end
R. altitude read by setting the altimeter at 29.92

DISCUSSION QUESTIONS AND EXERCISES   1-9

1. Define each of the following terms:

    a. atmospheric pressure

    b. International Standard Atmosphere (ISA)

    c. pressure altitude

    d. density altitude

2. State how each of the following affects density altitude:

    a. air pressure

    b. temperature

    c. humidity

3. How and why does density altitude affect the takeoff run?

4. Briefly describe how and why each of the following affects the takeoff run:

    a. headwind

    b. tailwind

    c. gross weight

    d. runway gradient

    e. CG

5.   The wind components
chart shown in Figure 6.1 has
several uses. First, for any
given wind report, you can use
it to determine the magnitude
of the crosswind component.
Second, you can determine the
maximum velocity for a given
crosswind angle. Finally, you
can use it to determine which
runway would be best suited for
a takeoff or landing without
exceeding the maximum crosswind
component. Use Figure 6.1 to
answer the following questions.

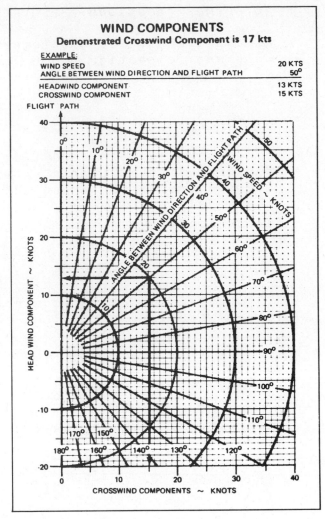

Figure 6.1

    a.   What is the crosswind
    component for a landing on
    Rwy 18 if the tower
    reports the wind at 220°
    at 30 kts.?

    b.   What is the headwind
    component for a landing on
    Rwy 18 if the tower
    reports the wind at 220°
    at 30 kts.?

    c.   Determine the maximum
    wind velocity for a 45°
    crosswind if the maximum
    crosswind component for
    your airplane is 25 kts.?

    d.   What is the maximum wind velocity for a 30° crosswind if
    the maximum crosswind component for an airplane is 12 kts.?

    e.   With a reported wind from the north at 20 kts., which
    runway (6, 14, 24, or 32) is appropriate for an airplane
    with a 13-kt. maximum crosswind component?

    f.   With a reported wind from the west at 25 kts., which
    runway (6, 14, 24, or 32) is appropriate for an airplane
    with a 13-kt. maximum crosswind component?

6.   To answer questions a-b, refer to Figure 6.2.

    a. If the outside air temperature is 90° F in Denver,
    Colorado (elevation 5,330 ft.), and the altimeter setting is
    30.30, what is the density altitude?

    b. Assume you are departing Rapid City (South Dakota)
    Regional Airport (field elevation 3,182 ft.), where the air
    temperature is 85° F and the altimeter is 29.60. What is
    the density altitude?

7.   When an airport tower controller reports that "the wind is
120 at 15," exactly what is the controller saying?

8.   Define **indicated airspeed**. How is it different from **true
airspeed**? How does the relationship between the two change as
altitude increases?

9.   What is the difference between the best angle-of-climb
airspeed and the best rate-of-climb speed?

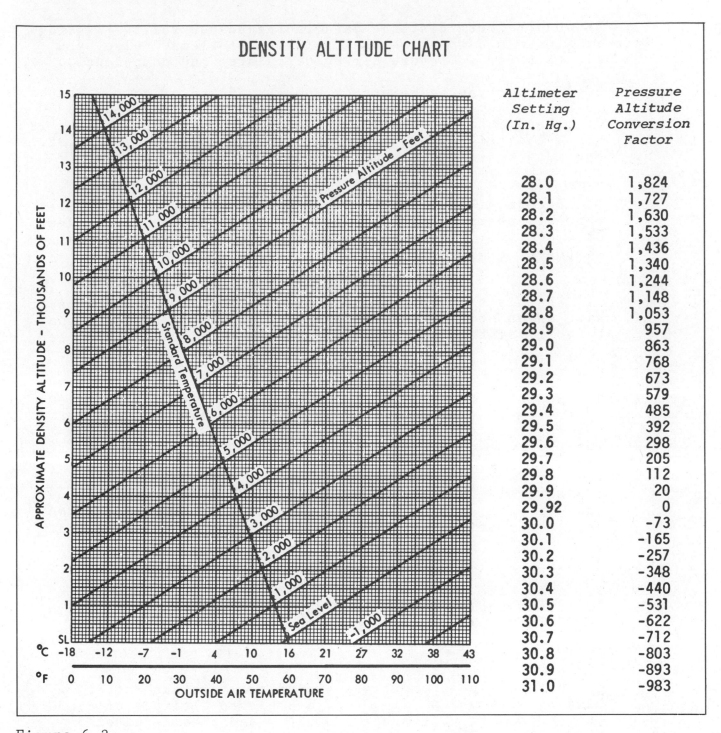

Figure 6.2

10. Use the takeoff distance data in Figure 6.3 and the landing data in Figure 6.4 to solve the following problems. Also, you may need to refer to Figure 6.1 to calculate wind components.

a. Suppose your airplane (2,800 lb. gross weight) is about to depart Rwy 31 at South Lake Tahoe (field elevation 4,870 ft., hard surface runway). Winds are reported at 340° at 18 kts., the temperature is 22° C, and the barometer is 29.32. How much runway will you need to get off the ground?

b. Determine that total takeoff distance necessary to clear a 50-ft. obstacle with a standard OAT, a pressure altitude of 4,000 ft., a takeoff weight of 2,800 lb., and a headwind component of 20 kts.

c. What is the approximate ground roll distance for takeoff with an OAT of 100° F, a pressure altitude of 2,000 ft., a takeoff weight of 2,950 lb., and calm wind conditions?

d. What is the total distance for a takeoff to clear a 50-ft. obstacle with an OAT of 59° F, a pressure altitude of sea level, a takeoff weight of 2,700 lb., and a tailwind of 10 kts.?

e. Determine the approximate ground roll distance for takeoff with an OAT of 90° F, a pressure altitude of 2,000 ft., a takeoff weight of 2,500 lb., and a 20-kt. headwind.

f. Determine the total distance to land the airplane over a 50-ft. obstacle with a standard OAT, a pressure altitude of 2,000 ft., a landing weight of 2,500 lb., and calm winds.

g. What is the approximate ground roll distance after landing with an OAT of 90° F, a pressure altitude of 4,000 ft., a landing weight of 2,800 lb., and a tailwind of 5 kts.?

h. What is the total landing distance over a 50-ft. obstacle with an OAT of 85° F, pressure altitude of 3,000 ft., landing weight of 2,850 lb., and a headwind of 25 kts.?

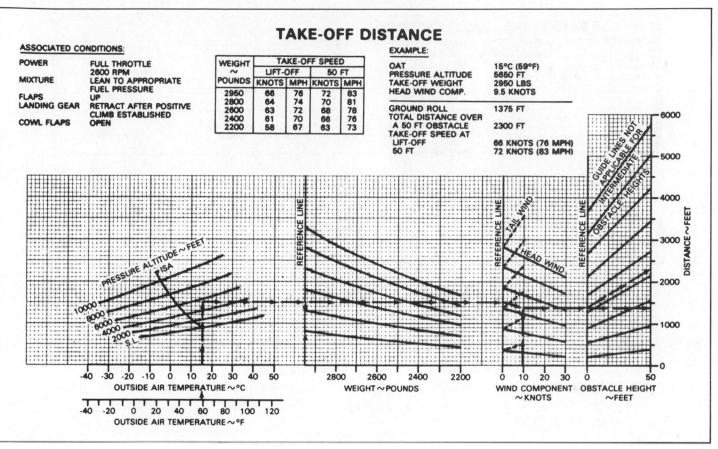

Figure 6.3

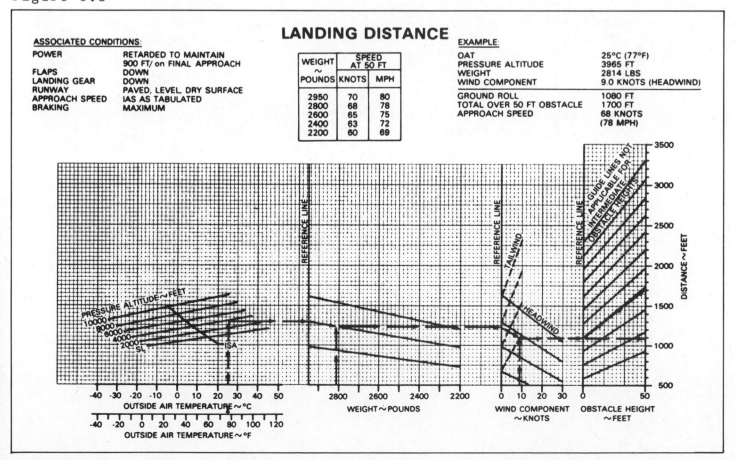

Figure 6.4

—LANDING DISTANCE—   FLAPS LOWERED TO 40° - POWER OFF
HARD SURFACE RUNWAY - ZERO WIND

GROSS WEIGHT LBS.	APPROACH SPEED, IAS, MPH	AT SEA LEVEL & 59° F.		AT 2500 FT. & 50° F.		AT 5000 FT. & 41° F.		AT 7500 FT. & 32° F.	
		GROUND ROLL	TOTAL TO CLEAR 50 FT. OBS	GROUND ROLL	TOTAL TO CLEAR 50 FT. OBS	GROUND ROLL	TOTAL TO CLEAR 50 FT. OBS	GROUND ROLL	TOTAL TO CLEAR 50 FT. OBS
1600	60	445	1075	470	1135	495	1195	520	1255

NOTES: 1. Decrease the distances shown by 10% for each 4 knots of headwind.
2. Increase the distance by 10% for each 60°F. temperature increase above standard.
3. For operation on a dry, grass runway, increase distances (both "ground roll" and "total to clear 50 ft. obstacle") by 20% of the "total to clear 50 ft. obstacle" figure.

Figure 6.5

11. Use the landing-distance data in Figure 6.5 to complete the following problems. You may also have to refer to Figure 6.1 to compute wind components.

a. Determine the landing ground roll with a pressure altitude at sea level, standard temperature, and an 8-kt. headwind.

b. What is the total landing distance required to clear a 50-ft. obstacle with a pressure altitude of 7,500 ft., an 8-kt. headwind, on a dry grass runway under standard temperature conditions?

c. Suppose you have been cleared to land on Rwy 7. The wind is reported at 100 at 28. What is the approximate landing roll distance under standard temperature conditions with a pressure altitude of 3,750 ft.?

d. Determine the total landing distance on a dry grass runway with a 50-ft. obstacle? The pressure altitude is 5,000 ft., the wind is calm, and the temperature is 101° F.

e. Determine the total distance required to land over a 50-ft. obstacle with a pressure altitude of 5,000 ft. and standard temperature conditions. You will be landing on Rwy 35. The wind is reported "010 at 34."

f. What is the approximate landing ground roll distance with a pressure altitude of 1,250 ft. under standard temperature conditions? You will be landing on Rwy 7. The winds are reported "360 at 24."

CRUISE PERFORMANCE
STANDARD DAY
AVERAGE CRUISE WEIGHT = 1600 POUNDS

ALTITUDE FEET	THROTTLE SETTING RPM	FUEL FLOW GPH	IAS KNOTS	TAS KNOTS
2500	2700	8.0	101	105
	2500	6.4	94	97
	2400	5.7	90	93
	2300	5.2	85	88
3500	2700	7.8	100	105
	2500	6.3	92	97
	2400	5.7	88	93
	2300	5.2	84	88
4500	2700	7.7	99	105
	2500	6.3	91	97
	2400	5.6	87	93
	2300	5.1	82	88
5500	2700	7.6	97	105
	2500	6.2	89	97
	2400	5.5	85	92
	2300	5.0	81	87
6500	2700	7.4	96	105
	2500	6.1	88	97
	2400	5.4	84	92
	2300	5.0	79	87
7500	2500	6.0	86	96
	2400	5.3	82	91
	2300	4.9	77	86
8500	2500	5.8	85	96
	2400	5.3	80	91
	2300	4.9	76	85
9500	2500	5.7	83	95
	2400	5.2	79	90
	2300	4.8	74	85
10500	2500	5.6	81	95
	2400	5.1	77	90
	2300	4.7	72	84
11500	2500	5.5	80	94
	2400	5.0	75	89
	2300	4.7	70	82

Cruise performance is based on best power mixture. Lean to maximum rpm for best performance.

Figure 6.6

12.   Refer to Figure 6.6. Suppose you are cruising at 6,500 ft. under standard temperature conditions. You have the throttle set at 2,500 RPM. What is your fuel flow, your indicated airspeed, and your true airspeed?

13. Refer to Figure 6.7. What is your maximum range at 2,400 RPM with a pressure altitude of 8,000 ft.?

14. Refer to Figure 6.8. What is your maximum endurance at 2,400 RPM with a pressure altitude of 8,000 ft.?

15. Name five things you can do as a pilot to conserve fuel.

16.   What is one of the most important factors that contributes to a safe and efficient landing?

17. T F When landing at an airport with a high density altitude, a pilot should add extra speed to the final-approach indicated airspeed to compensate for the thinner air.

Explain your answer.

18. An airplane takes off into a direct 20-kt. headwind. The stalling speed of the airplane is 50 kts., and its indicated airspeed in a climb is 65 kts. If the pilot turns the airplane downwind and the 20-kt. headwind becomes a 20-kt. tailwind, will the airplane stall? Explain why or why not.

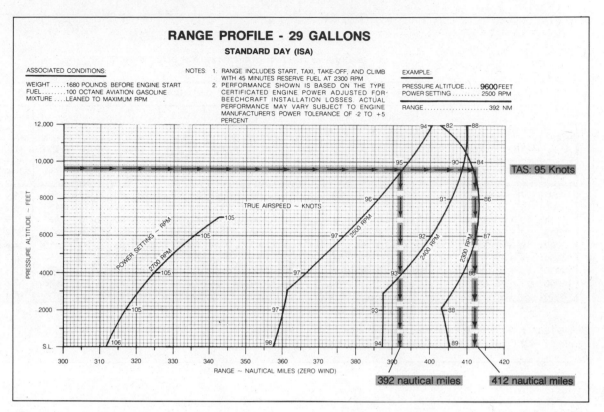

Figure 6.7

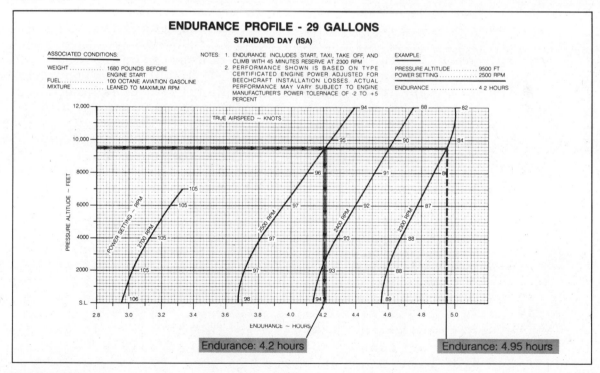

Figure 6.8

CRUISE POWER SETTINGS

75% MAXIMUM CONTINUOUS POWER (OR FULL THROTTLE)
2800 LBS

PRESS ALT.	ISA −20°C (−36°F)							STANDARD DAY (ISA)							ISA +20°C (+36°F)						
	IOAT		ENGINE SPEED	MAN. PRESS	FUEL FLOW PER ENGINE		TAS		IOAT		ENGINE SPEED	MAN. PRESS	FUEL FLOW PER ENGINE		TAS		IOAT		ENGINE SPEED	MAN. PRESS	FUEL FLOW PER ENGINE		TAS	
FEET	°F	°C	RPM	IN HG	PSI	GPH	KTS	MPH	°F	°C	RPM	IN HG	PSI	GPH	KTS	MPH	°F	°C	RPM	IN HG	PSI	GPH	KTS	MPH
SL	27	−3	2450	23.1	8.2	13.6	156	180	63	17	2450	23.8	8.2	13.6	159	183	99	37	2450	24.4	8.2	13.6	162	186
2000	19	−7	2450	22.8	8.2	13.6	158	182	55	13	2450	23.5	8.2	13.6	162	186	93	34	2450	24.2	8.2	13.6	165	190
4000	12	−11	2450	22.6	8.2	13.6	161	185	50	10	2450	23.3	8.2	13.6	165	190	86	30	2450	23.9	8.2	13.6	168	193
6000	7	−14	2450	22.3	8.2	13.6	164	189	43	6	2450	23.0	8.2	13.6	167	192	78	26	2450	23.4	8.0	13.4	170	196
8000	0	−18	2450	21.8	8.0	13.3	166	191	36	2	2450	21.8	7.8	12.8	167	192	72	22	2450	21.8	7.2	12.3	168	193
10000	−8	−22	2450	20.3	7.2	12.3	164	189	28	−2	2450	20.3	6.8	11.8	165	190	64	18	2450	20.3	6.6	11.4	166	191
12000	−15	−26	2450	18.8	6.4	11.3	162	186	21	−6	2450	18.8	6.2	10.9	163	188	57	14	2450	18.8	5.9	10.6	163	188
14000	−22	−30	2450	17.4	5.9	10.5	159	183	14	−10	2450	17.4	5.6	10.1	160	184	50	10	2450	17.4	5.3	9.8	160	184
16000	−29	−34	2450	16.1	5.3	9.7	156	180	7	−14	2450	16.1	5.1	9.4	156	180	43	6	2450	16.1	4.9	9.1	155	178

NOTES: 1. Full throttle manifold pressure settings are approximate.
2. Shaded area represents operation with full throttle.

Figure 6.9

19. Refer to Figure 6.9 above to answer the following questions.

a. What is the expected fuel consumption for a 1,000-mi. flight at a pressure altitude of 8,000 ft. with calm winds and a temperature of −18° C? Manifold pressure is 21.8" Hg.

b. What is the expected fuel consumption for a 500-mi. flight at a pressure altitude of 4,000 ft. with calm winds and a temperature of +30° C? Manifold pressure is 23.9" Hg.

c. What fuel flow should a pilot expect at 8,500 ft. on a standard day with 75 percent maximum continuous power?

d. Determine the approximate fuel flow at 75 percent maximum continous power at 6,500 ft. with a temperature of 36° F higher than standard.

REVIEW QUESTIONS

1.   (FAA 1010) What is the crosswind component for a landing on Rwy 18 if the tower reports the wind 220° at 30 kts.? (See Figure 6.10.)

1--19 kts.   3--30 kts.
2--23 kts.   4--34 kts.

2.   (FAA 1011) What is the headwind component for a landing on Rwy 18 if the tower reports the wind at 220° at 30 kts.? (See Figure 6.10.)

1--19 kts.   3--30 kts.
2--23 kts.   4--34 kts.

3.   (FAA 1012) Determine the maximum wind velocity for a 45° crosswind if the maximum crosswind component for the airplane is 25 kts. (See Figure 6.10.)

1--18 kts.   3--29 kts.
2--25 kts.   4--35 kts.

4.   (FAA 1013) What is the maximum wind velocity for a 30° crosswind if the maximum crosswind component for the airplane is 12 kts.? (See Figure 6.10.)

1--13 kts.   3--21 kts.
2--17 kts.   4--24 kts.

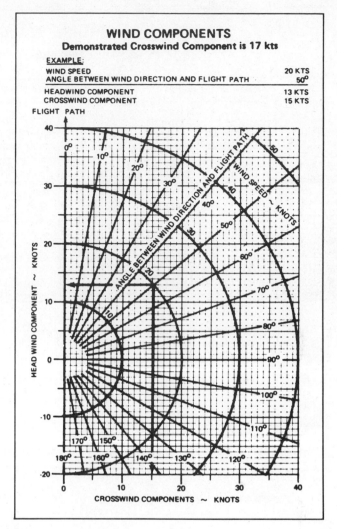

Figure 6.10

5.   (FAA 1014) With a reported wind of north at 20 kts., which runway (6, 14, 24, or 32) is appropriate for an airplane with a 13-kt. maximum crosswind component? (See Figure 6.10.)

1--Rwy 6.   3--Rwy 24.
2--Rwy 14.   4--Rwy 32.

6.   (FAA 1015) With a reported wind of south at 20 kts., which runway (6, 14, 24, or 32) is appropriate for an airplane with a 13-kt. maximum crosswind component? (See Figure 6.10.)

1--Rwy 6.   3--Rwy 24.
2--Rwy 14.   4--Rwy 32.

CRUISE POWER SETTINGS

75% MAXIMUM CONTINUOUS POWER (OR FULL THROTTLE)
2800 LBS

PRESS ALT.	ISA −20°C (−36°F)								STANDARD DAY (ISA)								ISA +20°C (+36°F)							
	IOAT		ENGINE SPEED	MAN. PRESS	FUEL FLOW PER ENGINE		TAS		IOAT		ENGINE SPEED	MAN. PRESS	FUEL FLOW PER ENGINE		TAS		IOAT		ENGINE SPEED	MAN. PRESS	FUEL FLOW PER ENGINE		TAS	
FEET	°F	°C	RPM	IN HG	PSI	GPH	KTS	MPH	°F	°C	RPM	IN HG	PSI	GPH	KTS	MPH	°F	°C	RPM	IN HG	PSI	GPH	KTS	MPH
SL	27	−3	2450	23.1	8.2	13.6	156	180	63	17	2450	23.8	8.2	13.6	159	183	99	37	2450	24.4	8.2	13.6	162	186
2000	19	−7	2450	22.8	8.2	13.6	158	182	55	13	2450	23.5	8.2	13.6	162	186	93	34	2450	24.2	8.2	13.6	165	190
4000	12	−11	2450	22.6	8.2	13.6	161	185	50	10	2450	23.3	8.2	13.6	165	190	86	30	2450	23.9	8.2	13.6	168	193
6000	7	−14	2450	22.3	8.2	13.6	164	189	43	6	2450	23.0	8.2	13.6	167	192	78	26	2450	23.4	8.0	13.4	170	196
8000	0	−18	2450	21.8	8.0	13.3	166	191	36	2	2450	21.8	7.8	12.8	167	192	72	22	2450	21.8	7.2	12.3	168	193
10000	−8	−22	2450	20.3	7.2	12.3	164	189	28	−2	2450	20.3	6.8	11.8	165	190	64	18	2450	20.3	6.5	11.4	166	191
12000	−15	−26	2450	18.8	6.4	11.3	162	186	21	−6	2450	18.8	6.2	10.9	163	186	57	14	2450	18.8	5.9	10.6	163	188
14000	−22	−30	2450	17.4	5.9	10.5	159	183	14	−10	2450	17.4	5.6	10.1	160	184	50	10	2450	17.4	5.3	9.8	160	184
16000	−29	−34	2450	16.1	5.3	9.7	155	180	7	−14	2450	16.1	5.1	9.4	156	180	43	6	2450	16.1	4.9	9.1	155	178

NOTES: 1. Full throttle manifold pressure settings are approximate.
2. Shaded area represents operation with full throttle.

Figure 6.11

7. (FAA 1017) What is the expected fuel consumption for a 1,000-mi. flight under the following conditions? (See Fig. 6.11.)

```
Pressure altitude ................. 8,000 ft.
Temperature ........................   −18° C
Manifold pressure ................. 21.8" Hg
Wind ..............................   Calm
```

1--41.9 gal.   3--71.4 gal.
2--69.6 gal.   4--73.8 gal.

8. (FAA 1018) What is the expected fuel consumption for a 500-mi. flight under the following conditions? (See Figure 6.11.)

```
Pressure altitude ................. 4,000 ft.
Temperature ........................   +30° C
Manifold pressure ................. 23.9" Hg
Wind ..............................   Calm
```

1--35.2 gal.   3--40.1 gal.
2--38.5 gal.   4--43.2 gal.

9. (FAA 1019) What fuel flow should a pilot expect at 8,500 ft. on a standard day with 75 percent maximum continuous power? (See Figure 6.11.)

1--12.05 gal./hr.   3--12.80 gal./hr.
2--12.55 gal./hr.   4--13.05 gal./hr.

10. (FAA 1020) Determine the approximate fuel flow at 75 percent maximum continuous power at 6,500 ft. with a temperature of 36° F higher than standard. (See Figure 6.11.)

1--12.4 gal./hr.   3--13.1 gal./hr.
2--12.7 gal./hr.   4--13.7 gal./hr.

11. (FAA 1031) Determine the total distance to land the airplane under the following conditions. (See Figure 6.12.)

    OAT ....   Std.
    Pressure altitude ....   2,000 ft.
    Weight ....   2,500 lb.
    Wind component ....   Calm
    Obstacle ....   50 ft.

1--850 ft.   2--1,250 ft.   3--1,450 ft.   4--1,700 ft.

12. (FAA 1032) What is the approximate ground roll distance after landing under the following conditions? (See Figure 6.12.)

    OAT ....   90° F
    Pressure altitude ....   4,000 ft.
    Weight ....   2,800 lb.
    Tailwind component ....   5 kts.

1--1,200 ft.   2--1,575 ft.   3--1,725 ft.   4--1,950 ft.

13. (FAA 1033) What is the total landing distance under the following conditions? (See Figure 6.12.)

    OAT ....   85° F
    Pressure altitude ....   3,000 ft.
    Weight ....   2,850 lb.
    Headwind component ....   25 kts.
    Obstacle ....   50 ft.

1--1,300 ft.   2--1,450 ft.   3--1,550 ft.   4--1,725 ft.

14. (FAA 1001) After takeoff, which airspeed would permit the pilot to gain the most altitude in a given period of time?

1--Cruising climb speed.
2--Best rate-of-climb speed.
3--Best angle-of-climb speed.
4--Minimum control speed.

15. (FAA 1002) Which would provide the greatest gain in altitude in the shortest distance during climb after takeoff?

1--Steepest pitch angle.
2--Cruising climb speed.
3--Best rate-of-climb speed.
4--Best angle-of-climb speed.

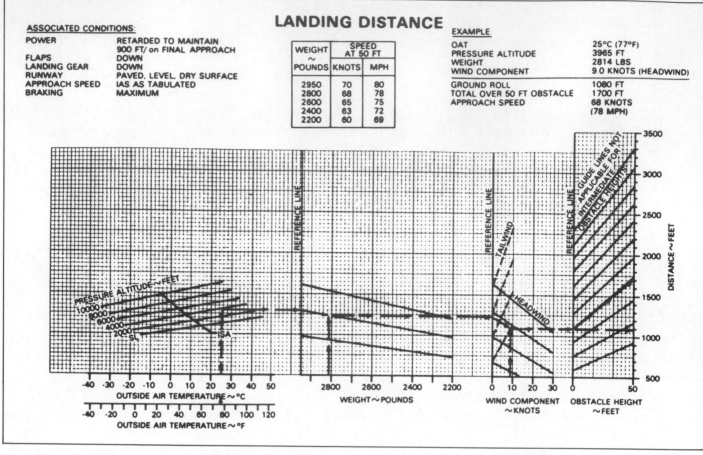

Figure 6.12

16. (FAA 1005) Density altitude is the

1--altitude reference to the standard datum plane.
2--pressure altitude corrected for non-standard temperature.
3--altitude read directly from the altimeter.
4--altitude above the surface.

17. (FAA 1008) What effect does high density altitude have on aircraft performance?

1--It increases engine performance.
2--It reduces an aircraft's climb performance.
3--It will decrease the runway length required for takeoff.
4--Lift increases; light air exerts less force on the airfoils.

18. (FAA 1045) What effect does higher density altitude have on propeller efficiency?

1--Increased efficiency due to less friction on the blades.
2--Reduced efficiency; the propellers exert less force.
3--Reduced efficiency due to the increased force of the thinner air on the propeller.
4--Increased efficiency; the propellers exert more force.

| —LANDING DISTANCE— | | FLAPS LOWERED TO 40° - POWER OFF HARD SURFACE RUNWAY - ZERO WIND | | | | | | | |
| GROSS WEIGHT LBS. | APPROACH SPEED, IAS, MPH | AT SEA LEVEL & 59° F. | | AT 2500 FT. & 50° F. | | AT 5000 FT. & 41° F. | | AT 7500 FT. & 32° F. | |
		GROUND ROLL	TOTAL TO CLEAR 50 FT. OBS	GROUND ROLL	TOTAL TO CLEAR 50 FT. OBS	GROUND ROLL	TOTAL TO CLEAR 50 FT. OBS	GROUND ROLL	TOTAL TO CLEAR 50 FT. OBS
1600	60	445	1075	470	1135	495	1195	520	1255

NOTES: 1. Decrease the distances shown by 10% for each 4 knots of headwind.
2. Increase the distance by 10% for each 60° F. temperature increase above standard.
3. For operation on a dry, grass runway, increase distances (both "ground roll" and "total to clear 50 ft. obstacle") by 20% of the "total to clear 50 ft. obstacle" figure.

Figure 6.13

19. (FAA 1034) With the following conditions, determine the landing ground roll from the Landing Distance Chart. (See Figure 6.13.)

        Pressure altitude .... Sea level
        Headwind .... 8 kts.
        Temperature .... Std.

1--356 ft.   2--401 ft.   3--490 ft.   4--534 ft.

20. (FAA 1035) What is the total landing distance required to clear a 50-ft. obstacle with the following conditions using the Landing Distance Chart? (See Figure 6.13.)

        Pressure altitude .... 7,500 ft.
        Headwind .... 8 kts.
        Temperature .... Std.
        Runway .... Dry grass

1--1,004 ft.   2--1,255 ft.   3--1,506 ft.   4--1,757 ft.

21. (FAA 1036) Determine the approximate landing roll distance from the Landing Distance Chart. (See Figure 6.13.)

        Pressure altitude .... 3,750 ft.
        Headwind .... 24 kts.
        Temperature .... Std.

1--193 ft.   2--338 ft.   3--628 ft.   4--772 ft.

22. (FAA 1037) What is the total landing distance required to clear a 50-ft. obstacle with the following conditions using the Landing Distance Chart? (See Figure 6.13.)

        Pressure altitude .... 5,000 ft.
        Headwind .... Calm
        Temperature .... 101° F
        Runway .... Dry grass

1--837 ft.   2--956 ft.   3--1,076 ft.   4--1,554 ft.

23. (FAA 1038) Determine the total distance required to land over a 50-ft. obstacle from the Landing Distance Chart. (See Figure 6.13.)

        Pressure altitude .... 5,000 ft.
        Headwind .... 32 kts.
        Temperature .... Std.

1--239 ft.   2--1,099 ft.   3--1,291 ft.   4--1,434 ft.

24. (FAA 1039) What is the approximate landing ground roll distance using the Landing Distance Chart? (See Figure 6.13.)

        Pressure altitude .... 1,250 ft.
        Headwind .... 8 kts.
        Temperature .... Std.

1--275 ft.   2--366 ft.   3--470 ft.   4--549 ft.

25. (FAA 1050) Which combination of atmospheric conditions will reduce aircraft takeoff and climb performance?

1--Low temperature, low relative humidity, and low density
   altitude.
2--High temperature, low relative humidity, and low density
   altitude.
3--High temperature, high relative humidity, and high density
   altitude.
4--Low temperature, high relative humidity, and high density
   altitude.

26. (FAA 1729) If the outside air temperature at a given altitude is warmer than standard, the density altitude is

1--lower than pressure altitude, but approximately equal to the
   true altitude.
2--higher than true altitude, but lower than pressure altitude.
3--higher than the pressure altitude.
4--lower than true altitude.

27. (FAA 1730) An aircraft is maintaining a constant pressure altitude and the outside air temperature is warmer than standard for that altitude. What is the density altitude with respect to pressure altitude?

1--Same.   3--Higher.
2--Lower.   4--Impossible to determine.

28. (FAA 1731) What are the standard temperature and pressure values for sea level?

1--15° C and 29.92" Hg.
2--59° C and 1013.2 mb.
3--59° F and 29.92 mb.
4--15° C and 1013.2" Hg.

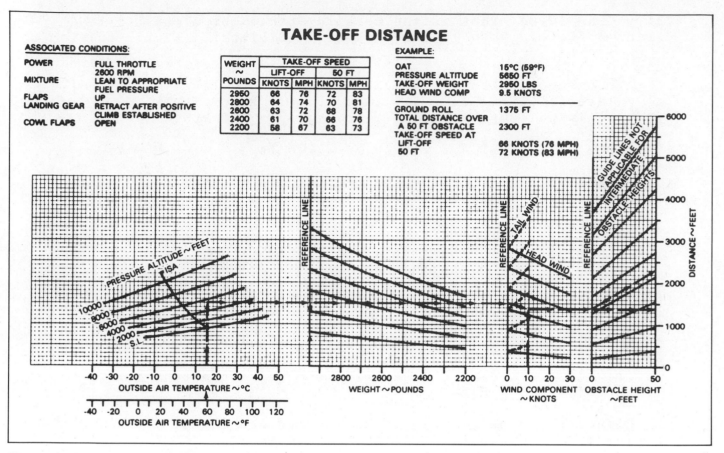

Figure 6.14

29. (FAA 1046) Determine the total distance for a takeoff to
clear a 50-ft. obstacle. (See Figure 6.14.)

        OAT ....   Std.
        Pressure altitude ....   4,000 ft.
        Takeoff weight ....   2,800 lb.
        Headwind component ....   20 kts.

1--1,250 ft.   3--1,750 ft.
2--1,500 ft.   4--1,900 ft.

30. (FAA 1047) What is the approximate ground roll distance for
takeoff under the following conditions? (See Figure 6.14.)

        OAT ....   100° F
        Pressure altitude ....   2,000 ft.
        Takeoff weight ....   2,950 lb.
        Headwind component ....   Calm

1--1,150 ft.   3--1,800 ft.
2--1,300 ft.   4--2,000 ft.

31. **(FAA 1048)** What is the total distance for a takeoff to clear a 50-ft. obstacle under these conditions? (Figure 6.14.)

```
        OAT ...................................     Std.
        Pressure altitude ..................  Sea level
        Takeoff weight .....................  2,700 lb.
        Tailwind component ................    10 kts.
```

1--1,000 ft.   2--1,400 ft.   3--1,700 ft.   4--1,900 ft.

32. **(FAA 1049)** Determine the approximate ground roll distance for takeoff under the following conditions. (See Figure 6.14.)

```
        OAT ...............................      90° F
        Pressure altitude ................   2,000 ft.
        Takeoff weight ...................   2,500 lb.
        Headwind component ...............     20 kts.
```

1--650 ft.   2--800 ft.   3--1,000 ft.   4--1,250 ft.

33. **(FAA 1732)** If, without adjusting the altimeter setting, a flight is made from an area of high pressure into an area of lower pressure and a constant altitude is maintained, the altimeter would indicate

1--higher than the actual altitude above sea level.
2--lower than the actual altitude above sea level.
3--the actual altitude above sea level.
4--the actual altitude above ground level.

34. **(FAA 1740)** Under which condition(s) will pressure altitude be equal to true altitude?

1--When the atmospheric pressure is 29.92" Hg.
2--When standard atmospheric conditions exist.
3--When indicated altitude is equal to pressure altitude.
4--When outside air temperature is standard for that altitude.

35. **(FAA 1741)** Under what condition will true altitude be lower than indicated altitude with an altimeter setting of 29.92 even with an accurate altimeter?

1--In colder than standard air temperature.
2--In warmer than standard air temperature.
3--When density altitude is higher than indicated altitude.
4--Under higher than standard pressure at standard air
    temperature.

36. **(FAA 1742)** Under what condition are pressure altitude and density altitude the same value?

1--At sea level, when the temperature is 0° F.
2--When the altimeter has no installation error.
3--When the altimeter setting is 29.92.
4--At standard temperature.

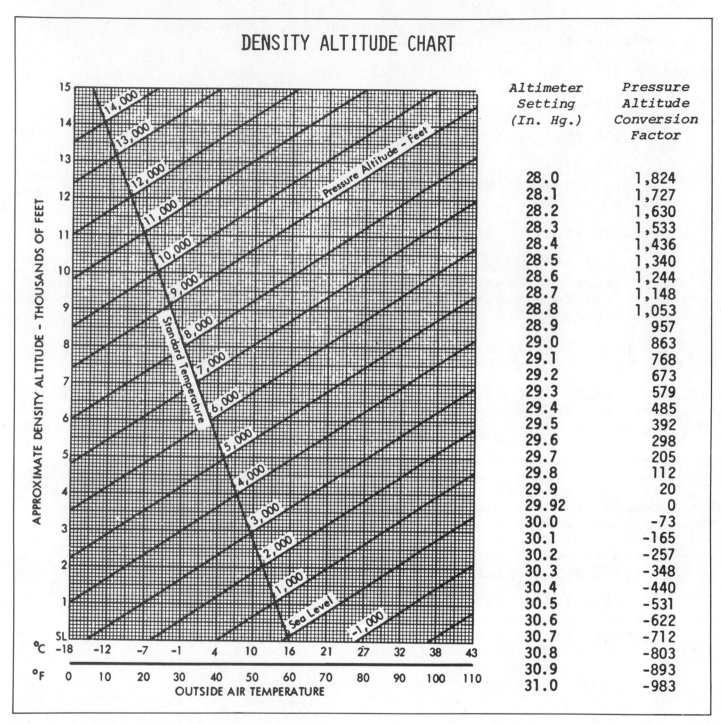

Figure 6.15

37. **(FAA 1733)** Determine the density altitude for the following conditions: (See Figure 6.15.)

        Altimeter setting 30.35
        Rwy temperature ....   +25° F
        Airport elevation .... 3,894 ft.

1--2,900 ft.   2--3,500 ft.   3--3,800 ft.   4--2,100 ft.

38. **(FAA 1734)** Determine the pressure altitude at an airport that is 3,563 ft. MSL with an altimeter setting of 29.96. (See Figure 6.15.)

1--3,527 ft.   2--3,556 ft.   3--3,639 ft.   4--3,507 ft.

39. **(FAA 1735)** What is the effect of a temperature decrease and a pressure increase on the density altitude from 90° F and 1,250 ft. pressure altitude to 60° F and 1,750 ft. pressure altitude? (See Figure 6.15.)

1--500-ft. increase.   3--1,300-ft. increase.
2--1,300-ft. decrease.   4--500-ft. decrease.

40. **(FAA 1736)** Determine the pressure altitude at an airport that is 1,386 ft. MSL with an altimeter setting of 29.97. (See Figure 6.15.)

1--1,451 ft.   2--1,562 ft.   3--1,684 ft.   4--1,341 ft.

41. **(FAA 1737 DI)** What is the effect of a temperature increase from 30° to 50° F on the density altitude if the pressure altitude remains at 3,000 ft.? (See Figure 6.15.)

1--900-ft. increase.   3--1,500-ft. increase.
2--1,100-ft. increase.   4--1,300-ft. decrease.

42. **(FAA 1738)** What is the effect of a temperature increase from 25° to 50° F on the density altitude if the pressure altitude remains at 5,000 ft.? (See Figure 6.15.)

1--1,650-ft. increase.   3--1,200-ft. increase.
2--1,400-ft. increase.   4--1,000-ft. increase.

43. **(FAA 1739)** Determine the density altitude for the following conditions. (See Figure 6.15.)

        Altimeter setting 29.25
        Rwy temperature ....   +81° F
        Airport elevation .... 5,250 ft.

1--4,600 ft.   2--5,877 ft.   3--8,400 ft.   4--7,700 ft.

44. (FAA 1743) Altimeter setting is the value to which the
scale of the pressure altimeter is set so the altimeter indicates

1--density altitude at sea level.
2--pressure altitude at sea level.
3--true altitude at field elevation.
4--pressure altitude at field elevation.

45. (FAA 1744) If a pilot changes the altimeter setting from
30.11 to 29.96, what is the approximate change in indication?

1--Altimeter will indicate 15 ft. higher.
2--Altimeter will indicate 15 ft. lower.
3--Altimeter will indicate 150 ft. lower.
4--Altimeter will indicate 150 ft. higher.

46. (FAA 1745) If a flight is made from an area of low pressure
into an area of high pressure without the altimeter setting being
adjusted and a constant indicated altitude is maintained, the
altimeter would indicate

1--the actual altitude above sea level.
2--higher than the actual altitude above sea level.
3--lower than the actual altitude above sea level.
4--the actual altitude above ground level.

47. (FAA 1746) If a flight is made from an area of high
pressure into an area of lower pressure without the altimeter
setting being adjusted and a constant indicated altitude is
maintained, the altimeter would indicate

1--lower than the actual altitude above sea level.
2--higher than the actual altitude above sea level.
3--the actual altitude above ground level.
4--the actual altitude above sea level.

48. (FAA 1747) What is the primary reason for computing density
altitude?

1--To determine pressure altitude.
2--To determine aircraft performance.
3--To establish FL's (flight levels) above 18,000 ft. MSL.
4--To ensure safe cruising altitude over mountainous terrain.

49. (FAA 1748) Which factor would tend to increase the density
altitude at a given airport?

1--Increasing barometric pressure.
2--Increasing ambient temperature.
3--Decreasing relative humidity.
4--Decreasing ambient temperature.

50. (FAA 1749) What condition would cause the altimeter to indicate a lower altitude than actually flown (true altitude)?

1--Air temperature lower than standard.
2--Atmospheric pressure lower than standard.
3--Pressure altitude the same as indicated altitude.
4--Air temperature warmer than standard.

51. (FAA 1750) If the altimeter indicates 1,380 ft. when set to 29.92, what is the pressure altitude?

1--1,280 ft.   3--1,480 ft.
2--1,380 ft.   4--1,580 ft.

52. Your altimeter reading when the Kollsman window is set to 29.92 in. of mercury is a measure of

1--atmospheric pressure.
2--density altitude.
3--International Standard Atmosphere (ISA).
4--pressure altitude.

53. Which of the following statements is correct?

1--Density altitude affects takeoff performance, but not landing
    performance.
2--Density altitude typically increases as altitude increases.
3--Extra groundspeed should be used to land at an airport with a
    high density altitude to compensate for the thinner air.
4--All of the above are correct.

54. Refer to Figure 6.15. Assuming an airport elevation of 3,165 feet, an outside air temperature of 93° F, and an altimeter setting of 30.10, what is the density altitude?

1--3,000 ft.   3--5,800 ft.
2--3,850 ft.   4--6,800 ft.

55. Given an airport elevation of 3,700 ft., an altimeter setting of 29.60, and an OAT of 75° F, determine the density altitude using Figure 6.15.

1--1,000 ft.   3--5,200 ft.
2--3,950 ft.   4--5,950 ft.

56. Which of the following conditions will increase the length of the takeoff roll?

1--Using a runway with a downhill gradient.
2--A headwind.
3--A soft field.
4--Increasing gross weight.

57. Actual speed through the air is defined as _____ airspeed.

1--calibrated   3--relative
2--indicated   4--true

58. If the barometer is 30.32 and you are at 120 MSL, what is
the pressure altitude?

1--280 ft.   3--100 ft.
2--80 ft.   4--520 ft.

59. Refer to Figure 6.6. Given a barometric setting of 29.22
and an indicated altitude of 5,800 ft., what is your true
airspeed and your fuel flow per hour if you maintain 2,500 RPM?

1--92 kts.; 5.4 gal./hr.
2--92 kts.; 6.1 gal./hr.
3--97 kts.; 6.1 gal./hr.
4--97 kts.; 5.4 gal./hr.

60. Refer to Figures 6.7 and 6.8. Assume a pressure altitude of
8,000 ft. and 2,400 RPM. What is your maximum range and maximum
endurance under these conditions?

1--385 NM; 4.1 hr.   3--425 NM; 4.7 hr.
2--405 NM; 4.5 hr.   4--450 NM; 4.1 hr.

61. Refer to Figure 6.7. How much could you extend your range
at a pressure altitude of 6,400 ft. if you were to reduce the
throttle from 2,400 RPM to 2,300 RPM?

1--6 NM   3--11 NM
2--8 NM   4--15 NM

ANSWERS

Key Terms and Concepts

    1. F   6. B   11. A   16. L
    2. Q   7. K   12. C   17. P
    3. I   8. J   13. M   18. N
    4. G   9. R   14. H
    5. E   10. O   15. D

Discussion Questions and Exercises

    5. a. 19 kts. The angle between the runway and wind is 40°.
        Read out the 40° line until you intersect the curved line
        that represents a wind speed of 30 kts. Now read down to
        the abscissa (crosswind component).

     b. 23 kts. The angle between the runway and wind is 40°. Read out the 40° line until you intersect the **curved line** that represents a wind speed of 30 kts. Now read **across** to the ordinate (headwind component).

     c. 35 kts. This will require some estimation on your part. First, find 25 kts. on the crosswind component scale. Go up until you intersect a point equidistant between 40° and 50° on the line that represents the angle between wind direction and flight path. Now you must **imagine** a curved line somewhere between 30 and 40 kts. As it so happens, the point is almost exactly equidistant between the two lines, or 35 kts.

     d. 24 kts. Use the same procedure outlined for answer 5c.

     e. Rwy 32 is the most appropriate, but there are two correct answers to this problem. First, logic tells us that we should take off **into** the wind, unless there is some extenuating circumstance that would dictate otherwise. The runway with a heading closest to north is Rwy 32; and, by checking the wind components in Figure 6.1 for a 40° angle at a wind velocity of 20 kts., we are barely within the airplane's crosswind limits. But, let's also look at Rwy 14. This would involve a departing tailwind; but, with a 40° crosswind (the difference between Rwy 14 and a tailwind at 180°), the airplane is still within limits.

     f. Rwy 24 would be the most appropriate. An analysis of Figure 6.1 also reveals that the airplane would be within its crosswind limits.

6.   a. 4,982 ft. Find the altimeter setting of 30.3 on the chart in Figure 6.2. Read across to find the altitude correction (-348 ft.). Subtract 348 from the field elevation of 5,330 ft. to arrive at 4,982 ft.

     b. 3,480 ft. Find the altimeter setting of 29.6 on the chart in Figure 6.2. Read across to find the altitude correction (+298 ft.). Add 298 to the field elevation.

7.   The wind is coming **from** 120° (magnetic) at 15 kts.

10. a. 1,200 ft. This is not an easy question. Nor is the chart particularly easy to read. I strongly suggest that you use a straightedge to help you read the chart accurately. If your answers are a little different from mine, it's probably due to reading errors. Differences of 100 ft. or so are common. Now to the details. First, you need to determine the pressure altitude. Go to Figure 6.2 and make the altitude correct for 29.30 (579 ft.), which must be added to the field elevation of 4,870 ft. The resulting pressure altitude is 5,449 ft. Second, you must use Figure 6.1 to calculate the headwind component. The angle between the runway and the wind is 30°. The headwind component is 14°. Third, refer to Figure 6.3. Read up the chart from 22° C until you intersect a pressure altitude corresponding to 5,449 ft. (estimate it). Now read **over** to the reference line. Read down, parallel to the shaded lines, until you intersect 2,800 lb. Read over to the next reference line. Again, staying parallel to the solid headwind

lines, read diagonally until you intersect 14 kts. Now
read over to the last reference line. If there were an
obstacle at the end of the runway, you would have to read
up, parallel to the diagonal lines. However, there is no
obstacle, so read directly across to the distance axis.

    b. 1,500 ft. Use the same procedure to read Figure 6.3 as
described in answer 10a.

    c. 1,300 ft. Use the same procedure to read Figure 6.3 as
described in answer 10a.

    d. 1,700 ft. Use the same procedure to read Figure 6.3 as
described in answer 10a.

    e. 650 ft. Use the same procedure to read Figure 6.3 as
described in answer 10a.

    f. 1,450 ft. Now it is time to use Figure 6.4. The
procedure for reading it is the same as the procedure you
just practiced on Figure 6.3. To review, read up from
the temperature scale until you intersect the pressure
altitude line. Read over to the reference line.
Parallel the weight line until you intersect the airplane
weight. Read over to the next reference line. Depending
upon whether you have a headwind or a tailwind, read down
or up the chart, parallel to the wind line. Read over to
the next reference line. At this point, if there is no
obstacle, read directly across to the distance scale. If
there is an obstacle, read up, parallel to the solid
lines, until you intersect the distance axis. If you are
still having trouble, follow through with the example
provided in the figure.

    g. 1,575 ft. See answer 10f.

    h. 1,300 ft. See answer 10f.

11. a. 356 ft. The ground roll will be reduced by 20 percent
due to the headwind. 445 - (.2 x 445) = 445 - 89 = 356.
Another way to compute this is to multiply 445 x 0.8.

    b. 1,255 ft. This requires you to reduce the landing roll
by 20 percent due to the headwind. But then, you must
add back in 20 percent of the "total to clear 50-ft.
obstacle" figure. The net effect is that the two 20
percent adjustments cancel each other out.

    c. 193 ft. In this answer you must interpolate between
5,000 and 2,500 ft. It just so happens that 3,750 is
exactly between them so you take half of the difference
between 495 and 470 (12.5 ft.) and add it to 470 to
arrive at a landing distance under standard conditions of
482.5 ft. The landing distance, however, will be reduced
by 60 percent due to the headwind component, or 289.5 ft.
The difference between 482.5 - 289.5 = 193 ft.

    d. 1,554 ft. In this answer you must add 20 percent to the
total to compensate for the dry grass runway and 10
percent more to compensate for 60° F above standard
conditions. So, you multiply 1,195 x 1.3 = 1,553.5 ft.

    e. 239 ft. First, you must compute the headwind component
from Figure 6.1. The angle is 20° and the headwind
component is 32 kts. Next, you must take the strong
headwind into account by subtracting 0.8 x 1,195 ft. =
956 ft. from 1,195. The answer is 239 ft.

f. 366 ft. Again, you must compute the headwind component from Figure 6.1. The angle is 70° and the calculated headwind component is 8 kts. Next, you must interpolate between sea level and 2,500 ft., which again happens to be exactly in between the two values on the chart (445 and 470). 445 + ([470 – 455]/2) = 457.5 ft. Next, correct for the 8 kts. headwind by subtracting 20 percent (91.5) to arrive at a landing distance of 366 ft.

12. Read directly from Figure 6.6. Fuel flow is 6.1 gal./hr. IAS is 88 kts. and TAS is 97 kts.

13. 406 mi. Read over on the chart from a pressure altitude of 8,000 ft. until you intersect the 2,400 RPM line. Now, read down to the range axis.

14. 4.5 hr. Read over on the chart from a pressure altitude of 8,000 ft. until you intersect the 2,400 RPM line. Now, read down to the endurance axis.

17. False. Indicated airspeed does not change with changes in altitude. True airspeed will be higher with higher density altitudes, but indicated airspeed remains the same.

18. No. Stalling speed has to do only with speed through the air. It is not affected by the speed of the wind over the ground.

19. a. 69.6 gal. First, determine the duration by dividing 1,000 mi. by 191 kts. (TAS). Find the ISA – 20° C portion of the table. Read over on the pressure altitude = 8,000 ft. line to find both TAS and fuel flow (13.3). Divide 1,000 by 191 to obtain 5.235 hr. Now, multiply 5.235 hr. x 13.3 = 69.6 gal.

b. 35.2 gal. Use the same logic as presented in answer 19a. Divide trip distance (500 mi.) by 193 to obtain 2.59 hr. Multiply 2.59 x 13.6 gal./hr. to arrive at 35.2 gal. In this answer, you will use the far right (ISA + 20° C) portion of the figure.

c. 12.55 gal./hr. In this answer, you will use the middle panel of the figure (Standard Day). Fuel flow **decreases** as you go higher, so the amount will be less than that required at 8,000 ft. (12.8). A pressure altitude of 8,500 ft. is 25 percent of the distance between 8,000 and 10,000 ft., so multiply the difference between 12.8 and 11.8 gal./hr x .25 = .25 gal./hr. Now, **subtract** .25 from 12.8 to arrive at 12.55 gal./hr.

d. 13.1 gal./hr. For this answer, you will use the right panel in Figure 6.9 (ISA + 36° F). 13.4 – 12.3 = 1.1 gal./hr. difference between 6,000 ft. and 8,000 ft. Take 25 percent of the difference to arrive at the value you must **subtract** from the 6,000 ft. value.

Review Questions

1. 1--See answer to Discussion Exercise 5a for computations.
2. 2--See answer to Discussion Exercise 5b for computations.
3. 4--See answer to Discussion Exercise 5c for computations.
4. 4--See answer to Discussion Exercise 5d for computations.

5. ?--Both Rwy 14 and Rwy 32 are within crosswind limits. See
    answer to Discussion Exercise 5e for an explanation. Rwy
    32 is the **best** choice since it is also a headwind, which
    makes alternative (4) the most appropriate choice.

6. ?--Both Rwy 14 and Rwy 32 are within crosswind limits. See
    answer to Discussion Exercise 5e for an explanation. Rwy
    14 is the **best** choice since it is also a headwind, which
    means alternative (2) is the most appropriate choice.

7. 2--See answer to Discussion Exercise 19a for computations.

8. 1--See answer to Discussion Exercise 19b for computations.

9. 2--See answer to Discussion Exercise 19c for computations.

10. 3--See answer to Discussion Exercise 19d for computations.

11. 3--See answer to Discussion Exercise 10f for computations.

12. 2--See answer to Discussion Exercise 10g for computations.

13. 1--See answer to Discussion Exercise 10h for computations.

14. 2--The best **rate** of climb, V_y, allows you to gain the most
    altitude in a given period of time.

15. 4--The best **angle** of climb, V_x, allows you to gain the most
    altitude in the shortest distance.

16. 2--This defines density altitude. Under standard
    conditions, pressure altitude and density altitude will
    be the same. Under non-standard conditions, you need to
    make adjustments.

17. 2--High density altitude means the air is less dense which
    in turn affects takeoff and landing characteristics as
    well as climb performance. Takeoff and landing distances
    increase while climb performance decreases.

18. 2--Density altitude decreases propeller efficiency; the air
    is less dense at high density altitudes.

19. 1--See answer to Discussion Exercise 11a for computations.

20. 2--See answer to Discussion Exercise 11b for computations.

21. 1--See answer to Discussion Exercise 11c for computations.

22. 4--See answer to Discussion Exercise 11d for computations.

23. 1--See answer to Discussion Exercise 11e for computations.

24. 2--See answer to Discussion Exercise 11f for computations.

25. 3--Decreases in air pressure (high density altitudes),
    increases in temperature, and increases in relative
    humidity reduce engine and propeller efficiency.

26. 3--Density altitude increases with increases in outside air
    temperature.

27. 3--See answer 25.

28. 1--Definition.

29. 2--See answer to Discussion Exercise 10b for computations.

30. 2--See answer to Discussion Exercise 10c for computations.

31. 3--See answer to Discussion Exercise 10d for computations.

32. 1--See answer to Discussion Exercise 10e for computations.

33. 1--Flying from an area of high pressure into an area of low
    pressure will cause the altimeter to **indicate** that you
    are higher than you actually are. I am **assuming** that the
    FAA means that you are maintaining a constant **indicated**
    altitude, as I see no way to maintain a constant altitude
    without adjusting the altimeter, which is precluded in
    the beginning of the question. This is simply a bad
    question.

34. 2--Pressure altitude and true altitude will be identical
     when standard atmospheric conditions exist. Remember,
     pressure altitude is defined in terms of standard
     atmospheric conditions.
35. 1--Colder air is more dense, so indicated altitude will be
     higher than true altitude.
36. 4--Density altitude is used to take into account **non-standard**
     temperatures. Pressure altitude equals density altitude
     only at standard temperature. Look at Figure 6.15 and
     read **up** the standard temperature line to determine the
     relationship between non-standard temperatures and
     pressure altitude.
37. 4--Refer to Figure 6.15. Use the right panel to figure the
     number of feet you will have to subtract from the airport
     elevation to determine the pressure altitude. Since
     30.35 falls midway between 30.3 (-348 ft.) and 30.4 (-440
     ft.), use the midway point, or -394 ft. Thus, pressure
     altitude is 3,500 ft. Read up the temperature axis along
     the 25° F line until you intersect the midway point
     represented by the diagonal pressure altitude lines of
     3,000 ft. and 4,000 ft. Read over to the density
     altitude line. From my reading of the chart, it appears
     to be about 2,000 ft. The closest answer available is
     2,000 ft.
38. 1--To figure pressure altitude, use the right panel of
     Figure 6.15. The difference between 29.92 and 29.96 is
     one-half the difference between 0 and -73 ft., or 36.5
     ft., which must be subtracted from the field elevation of
     3,563 ft., giving an answer of 3,526 ft.
39. 2--First, calculate the density altitude for 90° F by
     reading up the 90° F line until you intersect a pressure
     altitude of 1,250 ft., or roughly 3,600 ft. Next, read
     up the 60° F line until you intersect the 1,750 ft.
     pressure altitude line, or roughly 2,300 ft. Subtract
     these two figures and you will arrive at a 1,300-ft.
     decrease in density altitude.
40. 4--The difference between 29.92 and 29.97 is 0.05, which
     must be interpolated within the interval 29.92 to 30.00.
     Thus, you multiply 5/8 x 73 ft. = 45.62 ft., which must
     be **subtracted** from the field elevation of 1,386 ft. The
     answer is 1,340 ft. The closest alternative is 1,341 ft.
41. ?--This question has been designated as unusable by the FAA
     and removed from the FAA Question Selection Sheets.
42. 1--Follow the same steps used in answer 39. The numbers I
     computed by reading up the respective temperature lines
     were density altitudes of 5,500 ft. and 3,850 ft.
43. 3--First, correct for the altimeter setting of 29.25 by
     referring to the right panel of Figure 6.15. 29.25 falls
     midway between 29.20 and 29.30. The exact altitude that
     must be added to the field elevation is 626 ft. Thus,
     the pressure altitude in this problem is 5,876. Read
     vertically up the 81° line until you intersect a point
     slightly below the 6,000 ft. pressure altitude line.
     Read over to the density altitude axis to 8,500 ft.,
     which makes alternative (3) the best choice.

44. 3--When the scale of the pressure altimeter is set to local
     conditions, the altimeter will indicate true altitude at
     field elevation.

45. 3--Moving from higher pressure to lower pressure will mean
     that the altimeter will read lower. If you remember that
     each 1.0 in. of mercury represents a 1,000-ft. change in
     altitude, you can calculate the difference to be 150 ft.

46. 3--Moving from low pressure to high pressure will result in
     a lower reading from your altimeter than the actual
     altitude above sea level.

47. 2--Moving from high pressure to low pressure will result in
     a higher reading from your altimeter than the actual
     altitude above sea level.

48. 2--The primary reason for computing density altitude is to
     determine how your airplane will perform with respect to
     takeoff and landing distances and climb performance.

49. 2--Density altitude increases with increases in ambient
     temperature. In fact, density altitude takes into
     account non-standard temperatures.

50. 4--See answer 35.

51. 2--Pressure altitude is the altitude read from the altimeter
     when 29.92 is set in the Kollsman window.

52. 4--This defines pressure altitude, which must be corrected
     for non-standard temperatures to compute density altitude.

53. 2--As altitude increases, air pressure decreases, which in
     turn leads to a higher density altitude. IAS for
     approaches should stay the same. You should remember,
     however, that at high density altitudes, you will be
     moving faster relative to the ground at the same IAS.
     Finally, density altitude affects both takeoff and
     landing performance.

54. 3--First, correct for the altimeter setting by subtracting
     165 ft. from the field elevation. Use the right panel of
     Figure 6.15. Next, read up from the Degrees F axis at
     93° until you intersect the curved pressure altitude line
     at 3,000 ft. Now read over to the density altitude axis
     to find 5,800 ft.

55. 4--First, correct for the altimeter setting by adding 298
     ft. to the field elevation. Next, read up the 75° F line
     until you intersect the pressure altitude line that
     represents 4,000 ft. Now read over to the density
     altitude axis to find approximately 5,950 ft.

56. 3--A soft field provides more resistance for the wheels and
     thus will increase the takeoff roll, as will increasing
     the gross weight, taking off uphill, tailwinds, moving
     the CG forward, and increasing the density altitude.

57. 4--This defines true airspeed.

58. 1--([29.92 - 30.32] x 1,000) + true altitude = (-0.4 x 1,000)
     + 120 = -400 + 120 = -280 ft.

59. 3--First, figure the pressure altitude (29.92 - 29.22) = .7
     x 1,000 = 700 ft. + 5,800 ft. = 6,500 ft. Use a pressure
     altitude of 6,500 ft. as an approximation. Fuel flow is
     about 6.1 gal./hr. and TAS is 97 kts.

60. 2--On Figure 6.7 read over from a pressure altitude of 8,000
        ft. to the 2,400-RPM line and then down to the range,
        which in this question is 405 mi. On Figure 6.8 read
        over from a pressure altitude of 8,000 ft. until you
        intersect 2,400 RPM, then read down to endurance, which
        in this question is 4.5 hr. or 4 hr. and 38 min.
61. 3--Read over on the pressure altitude line representing
        6,400 ft. until you intersect each RPM line. The
        difference is approximately 11 mi.

7/AIRPORTS, AIRSPACE, AND LOCAL FLYING

MAIN POINTS

1.   Airports with towers to supervise ground and air traffic are called **controlled**; those without towers are called **uncontrolled**. The airport and its surrounding territory are called the **terminal area**. Controlled airports have an **airport traffic area (ATA)** where specific procedures exist. Finally, the local flying area is a zone up to 25 NM from a given airport.

2.   **Runways** are represented by one- or two-digit numbers corresponding to the nearest magnetic heading divided by 10. For example, a runway with a magnetic heading of 214° is Rwy 21, and its reciprocal (when used in the opposite direction) is Rwy 3 (a heading of 034°). L (left), R (right), and C (center) are used to denote parallel runways. A **basic runway** has a white number with a dashed centerline. A **nonprecision instrument approach runway** displays threshold markings (broad parallel stripes) before the runway number. **Precision approach runways** have more sophisticated markings, including side stripes and a touchdown zone. **Displaced thresholds** are marked by a series of chevrons at the end of the runway that create an area for clearance. This area may be used for rollout when landing in the opposite direction. A wide band of chevrons running up the center of the area before the threshold indicates an **overrun** or stopway that should not be used for any aircraft operation. An X at either end is used to mark a runway that is closed. **Taxiways** are marked with a yellow centerline. **Holding lines** are solid yellow lines perpendicular to the taxiway; they are not to be crossed at controlled airports until cleared by the tower. Under category II operations, the tower may tell you to stop at special Cat II holding lines (two lines connected by bars).

3.   Unless authorized by the FAA, the standard traffic pattern consists exclusively of left turns (except at entry). The legs of a pattern are **upwind** or **takeoff**, to about 300-400 ft. AGL; **crosswind**, 90° to the upwind leg and usually not to be

127

entered until crossing the departure end of the runway; **downwind**, parallel to the runway but in the opposite direction (typically flown at 1,000 ft. AGL, the **traffic pattern altitude**); **base**, parallel to the crosswind leg but at the opposite end of the field; and **final**, upwind, in line with the landing runway. The term **long final** refers to that portion just after turning from base; **short final** refers to the one-half-mile segment just prior to the threshold. One generally enters the pattern at an angle of 45° to the downwind leg, while traffic departures are typically made by executing a 45° turn away from the crosswind leg. At controlled airports, the tower will tell you what pattern restrictions, if any, apply.

4.   Preestablished traffic patterns exist for operating at all airports. This information is available from the tower at controlled airports and from an FSS or UNICOM frequency or from the **segmented circle** at uncontrolled airports. The segmented circle may have base to final approach indicators on the outside, and it may have a tetrahedron, wind sock, or wind tee in the middle to indicate the direction for landing. An amber light in the segmented circle, on the control tower, or on an adjoining building indicates that a right-hand pattern is in effect. An X in the circle means the field is closed.

5.   The normal **glide path** (glide slope) is about 3° to the horizon. Many airports have a **visual approach slope indicator** (**VASI**) consisting of two or three sets of colored lights (white, pink, and red). A safe, correct approach is one where red is maintained over white. Red/pink and red/red indicate that the airplane is too low, and white/pink and white/white indicate that it is too high. Red over white, just right!

6.   An **uncontrolled airport** may have a facility designated an **Aeronautical Advisory Station** (**ASS**) and operate a private radio service called UNICOM. Typical UNICOM frequencies are 122.8 and 123.0. Specific frequencies are shown on aeronautical charts and in the Airport/Facility Directory. UNICOM is not used for air traffic control, but rather for pilots to advise one another of taxiing, pattern, departure, and arrival intentions. When flying into airports with no control tower, FSS, or UNICOM, use MULTICOM (122.9) for communication at and around the airport. Private airports use either 122.725 or 122.75. **Flight Service Stations** (**FSS**) provide weather information, flight plan filing services, and airport advisory services if they are located at uncontrolled fields. When calling an FSS, it is referred to as "radio." For example, one might call Kansas City **Radio**.

7.   **Controlled airports** require two-way radio communication authorization for any movement on the ground or in the **airport traffic area** (**ATA**). The airport traffic area is an area extending from the ground up to, but not including, 3,000 ft. AGL, with a radius of 5 statute miles (SM) around the airport. The ATA is in effect only when the tower is operating. The speed limit in the ATA is 180 MPH (156 kts.). In the event of a radio failure, light signals are used to give the pilot directions. At

many busier airports, prerecorded information is available via **Automatic Terminal Information Service (ATIS)**.

8.   In good weather, all aircraft operate under the principle of **visual separation**. The FAA specifies minimum weather conditions that must be met for **visual flight rules (VFR)**. Poor weather, referred to as **instrument meteorological conditions (IMC)**, has a set of rules that govern flight called **instrument flight rules (IFR)**. The purpose of **controlled airspace** is to separate VFR and IFR traffic. Areas over which ATC has no control are called **uncontrolled airspace**. In VFR weather, also called **visual meteorological conditions (VMC)**, **most** controlled airspace below 18,000 ft. is treated as if it were uncontrolled. No contact with ATC is necessary to fly. In IFR weather, however, all aircraft must have clearance from ATC to fly in controlled airspace.

9.   There are several weather minimums that apply to VFR flights. In **uncontrolled airspace**, you need 1-mi. visibility and must remain clear of clouds **below 1,200 ft. AGL**. Between 1,200 ft. AGL and 10,000 ft. MSL, you must have 1-mi. visibility and you must stay 500 ft. below, 1,000 ft. above, or 2,000 ft. horizontally from clouds. **At or above 10,000 ft. MSL, and above 1,200 ft. AGL**, you must have 5-mi. visibility and remain at least 1,000 ft. below, 1,000 ft. above, or 1-mi. horizontally from clouds. In **controlled airspace**, there are two altitude divisions. **Below 10,000 ft. MSL**, you need 3-mi. visibility and must remain 500 ft. below, 1,000 ft. above, or 2,000 ft. horizontally from clouds. **At or above 10,000 feet MSL**, you need 5-mi. visibility and must remain at least 1,000 ft. below, 1,000 ft. above, or 1-mi. horizontally from clouds. To help you learn this, try drawing a picture for each type of airspace.

10.   There are a variety of types of **controlled airspace**. The **positive control area (PCA)** exists from 18,000 ft. MSL to FL 600, where FL means **flight level** and 600 refers to 60,000 ft. pressure altitude. It is IFR territory and requires transponder and two-way radio. The **continental control area** (14,500 ft. MSL and up) covers the 48 states and parts of Alaska. VFR flight is permitted in the continental control area, but added equipment is required. **Control zones** help separate VFR from IFR trafffic when instrument conditions prevail and exist around airports with instrument approach capability. They extend from the surface to the base of the continental control area or other controlled airspace. They typically extend outward at least 5-mi. from the airport and may have a keyhole appearance to accommodate instrument approaches. Flight under a ceiling in a control zone requires 3-mi. visibility and a ceiling of at least 1,000 ft. **Special VFR** clearance may be obtained in a control zone if you have 1-mi. visibility and can stay clear of clouds. Special VFR will be issued at night only if the pilot has an instrument rating. A control zone demarked by Ts on the chart means no special VFR is allowed. **Transition areas** around airports are indicated by magenta boundaries on sectional charts (a floor of 700-ft. AGL) and around airways by blue boundaries (a floor of

1,200-ft. AGL). They extend up to the base of the continental control areas and are also designed to help separate VFR and IFR aircraft. **Control areas** around Federal airways extend 4-mi. on either side of the airway and from 1,200 ft. AGL up to the overlying continental control area. Special-use airspace includes prohibited areas (forbidden to all aircraft), restricted area (marked on charts; may be flown over with permission), warning areas, military operation areas (MOAs), military training routes, and alert areas.

11. The distinctions between a **control zone** and an **airport traffic area** is important. While the size of a control zone can vary, an ATA is always 5 mi. in radius and 3,000 ft. tall. An ATA places a communication requirement on aircraft, whereas a control zone has a weather requirement. ATAs are in effect **only** when there is an operating control tower. On aeronautical charts, a control zone is depicted by dashed blue lines. An airport that can have an ATA has a blue symbol and has a control tower frequency in the data block.

12. Another service available at some airports is **radar**, whereby the controller can monitor progress and give specific approach and departure instructions. **Terminal Radar Programs** are available in two stages, II and III, depending upon the degree of control over IFR and VFR aircraft. At very busy airports, the control area is expanded to include a larger area of airspace called a **terminal control area (TCA)**. It is shaped like an upside-down wedding cake, and positive control is exercised when you are operating within it. TCAs are of two types, I and II. Group II TCAs require a 4096-code transponder, two-way radio, and navigation radio; you must also operate in accordance with instructions given by the controller. Group I TCAs are the same as Group II TCAs except that student pilots are not permitted to land at the primary airport in the TCA and an encoding altimeter is required.

13. Radio communications occur on VHF frequencies between 118.000 and 135.975 megahertz (MHz). Most modern VHF radios have the transmitter and receiver combined in one unit, a transceiver. Talking on the radio involves natural conversational tone and knowledge of the **phonetic alphabet**. There is no easy way to learn the phonetic alphabet except to use it. The quickest I have ever seen anyone learn it is two minutes. If you beat that record, let me know how you did it. Once you have learned it, practice it in other settings, such as asking for stock quotations or spelling your name over the telephone. In addition to the phonetic alphabet, some **numbers** are pronounced in special ways. For example, nine is pronounced "niner." Multidigit numbers are given by saying each number individually (for example, "heading two zero zero"). When a decimal is present, it is referred to as "point" (for example, UNICOM might be "one two three point zero"). One exception to the decimal rule is barometric pressure, which is read without a decimal (for example, "three zero zero six"). Altitudes below 10,000 are read

in thousands and hundreds (for example, "seven thousand two hundred" for 7,200). From 10,000 to FL 180, the thousand digit is stated separately, followed by the hundreds (for example, "one three thousand four hundred" for 13,400). **Time** is given with reference to the Greenwich meridian (zulu time) and follows military standards (0000-2400, or 24-hr. clock). **Airplane call signs** are used to identify each aircraft and are given without the prefix N (for example, Cessna N5009G would be identified "Cessna five zero zero niner gulf"). Always end each transmission with your call sign or abbreviated call sign to leave no doubt as to who is talking with ATC. Finally, ATC uses the **clock position reference system** to identify the position of airborne traffic relative to a given plane. The nose of the plane points toward 12 o'clock and the tail points toward 6 o'clock. A plane at 9 o'clock would be off your left wing.

14.   The following sequence is customary for initiating radio communication: (1) ground facility being called, (2) aircraft call number, (3) location and altitude, (4) intentions, and (5) other information (for example, ATIS received).

15.   **Night flying** has its own peculiar characteristics and considerations. Night vision is different from daylight vision. Your eyes require a considerable adaptation period (20-30 min.) to adjust to night vision, and this adjustment can easily be disrupted by sudden bursts of light. Cockpit illumination, whether red or white, should be kept at a minimum. Since night vision is concentrated in the eye's <u>peripheral</u> area, you may need to glance at objects quickly and indirectly several times to identify them correctly. Sometimes oxygen is recommended to maintain the pilot's visual acuity. You should also be aware that illusions are much more prevalent at night than during the day.

16.   Airplane lights include an **anticollision light** system (beacons or strobe lights), **position** or navigation lights (red, left wing; green, right wing; white, tail) required for night flight, and **landing lights**. When flying at night, you can tell if another airplane is approaching because green will be on the left and red on the right, thus the saying "red-right-returning." Landing or position lights can be used to signal the control tower at night in the event of a radio failure. **Airports** are identified by a rotating beacon featuring alternating green and white lights at night (a double-flashing white light alternating with green indicates a military airfield). If the beacon is rotating during the day, it means the weather is below VFR minimums. If present, green lights indicate the threshold, red lights the departure end, white lights side boundaries, and blue lights taxiways and taxiway turnoffs. When flying at night, it is wise to carry a flashlight in the event the lights fail. Finally, when flying at night, rely primarily on your instruments and not your senses for the airplane's attitude, altitude, vertical speed, and airspeed. Instruments seldom lie, except when they malfunction; your senses frequently do, particularly at night.

17. You should follow all published noise abatement procedures, unless compliance compromises the aircraft's safety. Also, there are areas over which you should excercise caution and stay within published limits such as parks and wildlife preserves.

18. Planning a flight, even one in which you simply practice takeoffs and landings, requires considerable attention to details, such as the airplane, the weather (required by the FAA), and runway conditions. That is why normal procedures are conducted by reference to checklists. The **preflight inspection** includes an examination of various parts of the airplane. **Ground operations** require coordinating your attention between events outside the cabin and indicators inside the cabin. Once you take off, the **in-flight checklist** becomes relevant. Next, there is a **landing checklist**. Finally, there is a **postflight checklist** to complete before leaving the aircraft.

19. **Taxiing** can be tricky, particularly if the wind is blowing. When surface winds are high, position the ailerons **into** the wind when taxiing into a quartering headwind and **away** from the wind when the wind is coming from the rear quarter. Keep the elevator neutral when the wind is coming from in front of you and down when it is coming from the rear.

20. At busy airports, you may encounter a **signaler** who will give you directions about where and how to park.

KEY TERMS AND CONCEPTS, PART 1

Match each term or concept (1-20) with the appropriate description (A-T) below. Each item has only one match.

___	1.	airport at night	___ 11.	crosswind
___	2.	departure	___ 12.	prohibited area
___	3.	words twice	___ 13.	flashing white
___	4.	prime meridian	___ 14.	control area
___	5.	ceiling	___ 15.	ATIS
___	6.	airport traffic area	___ 16.	vector
___	7.	IMC	___ 17.	156
___	8.	threshold	___ 18.	restricted airspace
___	9.	azimuth	___ 19.	VMC
___ 10.		Group II TCA	___ 20.	clear

A. pattern leg not to be entered before crossing the departure end of the runway
B. point beyond which landing aircraft can contact the runway
C. exists at controlled airports with specific flight procedures--5-mi. radius and up to 3,000 ft. AGL
D. alternating green and white beacon
E. red lights mark this end of the runway at night
F. repeat each key word or phrase twice
G. located in Greenwich, England
H. synonymous with direction

I.   airspace in which aircraft operation is forbidden
J.   airspace that may, for example, contain a high volume of
     military pilot training
K.   lowest layer of clouds beneath which you are flying that is
     classified as broken or overcast
L.   designated airspace such as a VOR Federal airway
M.   airspeed (in knots) not to be exceeded in an ATA
N.   instrument meteorological conditions
O.   two-way radio, 4096-code transponder, and mandatory radar
     service are required
P.   being given a heading and altitude to fly
Q.   prerecorded airport information available at busier
     controlled airports
R.   light signal used to indicate "return to starting point on
     airport"
S.   visual meteorological conditions
T.   word you should yell before engaging the starter

KEY TERMS AND CONCEPTS, PART 2

     Match each term or concept (1-20) with the appropriate
description (A-T) below. Each item has only one match.

___ 1. glide slope   ___ 11. control zone
___ 2. tetrahedron   ___ 12. holding line
___ 3. call sign   ___ 13. prohibited area
___ 4. taxiway turnoff   ___ 14. ARTCC
___ 5. Rwy 19   ___ 15. one mile
___ 6. terminal area   ___ 16. positive control area
___ 7. FSS   ___ 17. final
___ 8. vector   ___ 18. uncontrolled
___ 9. UNICOM   ___ 19. VHF communications
___ 10. verify   ___ 20. 3 o'clock

A.   where you file flight plans and receive weather information
B.   122.8 and 123.0 are typical VHF radio frequencies
C.   angle of descent on final approach
D.   portion of the pattern flown after the base leg
E.   three-dimensional triangle used as wind direction indicator
F.   place on taxiway where you must stop at a controlled airport
G.   reciprocal of Rwy 1
H.   controlled airport and its surrounding territory
I.   airport without an operating tower to control ground and air
     traffic
J.   horizontal distance you must maintain from clouds at or above
     10,000 ft. MSL in both controlled and uncontrolled airspace
K.   blue lights mark this point on a runway at night
L.   double-check the accuracy of the transmission
M.   falls within these radio frequencies: 118.00 to 135.975 MHz
N.   heading issued to an aircraft to provide navigational radar
     guidance
O.   airspace in which aircraft flight is not allowed
P.   provides air traffic control to IFR flight along controlled
     airways

Q. 18,000 ft. MSL to FL 600
R. airspace (typically at least 5-mi. radius from the airport)
    upward from the surface to 14,500 ft. MSL
S. aircraft identification number--for example, N7017G
T. where you would find an aircraft reported off your right wing

KEY TERMS AND CONCEPTS, PART 3

Match each term or concept (1-20) with the appropriate
description (A-T) below. Each term has only one match.

____ 1. wind T   ____ 11. read back
____ 2. Stage III radar   ____ 12. magenta
____ 3. upwind   ____ 13. transponder
____ 4. squelch   ____ 14. MULTICOM
____ 5. short final   ____ 15. transmitter
____ 6. rock your wings   ____ 16. X
____ 7. local flying area   ____ 17. three
____ 8. steady red   ____ 18. position lights
____ 9. 45   ____ 19. segmented circle
____ 10. flashing red   ____ 20. nonprecision instrument
                                              approach runway

A. positive radar separation of all **participating** VFR and IFR
    traffic
B. radio device that sends out a signal when it detects a radar
    wave
C. signal used to let the control tower know you have received a
    signal when your radio is inoperative
D. light signal used to indicate "give way to other aircraft;
    continue to circle"
E. light signal used to indicate "airport unsafe--do not land"
F. color used on aeronautical charts to depict uncontrolled
    airports and their legends
G. 122.9, one frequency used for air-to-air radio transmissions
H. a flight path parallel to the landing runway in the direction
    of landing
I. normal glide slope angle (in degrees)
J. portion of the traffic pattern one-half mile prior to the
    threshold
K. airport display used to indicate approach to final legs
L. markings that indicate a runway is closed
M. angle at which you should enter the downwind leg
N. runway with a heading number, dashed centerline, and broad
    parallel stripes
O. area up to approximately 25 mi. from a given airport
P. transceiver control that balances volume and static
Q. a type of landing direction indicator
R. lights on the wingtips and tail
S. repeat all the transmission that has just been received
T. portion of a radio that broadcasts

KEY TERMS AND CONCEPTS, PART 4

Match each term or concept (1-20) with the appropriate description (A-T) below. Each item has only one match.

___ 1. anticollision lights
___ 2. transition area
___ 3. chevrons
___ 4. downwind
___ 5. controlled airport
___ 6. overrun area
___ 7. basic runway
___ 8. base
___ 9. wind sock
___ 10. transceiver

___ 11. UHF communications
___ 12. engine starting checklist
___ 13. active
___ 14. Stage II radar service
___ 15. MOA
___ 16. reciprocal
___ 17. continental control area
___ 18. flashing green
___ 19. Group I TCA requirements
___ 20. peripheral vision

A. airspace designated for military operations
B. field of vision to either side of the eyes
C. controlled airspace extending upward from 700 ft. (from an airport with an instrument approach) or 1,200 ft. (from an airway) upward to the base of the next control area
D. airspace over the 48 states and parts of Alaska from 14,500 ft. MSL upward
E. two-way radio, an encoding altimeter, and mandatory radar service
F. radar sequencing of arriving VFR and IFR traffic; advisories for departing VFR traffic
G. light signal used to indicate "clear to taxi"
H. a direction 180° from a given direction
I. airport with an operating tower to control ground and air traffic
J. beacon or strobe system on fuselage
K. area of a runway marked by a wide band of chevrons indicating that this area should not be used for any aircraft operations
L. pattern leg parallel to the active runway but in the opposite direction from it
M. pattern leg parallel to crosswind but at the opposite end of the field
N. radio transmitter and receiver combined in a single unit
O. markings on a runway indicating a displaced threshold
P. runway with only a heading number and dashed centerline
Q. initial oil pressure check occurs with reference to this checklist
R. runway being used for takeoffs and landings
S. radio frequencies used mostly for military communications
T. wind indicator whose short end points **away** from the wind

DISCUSSION QUESTIONS AND EXERCISES

1. What is the difference between a controlled airport and an uncontrolled airport? What is the difference between an airport traffic area (ATA) and the local flying area?

2.   Briefly characterize each of the following:

   a. basic runway

   b. runway heading indicator

   c. nonprecision instrument approach runway

   d. displaced threshold

   e. closed runway

   f. taxiway

   g. holding line

   h. overrun area

3.   Draw Rwy 14. Indicate altitudes and directions of flight for a **right-hand** pattern with a pattern altitude of 800 ft. AGL.

4.   Draw a segmented circle for the following airport: Rwy 2-20 with left traffic for 2 and right traffic for 20; Rwy 10-28 with left traffic for 10 and right traffic for 28. The wind is 290° at 15. Put a tetrahedron in the circle. If you were landing here, what runway would you use and what would the traffic pattern be?

5.   Briefly describe the meanings of the color components of the
visual approach slope indicator (VASI). Be sure to describe both
types of systems.

6.   Briefly characterize each of the following. Include special
restrictions, communication functions, and communication
frequencies or procedures as appropriate.

   a.   UNICOM

   b.   MULTICOM

   c.   FSS

   d.   ATA

   e.   ATIS

   f.   Stage II radar service

   g.   Stage III radar service

   h.   Group I TCA

   i.   Group II TCA

7.   Briefly indicate what each of the following light signals
means for both ground and air operations:

                              Ground   Air

   a.   steady green

   b.   flashing green

   c.   steady red

     d.   flashing red

     e.   flashing white

     f.   alternating red and green

8.   Distinguish between visual meteorological conditions and instrument meteorological conditions. What type of rules apply to each condition?

9.   What is the difference between controlled and uncontrolled airspace?

10. Specify the weather minimums that apply to each of the following airspaces:

     a.   uncontrolled--below 1,200 ft. AGL

     b.   uncontrolled--between 1,200 ft. AGL and 10,000 ft. MSL

     c.   uncontrolled--at or above 10,000 ft. MSL and above 1,200 ft. AGL

     d.   controlled--below 10,000 ft. MSL

     e.   controlled--at or above 10,000 ft. MSL

11. Briefly describe each of the following.

     a.   positive control area

     b.   continental control area

    c.   control zone

    d.   transition area

    e.   prohibited area

    f.   control area around a Federal airway

12. You need to learn the phonetic alphabet, if you have not done so already. Refer to the text and study the letters and their names. Then reproduce as many as you can from memory in the space below. Once you have done as many as you can, complete the table, if necessary, by referring to the text. This exercise will help you learn the alphabet and will serve as a handy reference for review.

13. Describe exactly how you would say each of the following using standard radio communication language:

    a.   Tripacer N2497C

    b.   a heading of 290°

    c.   an altitude of 3,120 ft.

    d.   an altitude of 15,800 ft.

    e.   2240 zulu

    f.   Kansas City ARTCC

    g.   Tulsa FSS

h.   Repeat the transmission that was just received.

i.   Double-check the accuracy of the transmission.

j.   The message has been received and understood.

k.   123.6

l.   a barometer reading of 30.09

m.   position of an aircraft directly in front of you

14. List the five things you should state when initiating radio contact.

15. Suppose that you are flying in a Cessna 172 (N7280Q) equipped with a two-way radio and transponder (squawking 1200) over Marietta, Georgia, at 5,500 ft. on a head of 170°. You intend to land at Peachtree-De Kalb Airport and have carefully listened to the ATIS, information echo. Write out exactly what you would say when contacting Atlanta Approach Control.

16. Identify four extra precautions you should take when preparing for a night flight.

17. Briefly describe the function and location of each of the following airplane lights:

a.   anticollision lights

b.   position (navigation) lights

c.   landing lights

18.   What lights should you use to signal a control tower if your radio fails in flight during the day? at night?

19.   Briefly describe the standard runway lighting, taxi lighting, and beacon configuration at civilian and military airports.

20.   Suppose you are flying at night and in the distance you spot another aircraft at roughly the same altitude. How can you tell if the other aircraft is approaching you?

21.   Briefly describe the position the ailerons and elevator should be in when taxiing under the following conditions:

a.   strong headwind

b.   quartering headwind from the right

c.   quartering headwind from the left

d.   quartering tailwind from the right

e.   quartering tailwind from the left

f.   strong tailwind

REVIEW QUESTIONS

1.   (FAA 1353) How should the controls be held while taxiing a tricycle-gear equipped airplane into a left quartering headwind as depicted by A in Figure 7.1?

1--Left aileron up, neutral
     elevator.
2--Left aileron down, neutral
     elevator.
3--Left aileron up, down
     elevator.
4--Left aileron down, down
     elevator.

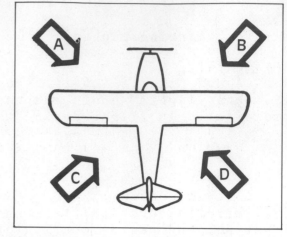

Figure 7.1

2.   (FAA 1354) How should the controls be held while taxiing a tricycle-gear equipped airplane into a left quartering tailwind as depicted by C in Figure 7.1?

1--Left aileron up, neutral elevator.
2--Left aileron down, neutral elevator.
3--Left aileron up, down elevator.
4--Left aileron down, down elevator.

3.   (FAA 1355) While taxiing a tricycle-gear equipped airplane into a right quartering headwind as depicted by B in Figure 7.1, the right aileron should be held

1--up and the elevator neutral.
2--down and the elevator neutral
3--up and the elevator down.
4--down and the elevator down.

4.   (FAA 1356) While taxiing a tricycle-gear equipped airplane in a right quartering tailwind as depicted by D in Figure 7.1, the right aileron should be held

1--up and the elevator neutral.
2--down and the elevator neutral.
3--up and the elevator down.
4--down and the elevator down.

5.   (FAA 1357) When taxiing with strong quartering tailwinds, which of the following aileron positions should be used?

1--Aileron parallel to the ground on the side from which the wind
     is blowing.
2--Neutral (streamlined position).
3--Aileron up on the side from which the wind is blowing.
4--Aileron down on the side from which the wind is blowing.

6.   (FAA 1358) Which of the following aileron positions should a
pilot generally use when taxiing in strong quartering headwinds?

1--Aileron up on the side from which the wind is blowing.
2--Aileron down on the side from which the wind is blowing.
3--Neutral.
4--Aileron parallel to the ground on the side from which the wind
   is blowing.

7.   (FAA 1359) Which wind condition would be most critical when
taxiing a nosewheel-equipped high-wing airplane?

1--Direct headwind.
2--Direct crosswind.
3--Quartering headwind.
4--Direct tailwind.

8.   (FAA 1419) What is the general direction of movement of the
other aircraft during a night flight if you observe a steady red
light and a flashing red light ahead at the same altitude?

1--The other aircraft is crossing to the left.
2--The other aircraft is crossing to the right.
3--The other aircraft is approaching head-on.
4--The other aircraft is headed away from you.

9.   (FAA 1420) What is the general direction of movement of the
other aircraft if during a night flight you observe a steady white
light and a flashing red light ahead and at the same altitude?

1--The other aircraft is crossing to the left.
2--The other aircraft is crossing to the right.
3--The other aircraft is approaching head-on.
4--The other aircraft is headed away from you.

10.   (FAA 1421) What is the general direction of movement of the
other aircraft if during a night flight you observe steady red and
green lights ahead and at the same altitude?

1--The other aircraft is crossing to the left.
2--The other aircraft is crossing to the right.
3--The other aircraft is approaching head-on.
4--The other aircraft is headed away from you.

11.   (FAA 1422) VFR approaches to land at night should be made

1--at a higher airspeed.
2--low and shallow.
3--with a steep descent.
4--the same as during daytime.

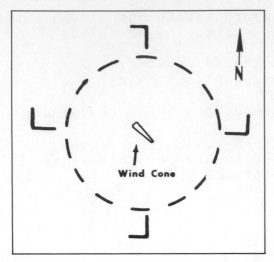

Figure 7.2

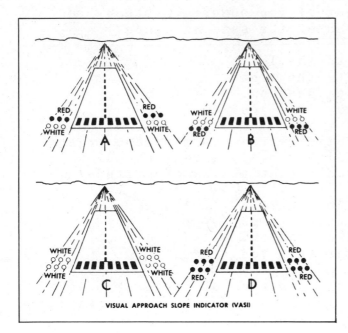

Figure 7.4

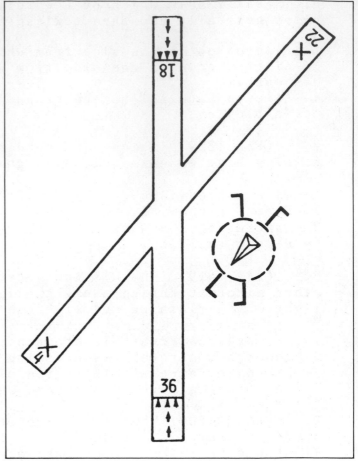

Figure 7.3

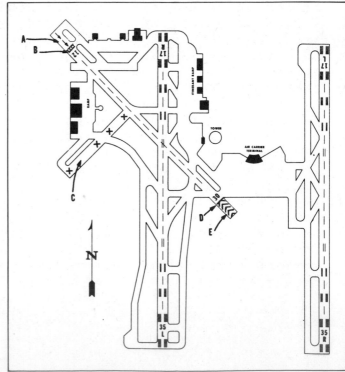

Figure 7.5

12. (FAA 1433) The segmented circle shown in Figure 7.2
indicates that the airport traffic is

1--left-hand for Rwy 17 and right-hand for Rwy 35.
2--right-hand for Rwy 9 and left-hand for Rwy 27.
3--right-hand for Rwy 35 and right-hand for Rwy 9.
4--left-hand for Rwy 35 and right-hand for Rwy 17.

13. (FAA 1434) The traffic patterns indicated in the segmented
circle depicted in Figure 7.2 have been arranged to avoid flights
over an area to the

1--south of the airport.
2--north of the airport.
3--southeast of the airport.
4--northwest of the airport.

14. (FAA 1435) The segmented circle depicted in Figure 7.2
indicates that a landing on Rwy 26 will be with a

1--right-quartering headwind.
2--left-quartering headwind.
3--right-quartering tailwind.
4--left-quartering tailwind.

15. (FAA 1436) Which runway and traffic pattern should be used
as indicated by the wind cone in the segmented circle depicted in
Figure 7.2?

1--Right-hand traffic on Rwy 35.
2--Right-hand traffic on Rwy 17.
3--Left-hand traffic on Rwy 35 or right-hand traffic on Rwy 26.
4--Left-hand traffic on Rwy 26 or Rwy 35.

16. (FAA 1437) Select the proper traffic pattern and runway for
a landing as indicated on the airport diagram in Figure 7.3.

1--Right-hand traffic and Rwy 4.
2--Right-hand traffic and Rwy 18.
3--Left-hand traffic and Rwy 22.
4--Left-hand traffic and Rwy 36.

17. (FAA 1438) If the wind is as shown by the landing direction
indicator in Figure 7.3, the pilot should land to the

1--north on Rwy 36 and expect a crosswind from the right.
2--south on Rwy 18 and expect a crosswind from the right.
3--southwest on Rwy 22 directly into the wind.
4--northeast on Rwy 4 directly into the wind.

18.   (FAA 1439) An airport's rotating beacon operated during the daylight hours indicates

1--there are obstructions on the airport.
2--that weather in the control zone is below basic VFR weather
    minimums.
3--parachute jumping is in progress.
4--the airport is temporarily closed.

19.   (FAA 1441) How can a military airport be identified at night?

1--Alternate white and green beacon light flashes.
2--Dual peaked (two quick) white flashes between intermittent
    green flashes.
3--White flashing beacon lights with steady green at the same
    location.
4--Alternate white and red beacon flashes.

20.   (FAA 1442) Illustration A in Figure 7.4 indicates that an aircraft is

1--off course.
2--below the glide slope.
3--on the glide slope.
4--above the glide slope.

21.   (FAA 1443) While on final approach to a runway equipped with a standard 2-bar VASI, the lights appear as shown by illustration D in Figure 7.4. This means that the pilot is

1--receiving an erroneous light indication.
2--above the glide slope.
3--below the glide slope.
4--on the glide slope.

22.   (FAA 1444) VASI lights appearing as shown by illustration C in Figure 7.4 would indicate that an airplane is

1--off course to the left.
2--on the glide slope.
3--below the glide slope.
4--above the glide slope.

23.   (FAA 1445) An on glide slope indication from a tri-color VASI is

1--a white light signal.
2--a green light signal.
3--an amber light signal.
4--a pink light signal.

24.   (FAA 1446)   An above glide slope indication from a tri-color
VASI is

1--a white light signal.
2--a green light signal.
3--an amber light signal.
4--a pink light signal.

25.   (FAA 1447)   A below glide slope indication from a tri-color
VASI is

1--a pink light signal.
2--a green light signal.
3--an amber light signal.
4--a red light signal.

26.   (FAA 1448)   When approaching to land on a runway served by a
VASI, the pilot shall

1--intercept and remain on the glide slope until touchdown only if
   the aircraft is operating on an instrument flight plan.
2--maintain an altitude that captures the glide slope at least 2
   mi. downwind from the runway threshold.
3--maintain an altitude at or above the glide slope.
4--remain on the glide slope and land between the two-light bar.

27.   (FAA 1449)   The airport taxiways are identified at night by

1--alternate red and green edge lights.
2--amber omnidirectional edge lights.
3--white directional edge lights.
4--blue omnidirectional edge lights.

28.   (FAA 1450)   The numbers 9 and 27 on a runway indicate that
the runway is oriented approximately

1--090° and 270° magnetic.
2--009° and 027° true.
3--090° and 270° true.
4--009° and 027° magnetic.

29.   (FAA 1451)   How is a runway recognized as being closed?

1--The letter C is painted in red after the runway number.
2--Red lights are placed at the approach end of the runway.
3--Yellow chevrons are painted on the runway beyond the threshold.
4--X is displayed on the runway.

30. (FAA 1452) According to the airport diagram shown in Figure 7.5

1--takeoffs and landings are permissible at position C since this
   is a short takeoff and landing runway.
2--the takeoff and landing portion of Rwy 12 begins at position
   B.
3--Rwy 30 is equipped at position E with emergency arresting gear
   to provide a means of stopping military aircraft.
4--takeoffs may be started at position A on Rwy 12, and the
   landing portion of this runway begins at position B.

31. (FAA 1453) What is the difference between area "A" and area
"E" on the airport depicted in Figure 7.5?

1--"A" may be used for taxi and takeoff; "E" may only be used as
   an overrun.
2--"A" may be used for all operations except heavy aircraft
   landings; "E" may only be used as an overrun.
3--"A" may only be used for taxi; "E" may be used for all
   operations except landings.
4--"A" may only be used for an overrun; "E" may only be used for
   taxi and takeoff.

32. (FAA 1454) Area C on the airport depicted in Figure 7.5 is
classified as

1--an STOL runway.
2--a parking ramp.
3--a multiple heliport.
4--a closed runway.

33. (FAA 1455) That portion of the runway identified by the
letter A in Figure 7.5

1--may be used for taxiing but should not be used for takeoffs or
   landings.
2--may be used for taxiing or takeoffs but not for landings.
3--may be used for taxiing, takeoffs, and landings.
4--may not be used except in an emergency.

34. (FAA 1456) The arrows that appear on the end of the
north/south runway as shown in Figure 7.3 indicate that the area

1--may be used for taxiing.
2--is usable for taxiing, takeoff, and landing.
3--cannot be used for landing, but may be used for taxiing and
   takeoff.
4--is available for landing at the pilot's discretion.

35. (FAA 1458) Under what condition, if any, may civil pilots
enter a restricted area?

1--For takeoff and landing to take care of official business.
2--With the controlling agency's authorization.
3--On airways with ATC clearance.
4--Under no condition.

36. (FAA 1459) What hazards to aircraft may exist in warning
areas and where are warning areas located?

1--Unusual, often invisible, hazards such as aerial gunnery or
    guided missiles over international waters.
2--Unusual, often invisible, hazards such as artillery firing or
    missiles over military reservations.
3--Known wind shear areas in canyons and mountainous areas during
    certain seasons.
4--Student training areas or parachute jumping areas in or near
    heavily populated areas.

37. (FAA 1460) When operating VFR in a MOA, a pilot

1--must obtain a clearance from the controlling agency prior to
    entering the MOA.
2--may operate only on the airways that transverse the MOA.
3--should exercise extreme caution when military activity is being
    conducted.
4--must operate only when military activity is not being
    conducted.

38. (FAA 1461) Who is responsible for collision avoidance in an
alert area?

1--The controlling agency.
2--All pilots, without exception.
3--Only the pilots transitioning the area.
4--All pilots except those participating in the training
    operation.

39. (FAA 1462) What minimum altitude is requested for aircraft
over national wildlife refuges?

1--500 ft. AGL   3--2,000 ft. AGL
2--1,000 ft. AGL   4--3,000 ft. AGL

40. (FAA 1463) Basic radar service in the terminal radar program
is best described as

1--traffic advisories and limited vectoring to VFR aircraft.
2--mandatory radar service provided by the ARTS program.
3--wind shear warning at participating airports.
4--sequencing and separation service to IFR aircraft.

41. (FAA 1464) Stage III service in the terminal radar program provides

1--IFR separation (1,000 ft. vertical and 3 mi. lateral) between all aircraft.
2--warning to pilots when their aircraft are in unsafe proximity to terrain, obstructions, or other aircraft.
3--sequencing and separation for participating VFR aircraft.
4--service to aircraft principally during the en route phase of flight.

42. (FAA 1465) From whom should a departing VFR aircraft request Stage II terminal radar advisory service during ground operations?

1--The nearest FSS.
2--Clearance delivery.
3--Tower controller, just before takeoff.
4--Ground control, on initial contact.

43. (FAA 1466) If radar traffic information is desired, which action should a pilot take prior to entering a TRSA?

1--Prior to entering the TRSA, contact approach control on the appropriate frequency.
2--Contact the tower and request permission to enter the TRSA.
3--Maintain an altitude below 2,000 ft. AGL prior to entering the TRSA.
4--Maintain an altitude at or below 3,000 ft. AGL until entering the airport control zone, then make initial contact with approach control for radar service.

44. (FAA 1467) Stage III service within TRSA utilize radar to provide

1--separation between all VFR aircraft operating within the TRSA.
2--radar vectoring if the weather minimums are below VFR conditions.
3--separation between IFR aircraft, because VFR aircraft are not permitted in the area.
4--separation between all participating VFR traffic and IFR aircraft operating within the TRSA.

45. (FAA 1468) What type airspace is designated as VOR Federal airways?

1--Control areas.
2--Positive control areas.
3--Transition areas.
4--Continental control areas.

46. (FAA 1469) If air traffic control advises that radar service
is being terminated when the pilot is departing a terminal radar
service area, that transponder should be set to code

1--0000.   3--4096.
2--1200.   4--7700.

47. (FAA 1477) What is the upper limit of a control zone in the
conterminous United States?

1--2,000 ft. above the surface.
2--Up to, but not including 3,000 ft. AGL.
3--The base of the continental control area.
3--There is no upper limit.

48. (FAA 1478) What are the horizontal limits of an airport
traffic area?

1--3 SM (statute mi.) from the airport boundary.
2--5 SM (statute mi.) from the airport boundary.
3--5 SM (statute mi.) from the geographical center of the airport.
4--3 SM (statute mi.) from the geographical center of the airport.

49. (FAA 1479) The vertical limit of an airport traffic area is
from the surface up to

1--and including 1,500 ft.
2--and including 2,000 ft.
3--but not including 3,000 ft.
4--the base of the overlying control area.

50. (FAA 1480) Prior to entering an airport advisory area, a
pilot

1--must obtain a clearance from air traffic control.
2--should monitor ATIS for weather and traffic advisories.
3--should contact approach control for vectors to the traffic
   pattern.
4--should contact the local FSS for airport and traffic
   advisories.

51. (FAA 1481) To determine if UNICOM is available at an airport
without a control tower, refer to the

1--ATIS.
2--appropriate Airport/Facility Directory.
3--Graphic Notices and Supplemental Data.
4--NOTAMS.

52. (FAA 1483) When landing at an airport that does not have a
tower, FSS, or UNICOM, broadcast your intentions in the blind on

1--123.0 MHz.   3--121.5 MHz.
2--123.6 MHz.   4--122.9 MHz.

53. (FAA 1484) After landing at a tower-controlled airport, when
should the pilot contact ground control?

1--Prior to turning off the runway.
2--After reaching a taxiway that leads directly to the parking
   area.
3--After leaving the runway and crossing the runway holding lines.
4--When advised by the tower to do so.

54. (FAA 1485) To obtain airport advisories a pilot should
contact an FSS on

1--123.6 MHz.   3--121.5 MHz.
2--122.8 MHz.   4--122.0 Mhz.

55. (FAA 1486) If instructed by ground control to taxi to Rwy 9,
the pilot may proceed

1--via taxiways and across runways to, but not onto, Rwy 9.
2--to the next intersecting runway where further clearance is
   required.
3--via taxiways and across runways to Rwy 9, where an immediate
   takeoff may be made.
4--via any route at the pilot's discretion onto Rwy 9 and hold
   until cleared for takeoff.

56. (FAA 1487) Absence of the sky condition and visibility on an
ATIS broadcast indicates that

1--the ceiling is at least 5,000 ft. and visibility is 5 mi. or
   more.
2--the sky condition is clear and visibility is unrestricted.
3--weather conditions are at or above VFR minimums.
4--the ceiling is at least 3,000 ft. and visibility is 7 mi. or
   more.

57. (FAA 1488) Pilots of aircraft arriving or departing certain
high activity areas can receive continuous broadcasts concerning
essential but routine information by monitoring

1--UNICOM.
2--ATIS.
3--aeronautical multicom service.
4--radar traffic information service.

58. **(FAA 1489)** ATIS is the continuous broadcast of recorded information

1--alerting pilots of radar-identified aircraft when their aircraft is in dangerous proximity to terrain or an obstruction.
2--concerning nonessential information to receive frequency congestion.
3--concerning noncontrol information in selected high-activity terminal areas.
4--concerning sky conditions limited to ceilings below 1,000 ft. and visibilities less than 3 mi.

59. **(FAA 1490)** While flying on an east heading, a radar ATC facility issues the following warning:

"TRAFFIC 3 O'CLOCK, 2 MILES, WESTBOUND. . . ."

The pilot should look for this traffic in the direction of the airplane's

1--left wingtip, and ahead of the wing.
2--left wingtip.
3--right wingtip.
4--nose and slightly to the right.

60. **(FAA 1491)** While flying north, a radar ATC facility issues the following warning:

TRAFFIC 10 O'CLOCK, 2 MILES, SOUTHBOUND. . . ."

The pilot would look for this traffic in a direction approximately

1--10° behind the left wing of the airplane.
2--60° to the left of the nose of the airplane.
3--30° to the right of the nose of the airplane.
4--20° to the left of the nose of the airplane.

61. **(FAA 1492)** During a flight, a radar controller advises:

"TRAFFIC 2 O'CLOCK, 5 MILES, NORTHBOUND. . . ."

Where should the pilot look for the traffic?

1--Just ahead of the right wingtip.
2--Slightly to the right of the airplane's nose.
3--Straight ahead.
4--Just ahead of the aircraft's left wingtip.

62. (FAA 1493) While flying on a north heading in calm wind, radar controller advises:

"TRAFFIC 9 O'CLOCK, 2 MILES, SOUTHBOUND. . . ."

The pilot should look for this traffic

1--behind the right wingtip.
2--off the left wingtip.
3--ahead of the left wingtip.
4--off the right wingtip.

63. (FAA 1496) If the aircraft's radio fails, what is the recommended procedure when landing at a controlled airport?

1--Observe the traffic flow, enter the pattern, and look for a
    light signal from the tower.
2--Enter a crosswind leg and rock the wings.
3--Flash the landing lights and cycle the landing gear while
    circling the airport.
4--Select 7700 on the transponder and fly a normal traffic
    pattern.

64. (FAA 1517) What minimum condition must exist to operate within a control zone special VFR?

1--2400 RVR.
2--500 ft. below clouds.
3--1,000 ft. above clouds.
4--1 mi. flight visibility.

65. (FAA 1518) Who should the pilot contact to receive a special VFR clearance in a control zone collocated within an airport traffic area?

1--FSS.
2--Airport control tower.
3--Air route traffic control center.
4--Approach control.

66. (FAA 1527) Where are airport traffic areas in effect?

1--At all airports that have a flight service station on the
    field.
2--Only at airports within a control zone.
3--Only at airports that have an operating control tower.
4--At all airports.

67. (FAA 1528) An airport traffic area is automatically in effect when

1--its associated control tower is in operation.
2--the weather is below VFR minimums.
3--nighttime hours exist.
4 radar service is available.

68. (FAA 1529) What are the horizontal limits of an airport
traffic area?

1--3 SM (statute mi.) from the airport boundary.
2--5 SM (statute mi.) from the airport boundary.
3--5 SM (statute mi.) from the geographical center of the airport.
4--3 SM (statute mi.) from the geographical center of the airport.

69. (FAA 1530) The vertical limit of an airport traffic area is
from the surface up to

1--and including 2,000 ft.
2--and including 3,000 ft.
3--but not including 3,000 ft.
4--the base of the overlying control area.

70. (FAA 1564) Unless otherwise specified, Federal airways
extend from

1--1,200 ft. above the surface upward to 14,500 ft. MSL and are 16
   NM wide.
2--700 ft. above the surface upward to the continental control
   area and are 10 NM wide.
3--1,200 ft. above the surface upward to 18,000 ft. MSL and are 8
   NM wide.
4--the surface upward to 18,000 ft. MSL and are 4 NM wide.

71. (FAA 1565) What type airspace is designated as VOR Federal
airways?

1--Control areas.
2--Positive control areas.
3--Transition areas.
4--Continental control area.

72. (FAA 1566) Control areas within the contiguous United States
extend upward from either 700 ft. or 1,200 ft. above the surface
to, but not including,

1--3,000 ft. MSL.
2--18,000 ft. MSL.
3--24,000 ft. MSL.
4--the base of the continental control area.

73. (FAA 1567) Within the contiguous United States, a control
zone extends from the surface upward to the base of

1--an airport traffic area.
2--the continental control area.
3--a transition area.
4--a terminal control area.

74. (FAA 1568) Control areas within the contiguous United States extend upward from either 700 ft. or 1,200 ft. above the surface to, but not including,

1--the base of the continental control area.
2--3,000 ft. MSL.
3--18,000 ft. MSL.
4--24,000 ft. MSL.

75. (FAA 1701) What is the minimum weather condition required for airplanes operating under special VFR in a control zone?

1--1-mi. flight visibility.
2--1-mi. flight visibility and 1,000-ft. ceiling.
3--3-mi. flight visibility and 1,000-ft. ceiling.
4--3-mi. flight visibility.

76. (FAA 1633) Unless otherwise authorized or required by air traffic control, the maximum indicated airspeed at which a reciprocating engine equipped aircraft should be flown within an airport traffic area is

1--156 kts.   2--180 kts.   3--200 kts.   4--288 kts.

77. (FAA 1644) An air traffic control clearance provides

1--authorization for flight in uncontrolled airspace.
2--priority over all other traffic.
3--adequate separation from all traffic.
4--authorization to proceed under specified traffic conditions in
   controlled airspace.

78. (FAA 1647) A flashing green air traffic control signal directed to an aircraft on the surface is a signal that the pilot

1--is cleared to taxi.
2--should exercise extreme caution.
3--should taxi clear of the runway in use.
4--should stop taxiing.

79. (FAA 1648) While on the final approach for landing, an alternating green and red light followed by a flashing red light is received from the control tower. Under these circumstances, the pilot should

1--land and clear the runway in use as safely and quickly as
   possible.
2--discontinue the approach, fly the same traffic pattern and
   approach again, and land.
3--abandon the approach, realizing the airport is unsafe for
   landing.
4--abandon the approach, circle the airport to the right, and
   expect a flashing white light when the airport is safe for
   landing.

80. (FAA 1649) A steady green air traffic control light signal
directed to an aircraft in flight is a signal that the pilot

1--is cleared to land.
2--should give way to other aircraft and continue circling.
3--should return for landing.
4--should exercise extreme caution.

81. (FAA 1650) What is the difference between a steady red and a
flashing red light signal from the tower to an aircraft
approaching to land?

1--Both light signals mean the same except the flashing red light
    requires a more urgent reaction.
2--A steady red light signals to continue circling and a flashing
    red light signals that the airport is unsafe for landing.
3--A steady red light signals to continue circling and a flashing
    red light signals to continue, but exercise extreme caution.
4--A steady red light signals that the airport is unsafe and a
    flashing red light signals to use a different runway.

82. (FAA 1651) A flashing white light signal from the control
tower to an aircraft taxiing is an indication

1--to taxi at a faster speed.
2--to taxi only on taxiways and not cross runways.
3--to return to the starting point on the airport.
4--that instrument conditions exist.

83. (FAA 1663) What is the purpose of an airport traffic area?

1--To provide for the control of aircraft landing and taking off
    from an airport with an operating control tower.
2--To provide for the control of all aircraft operating in the
    vicinity of an airport with an operating control tower.
3--To provide for the control of air traffic within a control zone
    that has an operating control tower.
4--To restrict aircraft without radios from operating in the
    vicinity of an airport with an operating control tower.

84. (FAA 1664) Unless otherwise authorized, two-way radio
communications with air traffic control are required for landings
or takeoffs

1--within control zones regardless of the weather conditions.
2--at all tower controlled airports regardless of the weather
    conditions.
3--at all tower controlled airports only when weather conditions
    are less than VFR.
4--at tower controlled airports within control zones only when
    weather conditions are less than VFR.

85. (FAA 1665) When is an airport traffic area in effect?

1--From 1 hr. before sunrise to 1 hr. after sunset.
2--When the associated control tower is in operation.
3--When the associated FSS is in operation.
4--From sunrise to sunset.

86. (FAA 1667) Which is the correct traffic pattern departure
procedure to use at a noncontrolled airport?

1--Depart in any direction consistent with safety, after crossing
   the airport boundary.
2--Make all turns to the left.
3--Comply with any FAA traffic pattern established for the
   airport.
4--Depart as prearranged with other pilots using the airport.

87. (FAA 1669) Regardless of weather conditions, air traffic
control authorization is required prior to operating an aircraft
within a

1--control zone.
2--terminal radar service area.
3--terminal control area.
4--transition area.

88. (FAA 1670) Which equipment is required when operating an
airplane within a Group II terminal control area?

1--A 4096 code transponder.
2--A 4096 code transponder with Mode C capability.
3--A VOR receiver with distance measuring equipment.
4--An automatic direction finder.

89. (FAA 1671) What minimum pilot certification is required for
operating in a Group I terminal control area?

1--Student pilot certificate with appropriate logbook
   endorsements.
2--Private pilot certificate.
3--Private pilot certificate with an instrument rating.
4--Commercial pilot certificate.

90. (FAA 1672 DI) What minimum navigation equipment is required
for operating in a Group I terminal control area?

1--VOR or TACAN receiver.
2--VOR or TACAN and a 4096 transponder.
3--VOR or TACAN and a 4096 transponder with Mode C capability.
4--DME, VOR or TACAN, and a transponder.

91. (FAA 1673) In which type of airspace is VFR flight prohibited?

1--Terminal control area.
2--Continental control area.
3--Control zone.
4--Positive control area.

92. (FAA 1674) The minimum flight visibility required for VFR flights above 10,000 ft. MSL and more than 1,200 ft. AGL is

1--1 SM.
2--3 SM.
3--5 SM.
4--not specified by regulation.

93. (FAA 1686) The minimum distance from clouds required for VFR operations on an airway below 10,000 ft. is

1--remain clear of clouds.
2--500 ft. below, 1,000 ft. above, and 2,000 ft. horizontally.
3--500 ft. above, 1,000 ft. below, and 2,000 ft. horizontally.
4--1,000 ft. above, 1,000 ft. below, and 1 mi. horizontally.

94. (FAA 1687) What minimum flight visibility is required for VFR flight operations on an airway below 10,000 ft.?

1--1 mi.   3--4 mi.
2--3 mi.   4--5 mi.

95. (FAA 1688) What minimum visibility and clearance from clouds are required for airplane VFR operations in uncontrolled airspace at 1,200 ft. AGL or below?

1--1 mi. visibility and clear of clouds.
2--3 mi. visibility and clear of clouds.
3--1 mi. visibility, 500 ft. below, 1,000 ft. above, and 2,000 ft. horizontal clearance from clouds.
4--3 mi. visibility, 500 ft. above, 1,000 ft. below, and 2,000 ft. horizontal clearance from clouds.

96. (FAA 1689) During operation outside controlled airspace at altitudes of more than 1,200 ft. AGL, but less than 10,000 ft. MSL, the minimum horizontal distance from clouds requirement for VFR flight is

1--500 ft.   3--1,500 ft.
2--1,000 ft.   4--2,000 ft.

97. (FAA 1690) According to FARs, VFR flight above 1,200 ft. AGL and below 10,000 ft. MSL requires a minimum visibility and vertical cloud clearance of

1--3 mi., and 500 ft. below or 1,000 ft. above the clouds in controlled airspace.
2--5 mi., and 1,000 ft. below or 1,000 ft. above the clouds at all altitudes.
3--5 mi., and 1,000 ft. below or 1,000 ft. above the clouds in the continental control area.
4--3 mi., and 1,000 ft. below or 2,000 ft. above the clouds at all altitudes within and outside of controlled airspace.

98. (FAA 1691) During operations within controlled airspace at altitudes of more than 1,200 ft. AGL, but less than 10,000 ft. MSL, the minimum distance above clouds requirement for VFR flight is

1--500 ft.   3--1,500 ft.
2--1,000 ft.   4--2,000 ft.

99. (FAA 1692) During operations at altitudes of more than 1,200 ft. AGL and at or above 10,000 ft. MSL, the minimum distance above clouds requirement for VFR flight is

1--500 ft.   3--1,500 ft.
2--1,000 ft.   4--2,000 ft.

100. (FAA 1693) Outside controlled airspace, the minimum flight visibility requirement for VFR flight above 1,200 ft. AGL and below 10,000 ft. MSL is

1--1 mi.   3--5 mi.
2--3 mi.   4--7 mi.

101. (FAA 1694) The minimum ceiling and visibility to operate an airplane VFR in a control zone are

1--500 ft. and 1 mi.
2--1,000 ft. and 3 mi.
3--1,400 ft. and 2 mi.
4--2,000 ft. and 3 mi.

102. (FAA 1695) During operations outside controlled airspace at altitudes of more than 1,200 ft. AGL, but less than 10,000 ft. MSL, the minimum distance below clouds requirement for VFR flight is

1--500 ft.   3--1,500 ft.
2--1,000 ft.   4--2,000 ft.

103. (FAA 1696) During operations within controlled airspace at altitudes of less than 1,200 ft. AGL, the minimum horizontal distance from clouds requirement for VFR flight is

1--500 ft.   3--1,500 ft.
2--1,000 ft.   4--2,000 ft.

104. (FAA 1697) During VFR operations outside controlled airspace at altitudes of less than 1,200 ft. AGL, the minimum flight visibility requirement when operating airplanes is

1--1 SM.
2--3 SM.
3--5 SM.
4--not specified by regulations.

105. (FAA 1679) For VFR flight operations above 10,000 ft. MSL and more than 1,200 ft. AGL, the minimum horizontal distance from clouds required is

1--1,000 ft.   3--1 mi.
2--2,000 ft.   4--5 mi.

106. (FAA 1698) A special VFR clearance authorizes the pilot of an aircraft to operate VFR while within a control zone

1--when the ceiling is less than 1,000 ft. and visibility less than 1 mi. if the aircraft does not exceed maneuvering speed.
2--if clear of clouds and the visibility is at least 1 mi.
3--with no minimum visibility requirements if clear of clouds.
4--at or below cloud base with a flight visibility of 1 mi. or less, provided the aircraft remains below 1,000 ft. above the surface.

107. (FAA 1699) No person may operate an airplane within a control zone at night under special VFR unless

1--the flight can be conducted 500 ft. below the clouds.
2--an instructor is aboard.
3--the airplane is equipped for instrument flight.
4--the flight visibility is at least 3 mi.

108. (FAA 1700) What are the minimum requirements for airplane operations under special VFR in a control zone at night?

1--The minimum visibility is 3 mi. and the airplane must operate clear of clouds.
2--The airplane must be under radar surveillance at all times while in the control zone.
3--The airplane must be equipped for IFR and with an altitude reporting transponder.
4--The pilot must be instrument rated and the airplane must be IFR equipped.

109. (FAA 1660) If an altimeter setting is not available before flight, to which altitude or setting should the pilot adjust the altimeter?

1--To 29.92" Hg for flight below 18,000 ft. MSL.
2--The elevation of the nearest airport corrected to mean sea level.
3--The elevation of the departure area.
4--Pressure altitude corrected for non-standard temperature.

110. (FAA 1661) When is it first required that the altimeter be set to 29.92" Hg, when climbing to cruising altitude or flight level?

1--12,500 ft. MSL.   3--18,000 ft. MSL.
2--14,500 ft. MSL.   4--24,000 ft. MSL.

111. (FAA 1662) Prior to takeoff, the altimeter should be set to

1--the current local altimeter setting, if available, or the departure airport elevation.
2--the corrected density altitude of the departure airport.
3--the corrected pressure altitude for the departure airport.
4--29.92" Hg.

112. (FAA 1501) In VFR conditions below 18,000 ft. MSL, what transponder code should be selected?

1--Code 1200, and the Ident feature should not be engaged.
2--Code 1200, and the Ident feature should be engaged.
3--Code 1400, and the Ident feature should not be engaged.
4--Code 1400, and the Ident feature should be engaged.

113. (FAA 1601) An operable transponder is required in which airspaces?

1--Group I TCAs, ADIZ, and positive control areas.
2--Group I and Group II TCAs only.
3--High-density airport traffic areas and Group I TCAs.
4--Group I TCAs and Group II TCAs and positive control areas.

114. (FAA 1598) Where, in the 48 contiguous United States and the District of Columbia, is a radar beacon transponder equipped with a Mode 3/A 4096 code capability required?

1--In Group I TCAs only.
2--In all TCAs only.
3--In all TCAs and in controlled airspace above 12,500 ft. MSL if more than 2,500 ft. AGL.
4--In controlled airspace above 12,500 ft. MSL if more than 2,500 ft. AGL only.

115. The runway being used for takeoffs and landings is referred
to as the _____ runway.

1--active   3--current
2--landing   4--authorized

116. Which of the following wind indicators has its small end
pointing into the direction of the wind?

1--Tetrahedron.
2--Tetrahedron, wind cone.
3--Tetrahedron, wind T.
4--Wind cone, wind T.

117. Solid yellow lines that run perpendicular to the taxiway are
called

1--displaced thresholds.   3--stopways.
2--holding lines.   4--thresholds.

118. Unless otherwise indicated, a magenta boundary around a
control area means that it has a floor of

1--700 ft. AGL.   3--3,000 ft. AGL.
2--1,200 ft. AGL.   4--14,500 ft. MSL.

119. What is the correct way to state a barometric reading of
29.32" Hg?

1--Two niner three two.
2--Two niner point three two.
3--Twenty-nine thirty-two.
4--Twenty-nine point thirty-two.

120. What is the correct way to state an altitude of 11,200 ft.?

1--One one thousand two hundred.
2--One one thousand two zero zero.
3--Eleven thousand two hundred.
4--Eleven thousand two hundred zero zero.

121. Which of the following terms is used to double-check the
accuracy of a transmission?

1--Acknowledge.   3--Roger.
2--Affirmative.   4--Verify.

122. Lights installed on the wingtips and tail of an airplane
are called

1--anticollision lights.
2--directional lights.
3--position lights.
4--taxiing lights.

ANSWERS

Key Terms and Concepts, Part 1

1. D	6. C	11. A	16. P
2. E	7. N	12. I	17. M
3. F	8. B	13. R	18. J
4. G	9. H	14. L	19. S
5. K	10. O	15. Q	20. T

Key Terms and Concepts, Part 2

1. C	6. H	11. R	16. Q
2. E	7. A	12. F	17. D
3. S	8. N	13. O	18. I
4. K	9. B	14. P	19. M
5. G	10. L	15. J	20. T

Key Terms and Concepts, Part 3

1. Q	6. C	11. S	16. L
2. A	7. O	12. F	17. I
3. H	8. D	13. B	18. R
4. P	9. M	14. G	19. K
5. J	10. E	15. T	20. N

Key Terms and Concepts, Part 4

1. J	6. K	11. S	16. H
2. C	7. P	12. Q	17. D
3. O	8. M	13. R	18. G
4. L	9. T	14. F	19. E
5. I	10. N	15. A	20. B

Discussion Questions and Exercises

7.   Takeoff leg: a heading of 280°, climb to at least 300
        ft. AGL (3,020 ft. on the altimeter), and make certain
        you are past the approach end of the runway. Crosswind:
        turn right to a heading of 010° and continue your climb.
        Downwind: turn right to a heading of 100° and continue
        your climb to 3,700 ft. MSL (1,000 ft. AGL). Base leg:
        turn right to a heading of 190° and continue your
        descent. Final: continue your descent and turn right to
        a heading of 280°. If you have set your altimeter (as
        required by the FAA) to 29.54, your altimeter will
        correspond to current conditions, and no further altitude
        corrections will have to be made.

13. a. "Tripacer two four niner seven charlie"
     b. "Heading two niner zero"
     c. "Three thousand one hundred twenty feet"
     d. "One five thousand eight hundred feet"
     e. "Two two four zero zulu"
     f. "Kansas City Center"
     g. "Birmingham Radio"
     h. "Say again"
     i. "Verify"
     j. "Acknowledge'
     k. "One two three point six"
     l. "Three zero zero nine" (no decimal)
     m. "12 o'clock"

Review Questions

1.   1--When there is a quartering headwind, the aileron should
        be up into the wind and the elevator in the neutral
        position. When there is a quartering tailwind, the
        aileron should be down into the wind and the elevator
        down.
2.   4--See answer 1.
3.   1--See answer 1.
4.   4--See answer 1.
5.   4--See answer 1.
6.   1--See answer 1.
7.   4--A quartering tailwind is most critical.
8.   1--The position lights are located as follows: left wing--
        red light; right wing--green light; tail--white light. The
        aircraft is crossing your flight path to the left.
9.   4--The aircraft is flying away from you. See answer 8.
10. 3--The aircraft is approaching head-on. See answer 8.
11. 4--VFR approaches should be done the same whether flying at
        night or during the day.
12. 4--It helps to label the runways and to position yourself as
        you would depart from each one, noting the traffic
        pattern in effect. The wind cone indicates a wind from
        the northwest (300° to 330°), so you would land on either
        Rwy 35 or Rwy 27. Please note that in some questions, it
        is Rwy 27 and in others, Rwy 26.
13. 3--See answer 12.
14. 1--See answer 12.
15. 3--See answer 12.
16. 2--The segmented circle indicates right-hand traffic for Rwy
        18. You would not use Rwy 22 since the X tells you that
        it is closed.
17. 2--See answer 16. Note that the tetrahedron tells you the
        wind is directly down Rwy 22, so there would be a right
        quartering crosswind.
18. 2--During daylight hours, it indicates weather below VFR
        minimums.
19. 2--Military airports have two quick white flashes between
        intermittent green flashes.

20.   3--With a dual-color VASI, white over white means you're too
      high, red over red means you're dead (too low), red over
      white is just right (on the glide slope).
21.   3--See answer 20.
22.   4--See answer 20.
23.   2--With a tri-color VASI, red indicates you are below the
      glide path, green means you are on the glide path, and
      amber means you are above it.
24.   3--See answer 23.
25.   4--See answer 23.
26.   3--See answers 20 and 23.
27.   4--Blue lights indicate taxiways and taxiway turnoffs.
28.   1--Runway headings are given relative to magnetic north to
      the nearest 10°. The last zero is dropped. Thus, a
      runway with a magnetic heading of 342° would be rounded
      to 340° and called Rwy 34.
29.   4--An X on the runway indicates that it is closed.
30.   4--The arrows leading to the band of markers at the approach
      end of Rwy 12 indicate that this portion of the runway
      may be used for taxiing and the takeoff roll, but you
      must land past the markers.
31.   1--See answer 30. The area of the runway covered with
      chevrons (Section E) is an overrun area that cannot be
      used for taxi, takeoff, or landing (except when the
      aircraft overruns Rwy 12).
32.   4--An X on the runway indicates that it is closed.
33.   2--See answer 30.
34.   3--See answer 30.
35.   2--Civil pilots must obtain authorization to enter a
      restricted area.
36.   1--Warning areas are typically located over international
      water and often contain unusual hazards.
37.   3--An MOA frequently has military training activity, during
      which the pilot should exercise extreme caution as these
      aircraft may operate at high speeds.
38.   2--In an alert area, all pilots are responsible for
      collision avoidance.
39.   3--The Fish and Wildlife Service requests pilots to fly at
      least 2,000 ft. above the terrain.
40.   1--Basic radar service in the terminal radar program
      involves traffic advisories and limited vectoring to VFR
      aircraft that request it.
41.   3--Stage III service in the terminal radar program provides
      sequencing and separation for **participating** VFR aircraft.
42.   4--Departing VFR pilots should contact ground control to
      request Stage II terminal radar advisory service.
43.   1--The pilot should contact approach control **prior** to
      entering a TRSA.
44.   4--Stage III service with a TRSA provides separation to all
      participating VFR traffic and to IFR traffic operating
      within the TRSA.
45.   1--A VOR Federal airway is defined as a control area.
46.   2--1200 is the VFR transponder code.
47.   3--The continental control area defines the upper limit of a
      control zone in the conterminous United States.

48. 3--An airport traffic area extends 5 SM from the
         geographical center of the airport and up to, but not
         including 3,000 ft. AGL.
49. 3--See answer 48.
50. 4--FSS provides airport and traffic advisories.
51. 2--UNICOM frequencies for airports with a control tower can
         be found on aeronautical charts and in the appropriate
         Airport/Facility Directory.
52. 4--Use the MULTICOM frequency, 122.9.
53. 4--You should contact ground control only after being
         advised to do so by the tower.
54. 1--123.6 is the standard FSS frequency on which to call for
         airport advisories.
55. 1--If you are cleared to taxi to the active runway, it means
         you may taxi via taxiways to the active runway, crossing
         all other runways in the process. You may not, however,
         cross the active runway at any point, including the
         departure end.
56. 1--When ATIS does not include sky condition and visibility
         information, it means the ceiling is at least 5,000 ft.
         and visibility is 5 mi. or greater.
57. 2--ATIS broadcasts essential but routine information at
         selected high-density airports. The information
         contained in ATIS broadcasts is **not** air traffic control
         directions.
58. 3--See answer 57.
59. 3--The nose of the plane points to 12 o'clock and the tail
         points to 6 o'clock. The right wing points to 3 o'clock
         and the left wing points to 9 o'clock. This is called
         the clock reference system.
60. 2--See answer 59.
61. 1--See answer 59.
62. 2--See answer 59.
63. 1--If your airplane fails, observe the traffic flow, enter
         the pattern, and look for a light signal from the tower.
64. 4--One mi. flight visibility is the minimum condition for
         special VFR within a control zone.
65. 2--The airport traffic area has an operating control tower,
         thus contact the airport control tower.
66. 3--An airport traffic area is in effect only when there is
         an operating control tower.
67. 1--See answer 66.
68. 3--See answer 48.
69. 3--See answer 48.
70. 3--Federal airways are control areas that exist 1,200 ft.
         above the surface upward to 18,000 ft. MSL and are 8 NM
         wide.
71. 1--See answer 70.
72. 4--See answer 47.
73. 2--See answer 47.
74. 1--See answer 47.
75. 1--See answer 64.
76. 1--156 kts. is the speed limit for reciprocating engine
         equipped aircraft that are operating within an airport
         traffic area.

77. 4--Air traffic control clearance provides authorization to
     proceed under specified traffic conditions in controlled
     airspace.
78. 1--See Figure 7.6.
79. 3--See Figure 7.6.
80. 1--See Figure 7.6.
81. 2--See Figure 7.6.
82. 3--See Figure 7.6.
83. 1--See answer 66.
84. 2--See answer 66.
85. 2--See answer 66.
86. 3--Each airport has a traffic pattern established by the FAA
     that is used to govern pattern procedures.
87. 3--Regardless of weather conditions, one must be authorized
     to operate within a terminal control area.
88. 1--A 4096 code transponder is required to operate an
     airplane within a Group II terminal control area.
89. 2--A pilot must hold at least a private pilot certificate to
     operate in a Group I terminal control area. Student
     pilots are permitted to fly through Group I TCAs, but not
     to land at them.
90. ?--This question has been designated as unusable by the FAA
     and removed from the FAA Question Selection Sheets.
91. 4--VFR flight is prohibited in a positive control area.
92. 3--See Figure 7.7.
93. 2--See Figure 7.7.
94. 2--See Figure 7.7.
95. 1--See Figure 7.7.
96. 4--See Figure 7.7.
97. 1--See Figure 7.7.
98. 2--See Figure 7.7.
99. 2--See Figure 7.7.
100. 1--See Figure 7.7.
101. 2--See Figure 7.7.
102. 1--See Figure 7.7.
103. 4--See Figure 7.7.
104. 1--See Figure 7.7.
105. 3--See Figure 7.7.
106. 2--See Figure 7.7.
107. 3--To fly special VFR at night, the pilot must be instrument
     rated and the airplane must be equipped for instrument
     flight.
108. 4--See answer 107.
109. 3--The altimeter should be set to the elevation of the
     departure area.
110. 3--29.92" Hg is used at 18,000 ft. MSL and.
111. 1--Prior to departure, the altimeter should be set to the
     current local altimeter setting, if available, or to the
     departure airport's elevation.
112. 1--Code 1200 should be used in VFR conditions; the Ident
     feature should not be engaged unless requested by air
     traffic control.

Type and color of signal	Meaning of the signal	
	Aircraft on the ground	**Aircraft in flight**
Steady green	Cleared for takeoff	Cleared to land
Flashing green	Cleared to taxi	Return for landing (followed by steady green at proper time)
Steady red	Stop	Yield to other aircraft and continue circling
Flashing red	Taxi clear of runway in use	Airport unsafe—do not land
Flashing white	Return to starting point on airport	——
Alternating red and green	General warning signal—exercise extreme caution	

Figure 7.6

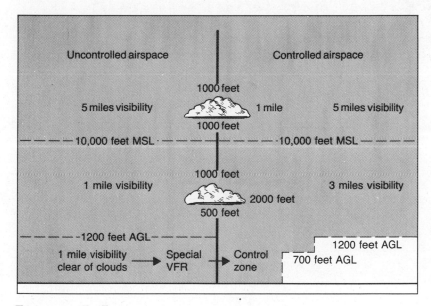

Figure 7.7

113. 4--An operable 3/A 4096 code transponder is required in Group I and II TCAs and in positive control space. In addition, an altitude encoding transponder (mode C) is required in all controlled airspace of the 48 contiguous states and the District of Columbia above 12,500 ft. MSL and in Group I TCAs.
114. 3--See answer 113.
115. 1--This is called the active runway.
116. 1--The tetrahedron's small end points into the wind; for the wind cone and the wind T, the large end points into the wind.
117. 2--These are called holding lines. At airports with an operating control tower, they indicate that you should stop and wait for further instructions.
118. 1--A magenta boundary indicates 700 ft. and a blue boundary indicates 1,200 ft. The upward boundary is 14,500 ft. MSL.
119. 1--Barometric pressure is stated without the decimal.
120. 1--The thousands are stated first (stated separately), followed by the hundreds.
121. 4--"Verify" is the word used to indicate that you should double-check the accuracy of your transmission.
122. 3--Position lights are located on the wingtips and tail.

8/METEOROLOGY: A PILOT'S VIEW OF WEATHER

MAIN POINTS

1.   **Air** consists primarily of nitrogen (78 percent), oxygen (21 percent), other gases, and varying amounts of water vapor. The **atmosphere** contains air and **particulate matter** such as pollen and pollutants that can affect precipitation as well as pilot visibility. The three levels of the atmosphere are the **troposphere** (surface to 29,000 ft. near the poles and 54,000 ft. near the equator), where most weather occurs; the **stratosphere** (29,000-54,000 ft. to 40-60 mi.), where the atmosphere is stable; and the **ionosphere** (upward to about 300 mi.).

2.   Atmospheric pressure can be measured: (1) in inches of mercury--the International Standard Atmosphere is the pressure at sea level, 29.92" Hg); (2) in pounds per square inch (psi)--the sea level standard is 14.7 in.; and (3) in millibars (mb--the standard is 1013.2). One psi equals 33.86 mb. Millibar pressures are common on weather charts. Atmospheric pressure decreases as altitude increases at a decreasing rate; that is, lower altitudes lose pressure more quickly than do higher altitudes.

3.   As water vapor increases, **condensation** (for example, clouds) or **precipitation** (discharge of moisture) may occur. Precipitation is affected by temperature (warm air can hold more moisture than can cold air) and the presence of particles on which moisture can condense. **Relative humidity** refers to the percentage of water vapor present compared to the amount it **could** hold under current conditions. When this value is 100 percent, the air is called **saturated**. The temperature at which saturation occurs is called the **dew point**. Because condensation is more likely to occur as the temperature of the air decreases, it is important to monitor both the current temperature and the dew point. When the spread between OAT and dew point is within 4° F and narrowing, condensation (for example, fog or low clouds) should be expected. Formation of water droplets also depends on

171

particles in the air called **condensation nuclei**. Two ways by which water changes to gas are **evaporation** (water to gas) and **sublimation** (solid to gas).

    4.   The surface of the earth absorbs and releases heat, depending upon its color and texture. For example, the land both absorbs and releases heat more quickly than water, which accounts for many weather patterns in coastal areas. The amount of radiation from the sun is less during the winter than it is during the summer.

    5.   Air loses temperature at an average rate of 2° C per 100 ft. in the troposphere (called the **standard lapse rate**), but this can be affected by many variables. It is important to know how the air and the surface below it are heated. **Radiation** heats the air and surface as it strikes them. **Conduction** refers to heating by direct contact, as when molecules collide with each other. **Convection** refers to currents set up as hot air rises or cool air falls. Horizontal transfer of heat (for example, a moving air mass) is called **advection**.

    6.   The principles of heat exchange affect the global circulation of the atmosphere, as does Earth's rotation. The apparent motion of the air is due to the **Coriolis force**, and it imparts a lateral component to the flow of winds. Generally, in the Northern Hemisphere, polar winds are from the east (**polar easterlies**); middle latitudes have **prevailing westerlies**; and, as you approach the equator, northeast winds (**northeast trades**) prevail. As polar easterlies meet prevailing westerlies, **circulation** begins, always in a counterclockwise direction around low-pressure regions and in a clockwise direction around high-pressure regions in the Northern Hemisphere.

    7.   Weather charts have lines of equal pressure called **isobars**. An elongated high-pressure area is called a **ridge**, and an elongated low-pressure area is called a **trough**. Winds above the friction level tend to flow parallel to the isobars. Rotational winds are called cyclones when associated with lows and anticyclones when associated with highs. How close together the bars are (called the **pressure gradient**) indicates the intensity of the winds. A steep gradient (bars close together) indicates higher winds.

    8.   The three **weather zones** are (a) the equator to 30° latitude--a belt of high pressure straddles the 30° line, (b) 30° to 60°--a belt of prevailing westerlies but much mixture of air from north and south, and (c) 60° to the pole--low pressure along 60° latitude. Other forces operating on the atmosphere include gravity, friction (for example, mountains), and centrifugal force.

    9.   The **ceiling** is defined as the height above the ground of the lowest layer of clouds, provided clouds cover more than half the sky. Sky cover is defined as **clear** (0 to 10 percent), scattered (10 to 50 percent), **broken** (60 to 90 percent), or

overcast (90 to 100 percent). **Visibility** refers to horizontal distance and is normally given in miles. Several conditions contribute to limited visibility. **Fog** is dangerous because it limits visibility severely. It is most likely to occur when relative humidity is high, condensation nuclei are present, and the air is cooling (as in the early evening after the sun sets). Four types of fog are **radiation fog**, or ground fog, which occurs under clear skies and calm winds; **advection fog**, which occurs when moist air moves over a cool surface, as in coastal areas; **upslope fog**, which occurs when moist air moves up to a cool region, as against a mountain; and **frontal fog**, which occurs when evaporating rain is lifted. **Precipitation** also affects visibility (snow more than light rain). Other **obscurations** (for example, dust and smog) limit visibility and may lead to reports of sky obscured or sky partially obscured since no ceiling can be accurately defined. The terms used to refer to weather conditions are **visual meteorological conditions (VMC)**, which means that flight can be maintained by the use of outside references, and **instrument meteorological conditions (IMC)**, which means that visual flight is not possible. These terms are not the same as VFR and IFR, both of which are a set of **rules**, not weather conditions. Finally, the most significant measure of visibility aloft is **slant range** ("over the nose").

   10. Ice that forms from a sublimation process (water vapor turns directly into a solid) is called **frost**. Frost on the airplane affects drag and lift and may prevent the airplane from becoming airborne. **Icing** is most likely to occur when there is visible moisture, OAT is between $-5°$ and $+3°$ C, and the aircraft skin is at or below freezing. VFR pilots should stay out of clouds and away from freezing rain. Structural ice comes in two varieties: **clear ice**, which is smooth-looking and clear; and **rime ice**, which is opaque and granular-looking. Ice reduces lift, increases the airplane's weight, decreases thrust, increases drag, and increases stalling speed.

   11. Vertical motion of the air can lead to unstable atmospheric conditions and turbulence. Dry air is more stable than moist air, and the **adiabatic lapse rate** is about $3°$ C per 1,000 ft. for dry air as compared to 1.1-$2.8°$ C for moist air. As mentioned earlier, the **standard lapse rate** is $2°$ C per 1,000 ft.; it is the average of lapse rates for dry and moist air. Convection currents (or thermals) develop as different shades and densities of the land differentially absorb and reflect heat, leading to vertical air movement. Around large bodies of water, convective currents produce **onshore winds** during the day as cool air over the ocean moves in to replace warm air over the land and **offshore winds** at night as this process reverses. Surface obstructions also affect turbulence. Even small buildings, hangars, and trees can affect the flow of the wind (referred to as land flow turbulence) and can be particularly disruptive as the wind increases in velocity. Flying near mountains can also be dangerous due to updrafts and downdrafts created by mountain flow. These vertical movements are also accentuated by an increase in the wind's velocity and can sometimes be identified

by **lenticular** (almond or lens-shaped) **clouds** on the leeward
(downwind) side of the mountain. **Wind shear**, common near
thunderstorms and mountains, is a sudden change in the wind's
horizontal or vertical movement. It is common when there is a
reversal of the lapse rate (**temperature inversion**) or when low-
altitude winds are different from surface winds. Finally, there
is **clear air turbulence (CAT)**, which is difficult to predict. It
typically occurs at higher altitudes.

12. Clouds are visible moisture and come in a variety of
different forms and sizes. They are classified by how they are
formed and by their altitudes. **Cumulus** (piled up) **clouds** are
formed by vertical currents of unstable air and have a fluffy
appearance. **Stratus** (layered) **clouds** are formed by the cooling
of a stable layer of air and have a sheet-like appearance.
Cirrus clouds are thin, wispy clouds that occur at high
altitudes. Additional descriptive words include **alto** (middle),
nimbo (rain), and **fracto** (broken). Low clouds typically have
bases less than 6,500 ft. AGL, middle clouds (alto) are found
between 6,500 ft. and 16,500 ft. AGL, and high clouds (cirriform)
occur between 16,500 ft. and 45,000 ft. AGL. The basic types of
cumulus clouds are cumulus, alto cumulus, stratocumulus, and
cumulonimbus. The stratus group includes stratus, altostratus,
and nimbostratus. The cirrus group includes cirrus,
cirrostratus, and cirrocumulus.

13. Of special note are **cumulonimbus** clouds, which are also
referred to as Cbs, TRWs, or thunderstorms. They are extremely
dangerous. A product of exaggerated vertical movement and
instability, Cbs come in a variety of types, such as **orographic**
(upslope lifting of moist air as it approaches a mountain range)
and **frontal** (one air mass lifts another). Thunderstorms develop
in three stages: **cumulus**, with rapid vertical development;
mature, marked by updrafts and heavy downdrafts out from the
center with rain and sometimes hail; and **dissipating**, identified
by the classic anvil-shaped cirriform top that develops at the
apex. Cbs can produce hail, lightning, updrafts and downdrafts,
and tornadoes. Navigating around Cbs requires great care: You
should give them a berth of at least 20 mi. Due to severe
vertical movement, flying under them is dangerous. If you are
caught in or near a thunderstorm, maintain maneuvering speed and
control the airplane's **attitude**. Attempt to keep pitch and bank
as level and constant as possible.

14. An **air mass** is an extensive body of air with fairly
consistent stability, temperature, and moisture content. As an
air mass begins to move along a pressure gradient, it behaves in
a predictable way. The line of discontinuity between two
differing air masses is called a **front**. The four main air mass
sources are: **arctic**, the coldest regions, from the poles; **polar**,
cold regions, high-pressure systems; **tropical**, a warm region,
low-pressure systems; and **equatorial**, the warmest region, low
pressure and calm winds. Masses that form over water are called
maritime and contain more moisture than those that form over
land, called **continental**. Air masses are classified as cold or

hot in reference to the ground over which they pass. A cold air
mass, designated k on a surface weather map, is a body of air
that is colder than the ground over which it passes, while a warm
air mass (w) is warmer than the surface over which it is moving.
Cold air moving over a warm surface produces convective current
(hot air rises) and is unstable, while warm air moving over cold
surfaces is stable. **Maritime Polar air masses** are moist,
unstable, and conducive to Cb buildups. **Continental Polar air
masses** tend to be stable, dry, and characterized by cirrus
clouds. **Maritime Tropical air masses** are typically humid,
internally stable, and characterized by stratiform clouds.
Continental Tropical air masses are dry, unstable, and not likely
to produce much precipitation. In the United States, polar and
arctic air moves from the northwest and tropical air from the
southwest. In general, cold air masses move faster than warm
ones do.

     15.   **Fronts** (lines of discontinuity between two differing
air masses) are named for the **advancing** air mass. A **warm front**
is a mass of warm air replacing cold air, and a **cold front** is a
mass of cold air supplanting warm air. An **occluded front** is one
in which a cold front outraces a slower-moving air mass and moves
it aloft. A **stationary front** is one in which the two fronts are
balanced and the zone of discontinuity remains relatively
constant geographically. **Frontogenesis** refers to the creation of
a front, as when one air mass overtakes another. When the air
masses normalize, the front dissipates (**frontolysis**). Frontal
passage is typically characterized by simultaneous changes in
temperature, barometric pressure, wind direction, and cloud
formation.

     16. **Warm air masses** tend to be moist and to climb over
retreating cold air masses. As a warm air mass climbs over the
cold air, its temperature decreases and condensation begins,
usually in the form of drizzle or light rain. The frontal zone
is usually stable and is preceded by high cirriform clouds. As
the front passes, the wind changes from southeast to southwest,
the barometer is steady or changes only slightly, the temperature
rises, and frontal weather dissipates. Warm fronts typically
have extensive cloud cover ahead of the front and low cloud cover
near the front, things important for VFR pilots to remember in
making flight plans.

     17. **Cold air masses** move faster than warm ones and tend to
push themselves under warm air masses. This gives rise to
cumulus clouds, turbulence, and--frequently--thunderstorms.
Squall lines, tightly knit lines of discontinuity, often precede
the front and are characterized by highly unstable conditions.
As the front passes, visibility improves, the wind shifts, the
temperature drops, and the barometer dips as the front passes,
then rises. Flying the front is not advisable.

     18. **Occluded air masses** are ones that have been overtaken
from behind by faster-moving cold fronts and forced aloft.
Thunderstorms almost always occur in the stratiform cloud layers;

the most activity is found in the area closest to the low,
typically the northernmost 50-100 mi.

    19. **Stationary air masses** move very slowly, if at all.
Winds associated with the frontal zone usually run parallel to
the line of discontinuity. Such fronts are characterized by low
cloud layers, drizzle, and an occasional thunderstorm. Sometimes
frontal waves develop as cold and warm air begin to rotate,
creating a miniature low-pressure cell (cyclogenesis).

KEY TERMS AND CONCEPTS, PART 1

    Match each term or concept (1-24) with the appropriate
description (A-X) below. Each item has only one match.

___	1. adiabatic lapse rate	___	13. mature
___	2. stratosphere	___	14. frontolysis
___	3. VMC	___	15. cirrus
___	4. troposphere	___	16. overcast
___	5. visibility	___	17. station pressure
___	6. temperature inversion	___	18. polar easterlies
___	7. clear	___	19. condensation nuclei
___	8. radiation	___	20. cumulonimbus
___	9. ridge	___	21. isobars
___	10. onshore	___	22. fog
___	11. maritime	___	23. cirrostratus
___	12. ceiling	___	24. sublimation

A. extends from 54,000 ft. at the equator to about 40-60 mi.
    above the earth
B. dissipation of a front
C. Cb stage with heavy downdrafts out of the center
D. reversal of the normal lapse rate
E. wind typically encountered on the beach during the day
F. flight can be maintained by use of outside references
G. horizontal distance one can see
H. lowest layer of clouds covering more than one-half of the sky
I. high-pressure area that takes an elongated form
J. winds flowing from the North Polar regions
K. temperature change of mechanically lifted dry air of about 3°
    C per 1,000-ft. change in elevation
L. particles such as pollen suspended in the air
M. pressure to which you set your altimeter before taking off
N. extends from the surface to 29,000 ft. over the poles
O. air mass that forms over water
P. clouds that have a wispy, horsehair appearance
Q. form of icing that is difficult to detect
R. condition of the sky when 95 percent covered by clouds
S. heat derived directly from the sun's rays
T. ice changes directly into water vapor
U. layer of cirrus clouds
V. clouds that are lying on or near the ground
W. continuous equal pressure lines on a weather chart
X. raining cumulus cloud

KEY TERMS AND CONCEPTS, PART 2

Match each term or concept (1-24) with the appropriate description (A-X) below. Each item has only one match.

___ 1. advection
___ 2. counterclockwise
___ 3. offshore
___ 4. anticyclones
___ 5. trough
___ 6. ionosphere
___ 7. scattered
___ 8. atmosphere
___ 9. radiation
___ 10. 14.7
___ 11. IMC
___ 12. convection
___ 13. 29.92" Hg
___ 14. evaporation
___ 15. squall line
___ 16. dissipating
___ 17. angle of incidence
___ 18. wind shear
___ 19. precipitation
___ 20. prevailing westerlies
___ 21. dew point
___ 22. northeast trades
___ 23. conduction
___ 24. relative humidity

A. direction wind moves around a low-pressure system
B. circulatory motion of the air caused by heat transfer
C. angle at which the sun's rays strike the earth
D. discharge of moisture from the atmosphere
E. ISA inches of mercury at sea level
F. air, water vapor, and particulate matter
G. outer layer of the atmosphere
H. tightly knit line of discontinuity that sometimes precedes a cold front
I. Cb stage in which an anvil-shaped cloud forms at the apex
J. sudden change in wind speed or direction
K. wind typically encountered on the beach at night
L. flight cannot be maintained by use of outside references
M. fog caused by the ground's cooling at night
N. condition of the sky when 30 percent covered by clouds
O. low-pressure area that takes an elongated form
P. winds flowing between 30° and 60° latitude
Q. high-pressure areas and their associated rotational winds
R. water changes from a liquid to a gas
S. ISA pounds per square inch (psi) at sea level
T. fog caused by moist air moving over a cold surface; common in coastal areas
U. ratio of water vapor in the air to the maximum amount it could hold
V. heat transferred directly from one molecule to another
W. winds flowing from 30° to the equator
X. temperature at which condensation or precipitation occurs

KEY TERMS AND CONCEPTS, PART 3

Match each term or concept (1-20) with the appropriate description (A-X) below. Each item has only one match.

___ 1. standard lapse rate
___ 2. rime
___ 3. 1013.2
___ 11. cold air mass
___ 12. slant range
___ 13. stratus

___ 4. front   ___ 14. clockwise
___ 5. upslope   ___ 15. occluded front
___ 6. pressure gradient   ___ 16. cyclones
___ 7. continental   ___ 17. obscuration
___ 8. broken   ___ 18. advection
___ 9. saturated   ___ 19. thermals
___ 10. cumulus   ___ 20. warm front

A. body of air that is colder than the surface it is passing over
B. clouds that are formed in layers
C. form of ice composed of small, brittle particles
D. fog caused when a moist air mass is lifted; common in and around mountains
E. condition of the sky when 75 percent covered by clouds
F. low-pressure areas and their associated rotational winds
G. direction wind moves around a high-pressure system
H. horizontal transfer of heat within an air mass as it passes over a surface
I. temperature decrease of 2° C per 1,000 ft. in elevation
J. air that has a relative humidity of 100 percent
K. ISA millibars (mb) at sea level
L. occurs when an advancing mass of warm air supersedes a cold air mass
M. air mass that forms over land
N. a rapidly moving cold front forces a slower moving air mass aloft
O. line of discontinuity between two differing air masses
P. clouds that appear to be piled up
Q. convective currents caused by differential heat reflection and absorption of the ground
R. "over-the-nose" visibility
S. condition that limits visibility, such as smog
T. pressure changes perpendicular to the isobars

DISCUSSION QUESTIONS AND EXERCISES

1. What is the difference between air and the atmosphere? Why is the distinction important?

2. State two characteristics of each of the three layers of the atmosphere identified below.

     a. stratosphere

   b. ionosphere

   c. troposphere

3.   What are the three common ways to report air pressure? For
each, state the value associated with standard conditions.

4.   How does the **rate of change** in air pressure change as
altitude increases? Give an example.

5.   What two things control precipitation in addition to the
water vapor itself? Explain how and why each affects
precipitation.

6.   Define each of the following terms:

   a. relative humidity

   b. dew point

   c. condensation nuclei

   d. evaporation

   e. sublimation

7.   Explain how each of the following affects air temperature:

   a.   the earth's rotation

   b.   unequal heating of land and water

   c.   the earth's revolution around the sun

8.   Characterize the following lapse rates:

   a.   standard lapse rate

   b.   adiabatic lapse rate (dry air)

   c.   adiabatic lapse rate (moist air)

9.   Briefly describe the Coriolis force. How is it important in understanding global circulation patterns?

10.   Briefly describe the three zones or cells that affect global circulation in the Northern Hemisphere and their associated weather conditions.

11.   State in what direction winds circulate around high-pressure and low-pressure systems in the Northern Hemisphere.

12. Define each of the following terms:

    a. ridge

    b. trough

    c. cyclone

    d. anticyclone

    e. pressure gradient

13. What is the relationship between the steepness of a pressure gradient and the stability of the air?

14. Briefly state how each of the following factors affects the wind:

    a. gravity

    b. friction

    c. centrifugal force

15. Define each of the following sky cover conditions and draw the symbol that is used to represent each one:

    a. clear

     b.   scattered

     c.   broken

     d.   overcast

16. What is fog? What three conditions are necessary for its formation?

17. Describe each of the following types of fog and state where you would be likely to encounter each:

     a.   radiation

     b.   advection

     c.   upslope

     d.   frontal

18. What is an obscuration? Give an example. Distinguish between reports of sky obscured and sky partially obscured.

19. Distinguish among VMC, VFR, IMC, and IFR.

20. What three conditions are necessary for the formation of
ice? Briefly distinguish between rime ice and clear ice. What
is the best way for VFR pilots to avoid icing conditions?

21.   State how turbulence is related to each of the following:

    a. lapse rate

    b. convective currents

    c. the coastlines

    d. land flow

    e. mountain flow

    f. wind shear

    g. temperature inversion

22. What is clear air turbulence? Why is it so difficult to
deal with?

23. Characterize each of the following cloud types, including a
description of what each looks like, what weather each is
associated with, and the altitudes at which each is likely to be
encountered:

    a. stratus

b. cumulus

c. cirriform

d. stratocumulus

e. nimbostratus

f. cirrocumulus

24. Why are cumulonimbus clouds considered particularly dangerous? Characterize the three stages of the Cb life cycle.

25. Why is it particularly dangerous to fly underneath a line of thunderstorms?

26. What is the most important thing you should do if you are caught in a thunderstorm cloud? Name three other things you should also do.

27. Characterize the four air mass source regions.

28. Distinguish between continental and maritime air masses.
How do they typically differ in temperature and moisture content?

29. Distinguish between warm and cold air masses.

30. Describe each of the following air mass types:

   a. maritime Polar

   b. continental Polar

   c. maritime Tropical

   d. continental Tropical

31. Characterize each of the following:

   a. warm front

   b. cold front

   c. occluded front

    d.   stationary front

    e.   cyclogenesis

32. In general, what should VFR pilots do when the weather begins to "close in"?

33. T F The layer of the atmosphere that gives rise to northern lights is the ionosphere.

34. T F The atmosphere gains pressure less quickly as altitude increases.

35. T F Air that has a relative humidity of 100 percent is called sublimated.

36. T F One factor in the intensity of solar radiation is the sun's rotation around the earth.

37. T F The primary process by which the sun heats the air is called radiation.

38. T F The direction of circulation around a region of low pressure in the Northern Hemisphere is counter-clockwise.

39. T F The steeper the pressure gradient is, the more likely high velocity winds are.

40. T F Wind speed along the isobars decreases as the center of a low-pressure area approaches and increases as the center of a high approaches.

41. T F Prevailing visibilities are given in thousands of feet when visibility is good.

42. T F Fog is a cloud that is lurking on or near the ground.

43. T F Fog caused by cooling of the ground at night (common under clear skies with calm winds) is called advection fog.

44. T F VMC and VFR refer to identical weather conditions.

45. T F Clear ice will accumulate rapidly in flight when the
           temperature is between freezing and -15° C and you are
           flying in cumuliform clouds.

46. T F Frost may form in flight when a cold aircraft descends
           from a zone of subzero temperatures to a zone of
           above-freezing temperatures and high relative
           humidity.

47. T F Moist air is typically less stable than dry air.

48. T F Onshore winds are common during the day, while
           offshore winds are common during the night.

49. T F Convection currents become less intense as the surface
           temperature increases.

50. T F An almond or lens-shaped cloud that appears stationary
           but that may reflect winds of 50 kts. or more is
           called a lenticular cloud.

51. T F Reversal of the normal lapse rate is called a
           temperature inversion.

52. T F A temperature inversion often develops near the ground
           on clear, cool nights in calm wind conditions.

53. T F The overhanging anvil of a thunderstorm points in the
           direction from which the storm has just moved.

54. T F Hail may be found in any level within a thunderstorm
           but not in the clear air outside the storm cloud.

55. T F The most important thing a pilot who enters a
           thunderstorm should do is keep a constant airspeed.

56. T F A cold air mass is colder than the surface over which
           it passes.

57. T F A squall line is typically associated with a fast-
           moving warm front.

58. T F Warm fronts are frequently associated with low
           ceilings and limited visibility.

REVIEW QUESTIONS

1.   (FAA 1723) What feature is associated with a temperature inversion?

1--A stable layer of air.
2--An unstable layer of air.
3--Chinook winds or mountain slopes.
4--Air mass thunderstorms.

2.   (FAA 1724) The most frequent type of ground or surface based temperature inversion is that produced by

1--terrestrial radiation on a clear, relatively still night.
2--warm air being lifted rapidly aloft in the vicinity of
   mountainous terrain.
3--the movement of colder air under warm air, or the movement of
   warm air over cold air.
4--widespread sinking of air within a thick layer aloft resulting
   in heating by compression.

3.   (FAA 1725) A temperature inversion would most likely result in which of the following weather conditions?

1--Clouds with extensive vertical development above an inversion
   aloft.
2--Good visibility in the lower levels of the atmosphere and poor
   visibility above an inversion aloft.
3--An increase in temperature as altitude is increased.
4--A decrease in temperature as altitude is increased.

4.   (FAA 1726) Which weather conditions should be expected beneath a low-level temperature inversion layer when the relative humidity is high?

1--Smooth air and poor visibility due to fog, haze, or low
   clouds.
2--Light wind shear and poor visibility due to haze and light
   rain.
3--Turbulent air and poor visibility due to fog, low stratus type
   clouds, and showery precipitation.
4--Updrafts and turbulence due to surface heating, fair
   visibility, and cumulus clouds developing at the top of the
   inversion.

5.   (FAA 1727) Every physical process of weather is accompanied by, or is the result of,

1--the movement of air.
2--a pressure differential.
3--a heat exchange.
4--moisture.

6.   (FAA 1728) What causes variations in altimeter settings between weather reporting points?

1--Unequal heating of the Earth's surface.
2--Variation of terrain elevation creating barriers to the movement of an air mass.
3--Coriolis force reacting with friction.
4--Friction of the air with the Earth's surface.

7.   (FAA 1751) Winds at 5,000 ft. AGL on a particular flight are southwesterly while most of the surface winds are southerly. This difference in direction is primarily due to

1--a stronger pressure gradient at higher altitudes.
2--friction between the wind and the surface.
3--stronger Coriolis force at the surface.
4--the influence of pressure systems at the lower altitudes.

8.   (FAA 1752 DI) In the Northern Hemisphere, what causes the wind to be deflected to the right?

1--The pressure gradient force.
2--Surface friction.
3--Centrifugal force.
4--Coriolis force.

9.   (FAA 1754) If the temperature/dew point spread is small and decreasing, and the temperature is 62°, what type of weather is most likely to develop?

1--Freezing precipitation.
2--Thunderstorms.
3--Fog or low clouds.
4--Rain showers.

10.   (FAA 1755) What is meant by the term dew point?

1--The temperature at which condensation and evaporation are equal.
2--The temperature at which dew will always form.
3--The temperature to which air must be cooled to become saturated.
4--The spread between actual temperature and the temperature during evaporation.

11.   (FAA 1756) The amount of water vapor which air can hold largely depends on

1--the dew point.
2--air temperature.
3--stability of air.
4--relative humidity.

12. (FAA 1757) The term dew point refers to the

1--temperature at which fog will form.
2--spread between actual temperature and temperature during
   evaporation.
3--temperature at which the evaporation and condensation points
   are equal.
4--temperature to which air must be cooled to become saturated.

13. (FAA 1758) What are the processes by which moisture is
added to unsaturated air?

1--Heating and sublimation.
2--Evaporation and sublimation.
3--Heating and condensation.
4--Supersaturation and evaporation.

14. (FAA 1759) The presence of ice pellets at the surface is
evidence that

1--there are thunderstorms in the area.
2--a cold front has passed.
3--there is freezing rain at a higher altitude.
4--the pilot can climb to a higher altitude without encountering
   more than light icing.

15. (FAA 1760) Clouds, fog, or dew will always form when

1--water vapor condenses.
2--water vapor is present.
3--relative humidity reaches or exceeds 100 percent.
4--the temperature and dew point are equal.

16. (FAA 1761) Which of the following measurements can be used
to determine the stability of the atmosphere?

1--Atmospheric pressure.
2--Actual lapse rate.
3--Surface temperature.
4--Wind velocity.

17. (FAA 1762) Which of the following would decrease the
stability of an air mass?

1--Warming from below.
2--Cooling from below.
3--Decrease in water vapor.
4--Sinking of the air mass.

18.   (FAA 1763)   The conditions necessary for the formation of cumulonimbus clouds are a lifting action and

1--unstable air containing an excess of condensation nuclei.
2--unstable, moist air.
3--either stable or unstable air.
4--stable, moist air.

19.   (FAA 1764)   What is a characteristic of stable air?

1--Stratiform clouds.
2--Unlimited visibility.
3--Fair weather cumulus clouds.
4--Temperature decreases rapidly with altitude.

20.   (FAA 1765)   Moist, stable air flowing upslope can be expected to

1--produce stratus type clouds.
2--produce a temperature inversion.
3--cause showers and thunderstorms.
4--develop convective turbulence.

21.   (FAA 1766)   If an unstable air mass is forced upward, what type clouds can be expected?

1--Layer-like clouds with a temperature inversion.
2--Layer-like clouds with little vertical development.
3--Layer-like clouds with considerable associated turbulence.
4--Clouds with considerable vertical development and associated
    turbulence.

22.   (FAA 1767)   What is the approximate base of the cumulus clouds if the temperature at 2,000 ft. MSL is 70° F and the dew point is 52° F?

1--3,000 ft. MSL.   3--6,000 ft. MSL.
2--4,000 ft. MSL.   4--8,000 ft. MSL.

23.   (FAA 1768)   At approximately what altitude above the surface would the pilot expect the base of cumuliform clouds if the surface air temperature is 82° F and the dew point is 55° F?

1--5,000 ft. MSL.   3--7,000 ft. MSL.
2--6,000 ft. MSL.   4--8,000 ft. MSL.

24.   (FAA 1770)   The suffix nimbus, used in naming clouds, means

1--a cloud with extensive vertical development.
2--a rain cloud.
3--a middle cloud containing ice pellets.
4--an accumulation of clouds.

25. (FAA 1771) An almond or lens-shaped cloud which appears
stationary, but which may contain winds of 50 kts. or more, is
referred to as

1--an inactive frontal cloud.
2--a funnel cloud.
3--a lenticular cloud.
4--a stratus cloud.

26. (FAA 1772) Clouds are divided into four families according
to their

1--origin.   3--height range.
2--outward shape.   4--composition.

27. (FAA 1773) Which clouds have the greatest turbulence?

1--Towering cumulus.
2--Cumulonimbus.
3--Nimbostratus.
4--Altocumulus castellanus.

28. (FAA 1774) Which of the following cloud types would
indicate convective turbulence?

1--Altocumulus standing lenticular clouds.
2--Nimbostratus clouds.
3--Towering cumulus clouds.
4--Cirrus clouds.

29. (FAA 1775) A moist, unstable air mass is characterized by

1--cumuliform clouds and showery precipitation.
2--poor visibility and smooth air.
3--stratiform clouds and continuous precipitation.
4--fog and drizzle.

30. (FAA 1776) What are characteristics of unstable air?

1--Turbulence and good surface visibility.
2--Turbuluence and poor surface visibility.
3--Nimbostratus clouds and good surface visibility.
4--Nimbostratus clouds and poor surface visibility.

31. (FAA 1777) A stable air mass is most likely to have which
of the following characteristics?

1--Showery precipitation.
2--Turbulent air.
3--Smooth air.
4--Cumuliform clouds.

32.   (FAA 1778)   An unstable air mass is most likely to have
which of the following characteristics?

1--Stratiform clouds and fog.
2--Turbulent air.
3--Continuous precipitation.
4--Fair to poor visibility in haze and smoke.

33.   (FAA 1779)   One weather phenomenon which will always occur
when flying across a front is

1--a change in the wind.
2--a large precipitation area, if the frontal surface is steep.
3--a large temperature change, especially at high altitudes.
4--the presence of clouds, either ahead of or behind the front.

34.   (FAA 1780)   The boundary between two different air masses is
referred to as a

1--foehn gap.   3--frontogenesis.
2--frontolysis.   4--front.

35.   (FAA 1781)   Steady precipitation, in contrast to showery,
preceding a front is an indication of

1--cumuliform clouds with moderate turbulence.
2--stratiform clouds with moderate turbulence.
3--cumuliform clouds with little or no turbulence.
4--stratiform clouds with little or no turbulence.

36.   (FAA 1783)   When flying at a low altitude across a mountain
range the greatest potential danger, caused by descending air
currents, will usually be encountered on the

1--leeward side when flying into the wind.
2--windward side when flying into the wind.
3--leeward side when flying with the wind.
4--windward side when flying with the wind.

37.   (FAA 1784)   Crests of standing mountain waves may be marked
by stationary, lens-shaped clouds known as

1--cumulonimbus mamma clouds.
2--standing lenticular clouds.
3--roll clouds.
4--rotor clouds.

38.   (FAA 1787)   A pilot can expect a wind shear zone in a
temperature inversion, whenever the wind speed at 2,000 to 4,000
ft. above the surface is at least

1--5 kts.   3--15 kts.
2--10 kts.   4--25 kts.

39. (FAA 1785) Hazardous wind shear is commonly encountered near the ground

1--near thunderstorms and during periods when the wind velocity is stronger than 35 kts.
2--during periods when the wind velocity is stronger than 35 kts. and near mountain valleys.
3--during periods of strong temperature inversion and near thunderstorms.
4--near mountain valleys and on the windward side of a hill or mountain.

40. (FAA 1786) Where does wind shear occur?

1--Only at higher altitudes, usually in the vicinity of jet streams.
2--At any level, and it can exist in both a horizontal and vertical direction.
3--Primarily at lower altitudes in the vicinity of mountain waves.
4--Only in the vicinity of thunderstorms.

41. (FAA 1788) An in-flight condition necessary for structural icing to form is

1--cumuliform clouds.   3--stratiform clouds.
2--cirrostratus clouds.   4--visible moisture.

42. (FAA 1789) In which environment is aircraft structural ice most likely to have the highest accumulation rate?

1--Cumulus clouds.   3--Stratus clouds.
2--Cirrus clouds.   4--Freezing rain.

43. (FAA 1793) Which conditions result in the formation of frost?

1--The freezing of dew.
2--The collecting surface's temperature is at or below freezing and small droplets of moisture fall on the collecting surface.
3--The temperature of the collecting surface is at or below the dew point of the adjacent air and the dew point is below freezing.
4--Small drops of moisture falling on the collecting surface when the surrounding air temperature is at or below freezing.

44. (FAA 1792) Why is frost considered hazardous to flight operation?

1--The increased weight requires a greater takeoff distance.
2--Frost changes the basic aerodynamic shape of the airfoil.
3--Frost decreases control effectiveness.
4--Frost causes early airflow separation resulting in a loss of lift.

45.   (FAA 1795) What conditions are necessary for the formation
of thunderstorms?

1--Lifting force, high humidity, and unstable conditions.
2--High humidity, high temperature, and cumulus clouds.
3--Low pressure, high humidity, and cumulus clouds.
4--Lifting force, high temperature, and unstable conditions.

46.   (FAA 1796) During the life cycle of a thunderstorm, which
stage is characterized predominantly by downdrafts?

1--Cumulus.   3--Mature.
2--Dissipating.   4--Anvil.

47.   (FAA 1797) Thunderstorms reach their greatest intensity
during the

1--updraft stage.   3--downdraft stage.
2--mature stage.   4--cumulus stage.

48.   (FAA 1798) What feature is normally associated with the
cumulus stage of thunderstorms?

1--Roll cloud.
2--Continuous updraft.
3--Frequent lightning.
4--Beginning of rain at the surface.

49.   (FAA 1799) Which weather phenomenon signals the beginning
of the mature stage of a thunderstorm?

1--The appearance of an anvil top.
2--The start of rain at the surface.
3--Growth rate of cloud is maximum.
4--Strong turbulence in the cloud.

50.   (FAA 1800) Thunderstorms which generally produce the most
intense hazard to aircraft are

1--air mass thunderstorms.
2--steady-state thunderstorms.
3--warm front thunderstorms.
4--squall line thunderstorms.

51.   (FAA 1801) Which thunderstorms generally produce the most
severe conditions, such as heavy hail and destructive winds?

1--Air mass thunderstorms.
2--Warm front thunderstorms.
3--Squall line thunderstorms.
4--Cold front thunderstorms.

52. (FAA 1802) Which weather phenomenon is always associated
with a thunderstorm?

1--Lightning.
2--Heavy rain showers.
3--Supercooled raindrops.
4--Hail.

53. (FAA 1803) If there is thunderstorm activity in the
vicinity of an airport at which you plan to land, which hazardous
and invisible atmospheric phenomenon might you expect to
encounter on the landing approach?

1--St. Elmo's fire.
2--Wind shear turbulence.
3--Tornadoes.
4--Virga.

54. (FAA 1804) A nonfrontal, narrow band of active
thunderstorms, that often develop ahead of a cold front, is known
as

1--an occlusion.   3--a squall line.
2--a prefrontal system.   4--a shear line.

55. (FAA 1805) Upon encountering severe turbulence, which
condition should the pilot attempt to maintain?

1--Constant altitude.
2--Constant airspeed (Va).
3--Level flight attitude.
4--Constant altitude and constant airspeed.

56. (FAA 1806) Fog associated with a warm front is generally
the result of saturation due to

1--evaporation of precipitation.
2--adiabatic cooling.
3--evaporation of surface moisture.
4--nocturnal cooling.

57. (FAA 1807) What situation is most conducive to the
formation of radiation fog?

1--Warm, moist air over low, flatland areas on clear, calm
   nights.
2--Moist, tropical air moving over cold, offshore water.
3--The movement of cold air over much warmer water.
4--Light wind moving warm, moist air upslope during the night.

58. (FAA 1808) What types of fog depend upon a wind in order to exist?

1--Radiation fog and ice fog.
2--Steam fog and downslope fog.
3--Precipitation-induced fog and ground fog.
4--Advection fog and upslope fog.

59. (FAA 1809) In which situation is advection fog most likely to form?

1--A warm, moist air mass on the windward side of mountains.
2--An air mass moving inland from the coast in winter.
3--A light breeze blowing colder air out to sea.
4--Warm, moist air settling over a warmer surface under no-wind conditions.

60. (FAA 1826) Ceiling, as used in aviation weather reports, is the height above the Earth's surface of the

1--highest layer of clouds located above the reporting station.
2--lowest layer of clouds or obscuration phenomena located above the reporting station.
3--highest layer of clouds that is reported as overcast and not classified as thin or partial.
4--lowest layer of clouds or obscuration phenomena that is reported as broken, overcast, or obscuration and not classified as thin or partial.

61. What gas accounts for the largest percentage of air?

1--Argon.   3--Oxygen.
2--Nitrogen.   4--Water vapor.

62. Which of the following numbers represents International Standard Atmosphere conditions in pounds per square inch (psi) at sea level?

1--14.7.   3--123.6.
2--29.92.   4--1013.2.

63. Water changes from a solid to a gas through a process called

1--condensation.   3--saturation.
2--evaporation.   4--sublimation.

64. Hot air rising is an example of

1--advection.   3--convection.
2--conduction.   4--radiation.

65. On a weather map, isobars that are close together represent a steep

1--anticyclone.   3--ridge.
2--pressure gradient.   4--trough.

66. If 80 percent of the sky is covered by clouds, it is classified as

1--broken.   3--overcast.
2--clear.   4--scattered.

67. Visibility "over the nose" of the airplane refers to the

1--longitudinal range minus the vertical range.
2--obscuration range.
3--slant range.
4--vertical range minus longitudinal range.

68. The type of ice that forms on an airplane surface depends on

1--an inversion aloft.
2--an increase in flight altitude.
3--the size of the water drops or droplets that strike the
   airplane surface.
4--the temperature/dew point spread.

69. What aerodynamic effects will structural icing have on an airplane?

1--Drag increases; thrust will not be affected.
2--Lift decreases; weight increases.
3--Stall speed decreases; thrust increases.
4--Weight increases; lift is not affected if drag and thrust
   remain constant.

70. Dry air tends to be _____ stable and to lose heat _____ rapidly than moist air.

1--less; less   3--more; less
2--less; more   4--more; more

71. Hail is most likely to be associated with

1--cirrocumulus clouds.   3--cumulus clouds.
2--cumulonimbus clouds.   4--stratocumulus clouds.

72. Gray, uniform, sheetlike clouds lurking at 2,000 ft. AGL would be classified as

1--altocumulus.   3--nimbostratus.
2--cumulus.   4--stratus.

73. _____ _____ air masses are typically dry, stable, and characterized by sparse precipitation and excellent visibilities.

1--Continental Polar   3--Maritime Polar
2--Continental Tropical   4--Maritime Tropical

74. When two air masses are so well balanced that neither prevails, this is called a/an

1--discontinuous front.   3--occluded front.
2--genetic front.   4--stationary front.

ANSWERS

Key Terms and Concepts, Part 1

1. K	7. Q	13. C	19. L
2. A	8. S	14. B	20. X
3. F	9. I	15. P	21. W
4. N	10. E	16. R	22. V
5. G	11. O	17. M	23. U
6. D	12. H	18. J	24. T

Key Terms and Concepts, Part 2

1. T	7. N	13. E	19. D
2. A	8. F	14. R	20. P
3. K	9. M	15. H	21. X
4. Q	10. S	16. I	22. W
5. O	11. L	17. C	23. V
6. G	12. B	18. J	24. U

Key Terms and Concepts, Part 3

1. I	6. T	11. A	16. F
2. C	7. M	12. R	17. S
3. K	8. E	13. B	18. H
4. O	9. J	14. G	19. Q
5. D	10. P	15. N	20. L

Discussion Questions and Exercises

33. T--This is due to the ionosphere's properties.
34. F--It loses pressure as altitude increases.
35. F--It is called saturated.
36. F--It depends upon the earth's rotation around the sun as well as the earth's daily rotation and the angle of incidence.
37. T--What more can one say?
38. T--And clockwise around a system of high pressure.

39. T--Steep pressure gradients mean rapid pressure changes that
     are typically associated with high winds.
40. T--Precisely!
41. F--It is reported in hundreds of feet when visibility is
     poor and in statute miles when it is good.
42. T--That is exactly what it is: moist and thick.
43. F--This is called radiation fog.
44. F--VMC specifies the weather and VFR specifies the rules
     that govern actual flight.
45. T--Cumuliform clouds produce large water droplets that are
     especially prone to produce clear ice.
46. T--Such conditions may lead to frost.
47. T--Mechanically lifted moist air can be particularly
     unstable.
48. T--At night, water retains heat while the land cools
     quickly, leading to offshore winds. During the day, the
     land warms more rapidly than the water, leading to
     onshore breezes as the cold air over the water moves in
     to replace the warm air.
49. F--It is just the opposite.
50. T--These are evidence of turbulent mountain waves on the
     leeward (downwind) side of the mountain.
51. T--That is the definition of an inversion.
52. T--This occurs because the ground cools at a rapid rate.
53. F--It points in the direction the storm is moving.
54. F--They may be thrown upward and outward as far as 5 mi.
55. F--One should first control the airplane's pitch and bank
     attitude.
56. T--Air masses are defined by the temperature of the mass in
     relation to the surface over which it travels.
57. F--It is associated with fast-moving cold fronts.
58. T--Warm fronts typically move slowly and cover a large area.

Review Questions

 1. 1--Inversions often develop in light wind conditions on
     clear, cool nights as the ground cools more quickly than
     the air above it. Thus, the temperature is higher above
     the ground than on the ground. Finally, such an
     inversion may trap fog or other particles.
 2. 1--See answer 1.
 3. 3--See answer 1.
 4. 1--See answer 1.
 5. ?--Air constantly moves, which causes changes in weather,
     making #1 a logical choice. But, the movement of air is
     the direct result of differences in heating and cooling
     in the lower atmosphere, which makes #3 a logical choice.
     I'm not sure what the FAA is driving at here.
 6. 1--Pressure differences are caused by unequal heating of the
     Earth's surface.
 7. 2--Pressure altitude is the altitude you read when 29.92"
     is set in the Kollsman window.
 8. ?--This question has been designated as unusable by the FAA
     and removed from the FAA Question Selection Sheets.

9. 3--Fog or low clouds are likely to form when the spread
    between the temperature and dew point is small and
    decreasing. The term "dew point" refers to that
    temperature to which the air must be cooled before it is
    fully saturated.
10. 3--See answer 9.
11. 2--See answer 9.
12. 4--See answer 9.
13. 2--Water changes to water vapor through the process of
    evaporation. Ice changes directly to gas through the
    process of sublimation. These are the two processes by
    which moisture is added to unsaturated air.
14. 3--Ice pellets on the surface mean freezing rain at higher
    altitudes.
15. 1--When water condenses, clouds, fog, or dew will form.
16. 2--Air stability is measured by the actual lapse rate.
17. 1--Warming from below will decrease air stability,
18. 2--Unstable, moist air with a vertical lifting action is
    necessary for the formation of a Cb.
19. 1--Stable air is characterized by stratiform clouds and fog,
    continuous precipitation, smooth air, and fair to poor
    visibility in haze and smoke. Unstable air is
    characterized by cumulus clouds, showery precipitation,
    turbulent air, and good visibility.
20. 1--Moist, stable air moving upslope is likely to produce
    stratus type clouds. Unstable, moist air moving upslope
    can be expected to produce clouds with considerable
    vertical development and associated turbulence.
21. 4--See answers 18 and 20.
22. 3--The convergence between the temperature and the dew point
    is about 4.4° F or 2.5° C per 1,000 ft. in a convective
    current. The difference in temperature (70 - 52) divided
    by 4.4 x 1,000 equals 4,090 ft., which--when added to
    2,000 ft. MSL--yields 6,090 ft. MSL. So, 6,000 ft. is
    the closest alternative.
23. 2--See answer 22. In this problem, the difference is
    (82 - 55)/4.4 x 1,000 = 6,136 ft., which must be added to
    2,000 ft. MSL. So, 8,000 ft. is the closest alternative.
24. 2--The suffix nimbus means a rain cloud, as in cumulonimbus.
25. 3--An almond or lens-shaped cloud which appears stationary,
    but which may contain winds of 50 kts. or more, is called
    a lenticular cloud.
26. 3--Clouds are divided into four families according to their
    height range.
27. 2--Cumulonimbus clouds have the greatest turbulence.
28. 3--Towering cumulus clouds are good evidence of considerable
    vertical movement that is associated with convective
    turbulence.
29. 1--See answer 19.
30. 1--See answer 19.
31. 3--See answer 19.
32. 2--See answer 19.
33. 1--Wind direction always changes when flying across a front.
34. 4--The boundary between two different air masses is called a
    front.

35. 4--See answer 19.
36. 1--Flying into the wind on the leeward side of a mountain
         produces the greatest potential danger when flying at low
         altitudes across a mountain range due to the downdrafts
         that may be present.
37. 2--See answer 25.
38. 4--Wind shear is common in a temperature inversion between
         2,000 and 4,000 ft. AGL when wind speed is stronger than
         25 kts.
39. 3--Ground wind shear is common during periods of strong
         temperature inversions and near thunderstorms.
40. 2--Wind shear can occur at any altitude, and it can exist in
         both vertical and horizontal directions.
41. 4--Visible moisture is necessary for structural icing to
         occur.
42. 4--Freezing rain produces the highest accumulation rate of
         structural ice.
43. 3--Formation of frost is similar to the formation of dew.
         The temperature must be lower than the dew point of the
         adjacent air and the dew point must be below freezing.
44. 4--Frost causes early airflow separation resulting in a loss
         of lift and may prevent the airplane from becoming
         airborne.
45. 1--Unstable air, high humidity, and a vertical lifting force
         are all necessary for the formation of cumulonimbus
         clouds.
46. 2--Downdrafts are characteristic of the dissipating cycle of
         a thunderstorm.
47. 2--Thunderstorms have their greatest intensity during their
         mature stage.
48. 2--Continuous updrafts are characteristic of the cumulus
         stage of a thunderstorm.
49. 2--The start of rain at the surface signals the beginning of
         the mature stage of a thunderstorm.
50. 4--Squall line thunderstorms generally produce the most
         severe weather conditions.
51. 3--See answer 50.
52. 1--Lightning is always associated with a thunderstorm.
53. 2--Wind shear is common near a thunderstorm.
54. 3--Narrow, nonfrontal bands of active thunderstorms that
         often develop ahead of a cold front are referred to as a
         squall line. They are among the most severe weather
         conditions and may produce hail and destructive winds.
55. 3--The foremost thing to maintain upon encountering severe
         turbulence is the airplane's attitude.
56. 1--Fog associated with a warm front, referred to as frontal
         fog, is generally the result of saturation due to
         evaporation.
57. 1--Radiation fog is most likely to form on clear, warm,
         moist nights over low, flatland areas. It does not
         depend on moving air.
58. 4--Advection and upslope fog depend upon wind. Advection
         fog is the movement of warm, moist air over a colder
         surface, and upslope fog is the movement of warm, moist
         air upslope.

59.   2--Advection fog is most likely to form when an air mass
      moves inland from the coast in winter. Also, see answer
      58.
60.   4--The term "ceiling," as used in aviation weather reports,
      is the height above the Earth's surface of the lowest
      layer of clouds or obscuration phenomena that is reported
      as broken (greater than 50 percent sky cover), overcast
      (greater than 90 percent sky cover), or obscuration and
      not classified as thin or partial.
61.   2--Air consists of 78 percent nitrogen, 21 percent oxygen,
      other gases, and varying amounts of water vapor.
62.   1--14.7 psi is the ISA standard at sea level.
63.   4--Sublimation is the process of changing a solid to a gas.
      Water changes from a liquid to a gas through a process
      called evaporation.
64.   3--Convective currents are established as hot air rises and
      cool air descends.
65.   2--The closeness of the isobars indicates the intensity of
      the pressure gradient.
66.   1--Clear (0-9 percent); scattered (10-49 percent); broken
      (50-89 percent); overcast (90-100 percent).
67.   3--The distance from the nose to the ground is a diagonal
      line known as the slant range.
68.   3--The type of droplet will determine what type of ice will
      form, clear or rime.
69.   2--Drag increases, weight increases, lift decreases, and
      thrust is reduced.
70.   4--Dry air is more stable than moist air, and it loses heat
      more rapidly than moist air.
71.   2--Cumulonimbus clouds are associated with all kinds of
      violent weather, including thunderstorms and hail.
72.   4--This defines stratus clouds.
73.   1--Continental Polar air masses have cold surface
      temperatures, low moisture content, and great stability
      in lower layers, especially in the source region.
74.   3--This defines a stationary front.

9/USING AVIATION WEATHER SERVICES

MAIN POINTS

1.   The National Weather Service (NWS) gathers information about the weather from such sources as direct observations, balloons, radar, and pilot reports.

2.   **Surface Analysis Charts** depict pressure patterns, fronts, surface winds, temperature and dew points, and visibility restrictions. Fortunately, Surface Analysis Charts have legends so you will not have to memorize the many symbols that appear on them. Cloud cover is represented by a circle that is progressively filled for greater amounts of cover. Wind direction is represented by a wind arrow whose tail points into the wind; wind velocity is indicated by feathers (10 kts. per) or a pennant (50 kts.). Isobars are represented in millibars (last two digits) and are spaced at 4-mb intervals. New surface analysis charts are issued every 3 hr., and the data displayed on them when they are first issued may be as much as 1 hr. old.

3.   **Weather Depiction Charts** shade all IFR areas (visibility below 3 mi. and/or a ceiling of less than 1,000 ft.). Marginal VFR areas (visibility 3-5 mi. and/or a ceiling of 1,000 to 3,000 ft.) are enclosed by a solid line.

4.   A **Significant Weather Prognosis Chart (Prog)** is a prognosis or forecast of how the weather may change. They are issued for low levels (surface to 24,000 ft.) and high levels (24,000 to 63,000 ft.). The typical low-level chart contains two successive 12-hr. estimates for surface and low-level conditions. Solid lines enclose IFR areas, scalloped lines enclose areas of marginal VFR, broken lines represent areas of moderate or greater turbulence, dotted lines depict surface freezing, and dashed lines represent freezing aloft. Shaded areas on the surface portion indicate areas of precipitation.

5.   The **Radar Summary Chart** shows areas of precipitation, not cloud cover, which typically is more widespread. The chart indicates precipitation trends and intensities.

6.   Weather reports are important in terms of both time frame and difference in relative accuracy between actual observations and future estimates based on those observations. Several reports can be used in flight planning.

7.   The **Surface Aviation Weather Report**, or sequence report, is given every hour and contains the following information:
   (1) station identifier, a three-letter code
   (2) type of report--for example, hourly or special
   (3) time of report, given in GMT or zulu
   (4) sky condition and ceiling: measured (M), estimated (E), indefinite (W), or variable (V)
   (5) visibility, given in statute miles
   (6) present weather
   (7) sea level pressure, given in millibars (no decimal, initial 9 or 10 omitted)
   (8) temperature, in Fahrenheit
   (9) dew point
   (10) wind direction (true) and speed in knots
   (11) altimeter setting, last three digits only, and
   (12) remarks--for example, pilot reports (PIREPs) and Notices to Airmen (NOTAMs).

PIREPs alert other pilots to in-flight situations, while NOTAMs alert pilots to closed runways, inoperable VOR stations, unusual airspace activity, and so forth. To file a PIREP, call FSS on the radio as soon as you detect an unusual or changing situation. Finally, the report may contain **runway visual range** data, the slant range visibility at airports equipped with instrument landing aids.

8.   **Terminal Forecasts (FTs)**, prepared three times daily for selected large airports, contain information similar to sequence reports. Winds are shown only if they are expected to be greater than 10 kts. and visibility only if it is expected to be 6 mi. or less. **Area Forecasts (FAs)** are released every 12 hr. for 18-hr. periods, with an additional 12-hr. outlook. **Winds And Temperatures Aloft Forecasts (FDs)** are important for performance and navigation calculations. Wind directions are reported with respect to true (not magnetic) north and temperatures are reported in Celsius.

9.   Several in-flight advisories are issued. Advisories are identified by a phonetic letter, and each area issuing these reports number them sequentially. A SIGMET (**significant meteorological** advisory) refers to severe weather developments such as turbulence or icing. A **convective SIGMET** refers to thunderstorm activity. An **AIRMET** (airmen's **meteorological** advisory) is important for all aircraft. Hurricane advisories, convective outlooks, severe weather watch bulletins, and special flight forecasts are also issued.

10.   **FAR 91.5** requires you to familiarize yourself with weather information prior to any flight outside the local area (typically defined as a 25-mi. radius). When you contact the weather facility (FSS, if one is available, or the NWS), give them your name, aircraft type and call numbers, destination, route of flight and altitude, type of flight (VFR), estimated time of departure (ETD), estimated time en route (ETE), estimated time of arrival (ETA), and any stops en route.

11.   Automated weather information is available from several sources. Pilot briefings are recorded on the **Pilot's Automatic Weather Answering Serivce (PATWAS)**, non-control airport information is available on the **Automatic Terminal Information Service (ATIS)**, and en route information is broadcast on the **Transcribed Weather Broadcasts (TWEB)** for 250 mi. around a navigational aid such as a VOR station. TWEBs begin at 15 min. past the hour, every hour, and are updated as changes occur. Finally, **En Route Flight Advisory Service (EFAS)** provides weather-related information on 122.0. To contact EFAS, call flight watch (for example, Kansas City Flight Watch) on 122.0 between the hours of 6 a.m. and 10 p.m. seven days a week.

KEY TERMS AND CONCEPTS

Match each term or concept (1-24) with the appropriate description (A-X) below. Each term has only one match.

___ 1.   TWEB
___ 2.   Winds Aloft Forecast
___ 3.   Area Forecasts
___ 4.   ceiling
___ 5.   PIREP
___ 6.   ATIS
___ 7.   severe weather watch
___ 8.   122.0
___ 9.   EFAS
___ 10. NWS
___ 11. NOTAMs
___ 12. Surface Aviation
         Weather Reports

___ 13. AIRMET
___ 14. PATWAS
___ 15. SIGMET
___ 16. station model
___ 17. echo
___ 18. true north
___ 19. RVR
___ 20. isobars
___ 21. convective SIGMET
___ 22. Terminal Forecast
___ 23. Radar Summary Chart
___ 24. Significant Weather
         Prognosis Chart

A.   height of the lowest layer of clouds above the Earth's surface when the total coverage is more than 50 percent
B.   in-flight pilot report on unusual weather conditions
C.   significant meteorological advisory pertaining to all aircraft
D.   forecasts of winds at selected altitudes; issued every 6 hr.
E.   hourly sequence report.
F.   short-term forecast of dangerous weather for small aircraft; issued by the NWS
G.   civilian agency that coordinates weather services in the United States
H.   chart that depicts precipitation through "echoes"
I.   significant meteorological advisory about thunderstorms

J. expected weather for the next 12 hr. for a geographical area; issued every 6 hr.
K. continuous recording of current and forecast weather along certain flight routes
L. continuous broadcast of meteorological information from a VOR facility.
M. forecasts for specific airports issued three times daily
N. wind directions are given with respect to this in Surface Analysis Weather Reports
O. appearance on a radar indicator of energy returned from a target such as a storm cell
P. EFAS radio frequency
Q. four-panel NWS charts for surface and significant weather estimates
R. publication containing current information essential to flight safety
S. slant range to runway at instrument-equipped airports
T. provides timely weather information at pilot request
U. lines on a surface analysis chart joining points of equal barometric pressure
V. in-flight advisory issued by National Severe Storms Forecast Center
W. grouping of weather information around a station on a surface analysis chart
X. continuous broadcast of recorded non-control information in selected terminal areas

DISCUSSION QUESTIONS AND EXERCISES

1. State what the initials below stand for and describe the weather advisory function each performs:

    a. NWS

    b. FSS

    c. PATWAS

    d. ATIS

    e. TWEB

   f. EFAS

   g. SIGMET

   h. AIRMET

   i. PIREP

   j. NOTAM

2.   Outline what you should tell a weather specialist when you
call for a briefing.

3.   What is a station model? How is it used on a Surface
Analysis Chart?

4.   What is the chief value to pilots of a Weather Depiction
Chart?

5.   How are the Significant Weather Prognosis Chart and the Surface Analysis Chart related? How are they different?

6.   Of what value to pilots is a Radar Summary Chart?

7.   Outline the information contained in a Surface Aviation Weather Report.

8.   What are the three ways to determine a station's ceiling? How are they coded on Surface Analysis Charts?

9.   How are Area Forecasts different from Terminal Forecasts?

10. Of what value to pilots are Winds And Temperatures Aloft
Forecasts?

11. Indicate how each of the following would be shown on a
surface weather map, including a statement as to how they are
color-coded:

  a. cold front aloft

  b. squall line

  c. ridge

  d. stationary front

  e. occluded front

  f. trough

  g. warm front

  h. high-pressure center

REVIEW QUESTIONS

1.   (FAA 1494) Below FL180, en route weather advisories should be obtained from an FSS on

1--122.1 MHz.   2--122.0 MHz.   3--123.6 MHz.   4--122.4 MHz.

2.   (FAA 1495) During what hours is EFAS (en route flight advisory service) normally available?

1--8:00 a.m. through 8:00 p.m. GMT.
2--7:00 a.m. through 12:00 p.m. local.
3--6:00 a.m. through 10:00 p.m. local.
4--Continuous operation.

[NOTICES TO AIRMEN]

NOTE: NOTICES ARE ARRANGED IN ALPHABETICAL ORDER BY STATE (AND WITHIN STATE BY CITY OR LOCALITY). NEW OR REVISED DATA ARE INDICATED BY BOLD AND UNDERSCORING THE AIRPORT NAME.

NOTE: ALL TIMES ARE LOCAL UNLESS OTHERWISE INDICATED.

NOTICE: OBSTRUCTION LIGHT OUTAGES NO LONGER REQUIRE DISTANT DISSEMINATION (NOTAM(D) OR CLASS II). CHECK LOCAL NOTAMS FOR OUTAGES.

DISTRICT OF COLUMBIA

WASHINGTON NATIONAL ARPT: NW thr Rwy 15/33 relctd 200' and an additional 300' dsplcmt to Rwy 15 until 6/1. REIL and VASI Rwy 15 OTS until 6/1. Rwy 15 Clsd to B727 acft and larger until 6/1. (1/81) LGTD crane 396 ft MSL 1.8 miles west Rwy 15 until 8/3. (3/81) Helipad Clsd UFN. (3/81)

FLORIDA

FT LAUDERDALE EXECUTIVE ARPT: TPA 800 ft. TPA for jets 1300 ft. 4 Box VASI cmsnd left side Rwy 26. 2 Box VASI cmsnd left side Rwy 13. (3/81)
FT LAUDERDALE-HOLLYWOOD INTL ARPT: TPA 1000 ft. Thr Rwy 27L dsplcd 401 ft. Thr Rwy 09L dsplcd 609 ft. Thr Rwy 27R dsplcd 599 ft. (3/81)
FT MYERS PAGE FIELD ARPT: Unicom Freq 123.05. (3/81)
MIAMI INTL ARPT: Wide Body & DC-8 acft lndg Rwy 09L & desiring to taxi west on Twy M are advised to use Twys M-8, M-9, M-10, M-11 or the end. Twys are numbered W to E. (3/81) RVR Rwy 27L OTS UFN. (3/81)
NEW SMYRNA BEACH MUNI ARPT: Rwy 15/33 Rwy lgts oper dusk-dawn. Rwy 11/29 Rwy lgts oper dusk-2400. For VASI and Rwy 11/29 Rwy lgts after 2400 key 122.8 3 times for low, 5 times for med, and 7 times for high. (1/81)

Figure 9.1

3.   (FAA 1504 DI) What is the status of Rwy 15/33 at Washington National Airport? (See Figure 9.1.)

1--Rwy 15 threshold displaced 300 ft. Rwy 33 threshold 200 ft., REIL and VASI out of service, Rwy 15 restricted for B727 and heavier aircraft.
2--Rwy 15/33 closed until further notice to all aircraft except B727 and heavier, REIL and VASI out of service.
3--Rwy 15/33 relocated, former Rwy 15 closed until further notice, REIL and VASI out of service, Rwy 15 threshold displaced.
4--Rwy 15/33 relocated a total of 500 ft., Rwy 15 closed to B747 aircraft and heavier, REIL and VASI out of service.

4.   (FAA 1505) The traffic pattern altitude for light airplanes
and gyroplanes at Ft. Lauderdale-Hollywood International Airport
(Figure 9.1) is changed to

1--800 ft.   3--3,000 ft.
2--1,000 ft.   4--an additional 1,000 ft.

5.   (FAA 1506) What is the status of the runway lights for a
landing at New Smyrna Beach Muni. on Rwy 15 after 2400? (See
Figure 9.1.)

1--For runway lights, key the transmitter the proper number of
   times on 122.8 MHz.
2--The lights on this runway are not operated at night.
3--The runway lights operate from dark to dawn.
4--Runway lights are on request.

6.   (FAA 1531) Ceiling, as used in weather reports, is defined
as the height above the Earth's surface of the

1--lowest reported obscuration and the highest layer of clouds
   reported as overcast.
2--lowest layer of clouds reported as broken or overcast and not
   classified as thin.
3--lowest layer of clouds reported as scattered, broken, or thin.
4--highest layer of clouds reported as broken or thin.

7.   (FAA 1782) What is indicated when a current SIGMET
forecasts embedded thunderstorms?

1--Thunderstorms have been visually sighted.
2--Severe thunderstorms are embedded within a squall line.
3--Thunderstorms are dissipating and present no serious problem
   to IFR flight.
4--Thunderstorms are obscured by massive cloud layers and cannot
   be seen.

8.   (FAA 1831) Radar weather reports are of special interest to
pilots because they report

1--large areas of low ceilings and fog.
2--location of precipitation along with type, intensity, and
   trend.
3--location of broken to overcast clouds.
4--icing conditions.

9.   (FAA 1832) From which primary source should information be
obtained regarding expected weather at your destination and
estimated time of arrival?

1--Low Level Prog Chart.
2--Weather Depiction Chart.
3--Terminal Forecast.
4--Radar Summary and Weather Depiction Chart.

```
INK SA 1854 CLR 15 106/77/63/1112G18/000
BOI SA 1854 150 SCT 30 181/62/42/1304/015
LAX SA 1852 7 SCT 250 SCT 6HK 129/60/59/2504/991
MDW RS 1856 -X M7 OVC 11/2R+F 990/63/61/3205/980/RF2 RB12
JFK RS 1853 W5 X 1/2F 180/68/64/1804/006/RO4RVR22V30 TWR VSBY 1/4
```

Figure 9.2

10. (FAA 1827) What is the sky condition depicted for Chicago
Midway Airport (MDW) in Figure 9.2?

1--Thin overcast measured ceiling 700 ft., overcast 1,100 ft.,
   visibility 2 mi. in rain plus fog.
2--Sky partially obscured, measured ceiling 700 overcast,
   visibility 1-1/2, heavy rain, fog.
3--Thin overcast measured 700 ft. overcast, visibility 1-1/2,
   heavy rain, fog.
4--Sky partially obscured, measured ceiling 700 overcast,
   visibility 11, occasionally 2, with rain and heavy fog.

11. (FAA 1828) Which of the reporting stations in Figure 9.2
have VFR weather?

1--All
2--All except JFK.
3--All except JFK and MDW.
4--INK only.

12. (FAA 1829) The wind direction and velocity at JFK (see
Figure 9.2) is from

1--180° at 40 kts.
2--040° variable 22 to 30 kts.
3--180° at 4 kts.
4--018° at 4 to 6 kts.

13. (FAA 1830) What are the wind conditions at Wink, Texas
(INK)? (See Figure 9.2.).

1--Calm.
2--111° gusting to 18 kts.
3--011°, 12 kts. gusting to 18 kts.
4--110°, 12 kts. gusting to 18 kts.

14. (FAA 1833) According to the Terminal Forecast for Oklahoma
City (OKC) in Figure 9.3, the cold front should pass through

1--between 21Z and 02Z the next day.
2--between 18Z and 21Z with heavy thunderstorms.
3--between 1515Z and 18Z.
4--after 09Z the next day.

```
OK FT 011447

GAG FT 011515 100 SCT 250 SCT 2610. 16Z 60 SCT C100 BKN 3315G22 CHC C50
  BKN 5TRW. 01Z 250 SCT 3515G25. 09Z VFR WIND..

HBR FT 011515 C120 BKN 250 BKN 3010. 17Z 100 SCT C250 BKN 3215G25 CHC C30
  BKN 3TRW. 00Z 250 SCT 3515G25. 09Z VFR WIND..

MLC FT 011515 C20 BKN 1815 BKN OCNL SCT. 20Z C30 BKN 1815G22 CHC C20 BKN
  1TRW. 03Z C30 BKN 2015 CHC C7 X 1/2TRW+G40. 09Z MVFR CIG TRW..

OKC FT 011515 C12 BKN 140 BKN 1815G28 LWR BKN V SCT. 18Z C30 BKN 250 BKN
  2315G25 LWR BKN OCNL SCT CHC C7 X 1/2TRW+G40. 21Z CFP 100 SCT C250
  BKN 3315G25 CHC C30 BKN 5TRW-. 02Z 100 SCT 250 SCT 3515G25. 09Z VFR
  WIND..

PNC FT 011515 C100 BKN 250 BKN 1810. 16Z CFP 20 SCT C100 BKN 3115 SCT V
  BKN. 00Z 250 SCT 3515G25.  09Z VFR WIND..

TUL FT 011515 C20 BKN 1915G22. 19Z C30 BKN 1815G25 CHC 3TRW. 23Z CFP C100
  BKN 250 BKN 3215G25 CHC C30 BKN 5TRW. 09Z VFR WIND..
```

Figure 9.3

15. (FAA 1834) What wind conditions are expected at Hobart
(HBR) at 16Z? (See Figure 9.3.)

1--Calm.
2--115° at 15 kts.
3--300° at 10 kts.
4--320° at 15 kts. gusting to 25 kts.

16. (FAA 1835) What ceiling is forecast for Gage (GAG) between
16Z and midnight Z? (See Figure 9.3.)

1--100 scattered.   3--1,000 broken.
2--6,000 scattered.   4--10,000 broken.

17. (FAA 1836) What is the outlook for weather conditions at
McAlester (MLC)? (See Figure 9.3.)

1--Ceilings 2,000-3,000 ft. with southerly winds.
2--Ceilings 700 ft., sky obscured, visibility 1/2 mi. in the
   thundershowers.
3--VFR except in the thundershowers, peak wind gusts 40 kts.
4--Marginal VFR due to low ceilings and thundershowers.

18. (FAA 1837) The wind condition in the Terminal Forecast
Outlook for Ponca City (PNC) in Figure 9.3 is

1--missing.
2--for velocities of 25 kts. or stronger.
3--for a wind shift from south to northwest.
4--for the wind to change from a gusty condition to calm.

```
   FA 011240
DFW FA 011240
VALID 011300Z-020700Z
OTLK 020700Z-021900Z

NM OK TX AND CSTL WTRS...

HGTS MSL UNLESS NOTED...

TSTMS IMPLY PSBL SVR OR GTR TURBC..SVR ICG..AND LOW-LVL WIND SHEAR...

FLT PRCTN...SWRN TX W PECOS RVR AND NM...OCNL MDT TURBC BLO 150 WITH
STG UDDFS VCNTY MTNS.

S CNTRL TX SERN TX...PATCHY CIGS AOB 010 AND VSBY BLO 3 MI IPVG AFT
15Z.

TX AND OK ALG AND WITHIN 100 MI OF CDFNT...OCNL MDT TURBC BLO 100 AND
W OF FNT LLWS.

SYNS... CDFNT VCNTY ICT-LBB-HOB LN SWWD WL MOV EWD TO ABT
FYV-BIG BEND LN BY 07Z.

SIGCLD AND WX...
NM AND PTN OF W TX W OF INK-BIG BEND LN...
GENLY 150-200 SCT LCLY BKN WITH LWR 80-120 SCT NERN NM SPRDG OVR AREA
AFTN AND CLRG AFDK. PATCHY 40-50 SCT NERN NM AFDK. OTLK... VFR.

OK AND TX W OF ICT-CDS-INK LN...
CIG 100-150 BKN TOPS LYRS 300. OTLK... VFR CIG ABV 100.
OK AND TX E OF ICT-CDS LN...
CIG 12-25 BKN V SCT 60 120 BKN V SCT TOPS LYRS 300 WITH CIG LCLY LWR
OVR ERN OK LCLY CIG 10-14 OVC. CIG GRDLY LFTG TO 20-30 BKN V SCT 80
BY NOON WITH SCT TSTMS LCLY LWRG CONDS BLO CIG 10 X 2TRW TOPS 400.
CONDS WL LFT ABT 50-100 MI BHD CDFNT TO CIG 100 BKN. OTLK... MVFR CIG
TRW BCMG VFR 100 MI W OF CDFNT.

CSTL WTRS...
GENLY 25 SCT. WDLY SCT SHWRS DVLPG AFT 18Z. OTLK...VFR BCMG MVFR TRW
BY MID MRNG.

ICG AND FRZLVL...NONE OF CONSEQUENCE OUTSIDE SHWRS AND TSTMS. FRZLVL
090 NRN NM SLPG TO 145 SRN TX.

TURBC...SWRN TX W PECOS RVR AND NM...OCNL MDT TURBC BLO 120 WITH STG
UDDFS VCNTY MTNS. OCNL MDT TURBC WITHIN 100 MI OF FNT BLO 100.

THIS FA ISSUANCE INCORPORATES THE FOLLOWING AIRMETS STILL IN EFFECT...
NONE.
```

Figure 9.4

19. (FAA 1838) When is the wind forecast to shift at Tulsa
(TUL)? (See Figure 9.3.).

1--15Z.
2--Between 19Z and 23Z.
3--Between 23Z and 09Z the next day.
4--After 09Z the next day.

20. (FAA 1839) Hazards to light aircraft associated with the
front as forecast in Figure 9.4 are

1--shifting winds, moderate turbulence, and ceilings as low as
   1,000 ft.
2--moderate turbulence, strong updrafts and downdrafts near the
   mountains, and icing conditions near the thunderstorms.
3--lowering ceilings, icing conditions, and moderate turbulence
   in and around thunderstorms.
4--low-level wind shear in the vicinity of the mountains and
   moderate turbulence in and around thunderstorms.

21. (FAA 1840) What general cloud conditions are forecast in
the Area Forecast for the day? (See Figure 9.4.)

1--Scattered to broken ahead of the front and scattered to clear
   behind the front.
2--Embedded thunderstorms within 100 NM of the front and on both
   sides.
3--Scattered to broken ahead of the front, broken to overcast
   along the front, lifting to 10,000 ft. 50-100 mi. behind the
   front.
4--Broken ahead of the front, widely scattered thunderstorms
   along the front, clear after frontal passage.

22. (FAA 1841) According to the Area Forecast in Figure 9.4,
the overall outlook is for

1--clearing over the entire area.
2--VFR conditions with ceilings above 1,000 ft.
3--VFR in the western portion and marginal VFR in the eastern
   portion and coastal waters.
4--scattered thunderstorms west of the front and marginal VFR
   changing to VFR immediately behind the front.

23. (FAA 1842) What is the forecast for southern Texas and the
coastal waters? (See Figure 9.4.)

1--Low ceilings and visibility moving from land to coastal waters
   after 15Z becoming VFR over land and MVFR over water.
2--Patchy ceilings at or below 1,000 ft. and visibility below 3
   mi. improving after 15Z.
3--Generally 2,500 scattered, widely scattered showers developing
   after 18Z becoming MVFR by midmorning.
4--Moderate turbulence, patchy ceilings below 1,000 ft. with
   visibilities below 3 mi.

24. (FAA 1843 DI)   What is the forecast for northern New Mexico?
(See Figure 9.4.)

1--Occasionally moderate turbulence below 15,000 ft. with strong
    updrafts and downdrafts in the vicinity of the mountains with
    low ceilings and visibility.
2--Scattered clouds 8,000 to 12,000 ft. spreading over the area
    afternoon and clearing after dark to patchy scattered clouds
    at 4,000 to 5,000 ft.
3--Locally broken to scattered clouds at 15,000 to 20,000 ft.
    lowering to 8,000 to 12,000 ft. scattered in the morning and
    clearing by noon.
4--Ceilings 10,000 to 15,000 ft. broken, tops of layers 30,000
    ft. and clearing after 07Z above 10,000 ft.

```
      FDUS2 KWBC 011644
DATA BASED ON 011200Z

VALID 020000Z   FOR USE 2100-0600Z. TEMPS NEG ABV 24000

FT  3000    6000    9000    12000   18000   24000   30000   34000   39000

ABI       2512+14 2519+09 2426+02 2438-14 2345-26 234840 234546 254152
ABQ               3115+02 3125-04 2934-18 2944-31 285240 285145 295151
AMA       3013    3120+03 2823-03 2536-16 2447-30 245642 255447 265351
ATL 1510 1811+13 1909+08 2107+02 2708-12 2710-24 271339 271648 252057
BNA 1920 2220+13 2220+08 2319+02 2421-12 2426-24 243240 243449 243659
BRO 1811 1917+17 2015+11 2214+05 2517-09 2626-21 273937 274845 276155
DAL 2115 2322+15 2325+09 2328+02 2335-13 2342-25 234940 244946 254754
```

Figure 9.5

25. (FAA 1844)   Which altitude should provide the best
groundspeed advantage for a westerly flight from Brownsville
(BRO)?   (See Figure 9.5.)

1--3,000 ft.   2--12,500 ft.   3--32,000 ft.   4--36,000 ft.

26. (FAA 1845)   Which altitude should provide the best
groundspeed advantage for an easterly flight from Atlanta (ATL)?
(See Figure 9.5.)

1--3,000 ft.   2--19,500 ft.   3--25,500 ft.   4--34,000 ft.

27. (FAA 1846)   What wind is forecast for Atlanta (ATL) at
24,000 ft.?   (See Figure 9.5.)

1--270° at 10 kts.
2--270° at 10 kts. gusting to 24 kts.
3--027° at 10 kts.
4--027° at 10 kts. gusting to 24 kts.

28.   (FAA 1847)   Which altitude should provide the best
groundspeed advantage for a flight from Nashville (BNA) on a 360°
course?   (See Figure 9.5.)

1--3,000 ft.   3--23,500 ft.
2--9,500 ft.   4--38,000 ft.

29.   (FAA 1848)   What wind is forecast for 6,000 ft. at Abilene
(ABI)?   (See Figure 9.5.)

1--250° at 12 kts. gusting to 14 kts.
2--251° at 1 kt. gusting to 14 kts.
3--250° at 12 kts.
4--025° at 12 kts.

30.   (FAA 1849)   The wind and temperature for Albuquerque (ABQ)
as depicted in Figure 9.5 at 30,000 ft. are forecast to be from

1--285° at 24 kts. and 0° C.
2--280° at 52 kts. and -40° C.
3--280° at 52 kts. and 40° C.
4--280° at 52 MPH and 40° C.

31.   (FAA 1850)   Of what value is the depiction chart in Figure
9.6. to the pilot?

1--To determine general weather conditions on which to base
    flight planning.
2--For a forecast of cloud coverage, visibilities, and frontal
    activity.
3--To determine the frontal trends and air mass characteristics.
4--For an overall view of thunderstorm activity and forecast
    cloud heights.

32.   (FAA 1851)   The marginal weather in southeast New Mexico as
depicted in Figure 9.6 is due to

1--reported thunderstorms.
2--600-ft. overcast ceilings.
3--rain showers.
4--low visibility.

33.   (FAA 1852)   What weather phenomenon is causing IFR
conditions along the coast of Oregon and California?   (See Figure
9.6.)

1--Squall line activity.   3--Heavy rain showers.
2--Low ceilings.   4--Drizzle.

34.   (FAA 1853)   What is the status of the front that extends
from New Mexico to Indiana as depicted in Figure 9.6?

1--Stationary.   3--Retreating.
2--Occluded.   4--Dissipating.

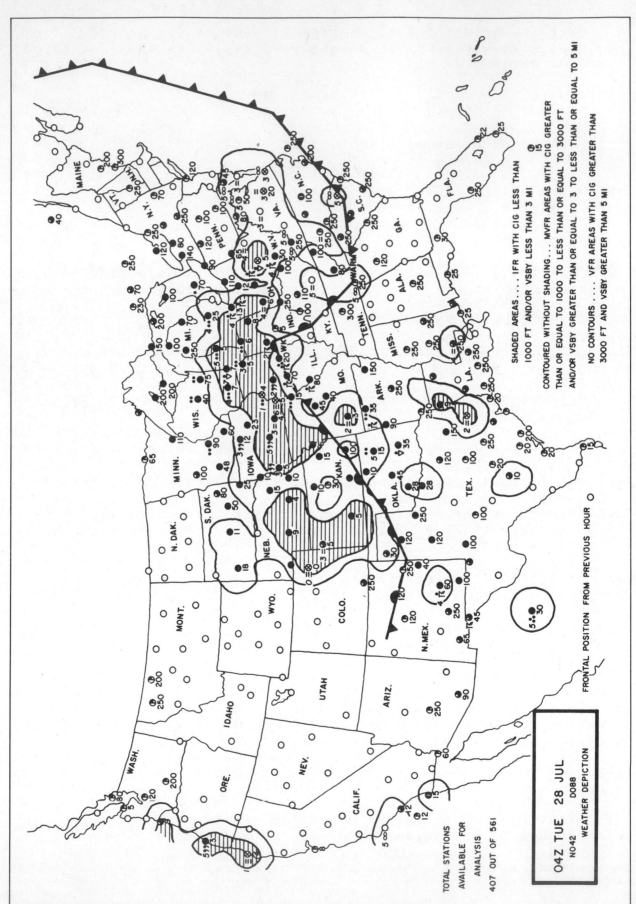

Figure 9.6

35. (FAA 1854) According to the depiction chart in Figure 9.6.,
the current weather for a flight from central Arkansas to
southeast Alabama is

1--broken clouds at 2,500 ft.
2--visibility from 3 to 5 mi.
3--broken to scattered clouds at 25,000 ft.
4--for and low ceilings to move in from the gulf.

36. (FAA 1855) What weather is forecast for the Gulf Coast area
just ahead of the cold front during the first 12 hr.? (See
Figure 9.7.)

1--Marginal VFR to IFR with intermittent thundershowers and rain
     showers.
2--IFR with moderate or greater turbulence over the coastal
     areas.
3--Thunderstorm cells moving northeastward ahead of the front.
4--Rain and drizzle dissipating, clearing along the front.

37. (FAA 1856) At what level are icing conditions forecast for
northern California during the first 12 hr. as depicted in Figure
9.7?

1--2,000 ft.   3--Surface to 8,000 ft.
2--3,200 ft.   4--8,000 to 12,000 ft.

38. (FAA 1857) For what phase of flight planning are the
Significant Weather Progs designed? (See Figure 9.7.)

1--A complete set of weather forecasts for overall planning.
2--Information to avoid areas and/or altitudes of most
     significant icing and turbulence.
3--An anlysis of frontal activity, cloud coverage, and areas of
     precipitation.
4--Overall depiction of current ceilings and visibilities with
     fronts and icing levels.

39. (FAA 1858) Interpret the weather symbol depicted in lower
California on the 12-hr. Significant Weather Prog in Figure 9.7?

1--Moderate turbulence, surface to 18,000 ft.
2--Thunderstorm tops at 18,000 ft.
3--Base of clear air turbulence, 18,000 ft.
4--Moderate turbulence, 180 mb level.

40. (FAA 1859) The band of IFR weather associated with the cold
front in the western states (Figure 9.7) is forecast to move

1--southeast at 30 kts. with moderate snow showers.
2--northeast at 12 kts. with the front and producing snow
     showers.
3--eastward at 30 kts. with the low and producing snow showers.
4--eastward at 30 kts. with continuous snow.

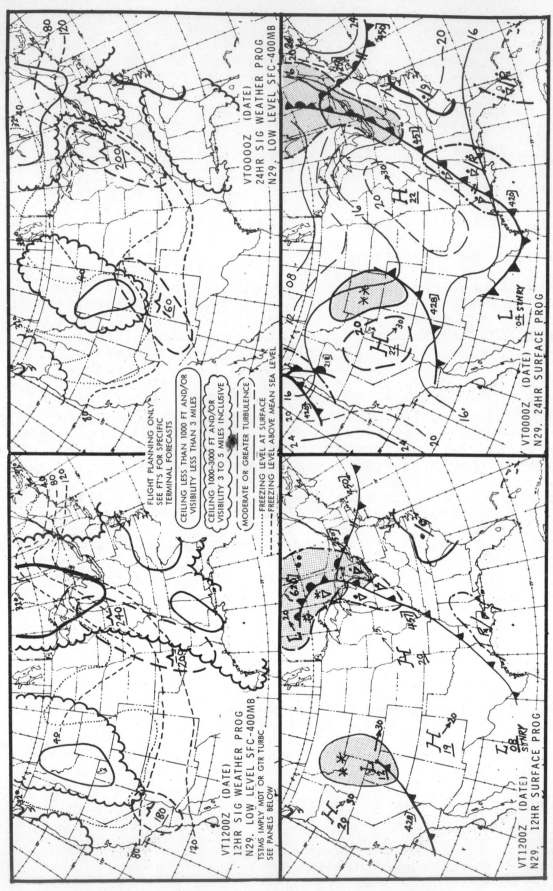

Figure 9.7

41. (FAA 1860) The section of the Area Forecast entitled FLT PRCTN contains a

1--summary of the expected hazardous weather.
2--statement listing those AIRMETS still in effect.
3--summary of general weather conditions over several states.
4--statments TSTME IMPLY PSBL SVR OR GTR TURBC, SVR ICG, AND LOW-
   LVL WIND SHEAR.

42. (FAA 1861) The section of the Area Forecast entitled SIG CLDS AND WX contains a

1--summary of cloudiness and weather significant to flight
   operations broken down by states or other geographical areas.
2--summary of forecast sky cover, cloud tops, visibility, and
   obstructions to vision along specific routes.
3--statement of AIRMETS and SIGMETS still in effect at the time
   of issue.
4--summary of only those clouds and weather considered adverse to
   safe flight operation.

43. (FAA 1862) To best determine forecast weather conditions between weather reporting stations, the pilot should refer to the

1--pilot reports.   3--weather maps.
2--prognostic charts.   4--area forecasts.

44. (FAA 1863) To determine the freezing level and areas of probable icing aloft, the pilot should refer to the

1--Radar Summary Chart.   3--Area Forecast.
2--Weather Depiction Chart. 4--Surface Analysis.

45. (FAA 1864) To obtain a continuous transcribed weather briefing including winds aloft and route forecasts for a cross-country flight, a pilot should monitor

1--a TWEB on a low-frequency radio receiver.
2--a VHF radio receiver tuned to an ATIS frequency.
3--the regularly scheduled weather broadcast on a VOR frequency.
4--a high-frequency radio receiver tuned to the En Route Flight
   Advisory Service (Flight Watch) station.

46. (FAA 1865) Individual forecasts for specific routes of flight can be obtained from which weather service?

1--Transcribed Weather Broadcasts.
2--Terminal Forecasts.
3--Area Forecasts.
4--In-flight advisories.

47. (FAA 1866) TWEBs (transcribed weather broadcasts) may be
monitored by tuning the appropriate radio receiver to certain

1--FSS communication frequencies.
2--airport advisory frequencies.
3--VOR and NDB frequencies.
4--NDB frequencies only.

48. (FAA 1867) SIGMETs are issued as a warning of weather
conditions hazardous

1--particularly to light aircraft.
2--to all aircraft.
3--only to light aircraft operations.
4--particularly to heavy aircraft.

49. (FAA 1868) AIRMETs are issued as a warning of weather
conditions hazardous

1--to all airplanes.
2--particularly to light airplanes.
3--to VFR operations only.
4--particularly to heavy airplanes.

50. (FAA 1869) Which in-flight advisory would contain
information on severe icing?

1--Convective SIGMET.
2--SIGMET.
3--AIRMET.
4--PIREP.

51. (FAA 1870) What information is contained in a convective
SIGMET in the contermininous United States?

1--Tornadoes, embedded thunderstorms, and hail 3/4 in. or greater
   in diameter.
2--Severe icing, severe turbulence, or widespread dust storms
   lowering visibility to less than 3 mi.
3--Weather less than basic VFR and sustained winds of 30 kts. or
   greater at the surface.
4--Ceilings less than 500 ft. and visibility less than 1 mi.

52. (FAA 1871) What information is provided by the Radar
Summary Chart that is not shown on other weather charts?

1--Lines and cells of hazardous thunderstorms.
2--Ceilings and precipitation between reporting stations.
3--Types of precipitation between reporting stations.
4--Areas of cloud cover and icing levels within the clouds.

This is a telephone weather briefing from the FSS at New Orleans for a flight from New Orleans to Jacksonville, Florida. The pilot calls at 17Z and advises that the departure time is 18Z with an ETE of about 3 hours.

'We have some adverse weather conditions today. Convective Sigmet 8 EASY calls for a broken line of very heavy thunderstorms currently extending from a Birmingham, Alabama - Pensacola line and southward into the Gulf of Mexico, few tops to 45 thousand, possible surface wind gusts to 60 knots and large hail. The line is moving east at 20 knots. A severe thunderstorm watch is in effect for the area enclosed by a Birmingham, Pensacola, Jacksonville, Augusta, Birmingham line until 23Z.'

'On the synopsis, a cold front extends from a low-pressure center in Indiana southward through the Florida Panhandle near Pensacola into the Gulf of Mexico and is moving east at 20 knots.'

'Last hour here at New Orleans we had 4 thousand scattered estimated ceiling 8 thousand broken, 25 thousand broken, visibility 7, temperature 63, dew point 55, wind 310 at 12 gusting to 19, altimeter 29 point 98. En route there were scattered clouds at 4 thousand AGL with ceilings 8 thousand from New Orleans to Mobile becoming 4 thousand the rest of the route. There is a broken line of very heavy thunderstorms, tops 45 thousand from Birmingham to Pensacola and southward into the Gulf of Mexico moving east at 20 knots. Jacksonville reported 15 hundred scattered, estimated ceiling 4 thousand broken, 8 thousand overcast, visibility 7, temperature 75, dew point 70, wind 160 at 15 gusting to 25, altimeter 30 point 02, cumulonimbus to the southwest moving northeast.'

'The terminal forecast for New Orleans for 18Z calls for 8 thousand scattered, 25 thousand scattered, visibility 7, wind 310 at 10.'

'En route you will have scattered clouds at or above 8 thousand from Louisiana to Mississippi and ceilings 4 thousand in extreme southern Alabama through the rest of the route with clouds layered to above 20 thousand. A line of very heavy thunderstorms moving east through the Florida Panhandle along your route of flight. Convective Sigmet 8 EASY is in effect. Also in effect is the severe thunderstorm watch from Pensacola to Jacksonville. The freezing level is at 13 thousand 5 hundred MSL along the entire route. Severe icing, turbulence, and large hail can be expected within the very heavy thunderstorms.'

'At 21Z, Jacksonville will have 15 hundred scattered, ceiling 4 thousand broken, 8 thousand overcast, visibility 7, wind 160 at 15 gusting to 20. They will have thundershowers in the vicinity and a chance of 5 hundred overcast, visibility 1 mile, heavy thundershowers, hail, and wind gusting to 50 after 21Z.'

'Your winds aloft for the route are: from New Orleans to Mobile, 3 thousand 260 at 25, 6 thousand 270 at 36, 9 thousand 270 at 48, 12 thousand 270 at 54, 15 thousand and above 270 at 55. From Mobile to Jacksonville, 3 thousand 180 at 30, 6 thousand 190 at 45, 9 thousand 210 at 55, 12 thousand 220 at 63, 15 thousand 230 at 67, 18 thousand and above 230 at 70.'

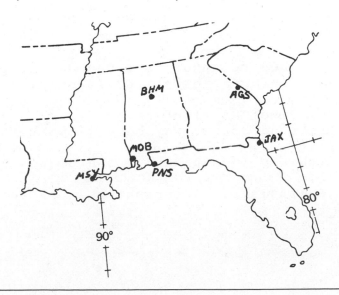

Figure 9.8

53. (FAA 1874) When telephoning a weather briefing facility for preflight weather information, you should state

1--that you possess a current medical certificate.
2--your intended route, destination, and type of aircraft.
3--the color of the aircraft and number of occupants on board.
4--your total flight time.

54. (FAA 1875) What ceilings should the flight encounter except in the area of the line of thunderstorms? (See Figure 9.8.)

1--1,500 ft.   3--8,000 ft.
2--4,000 ft.   4--25,000 ft.

55. (FAA 1876) At what time should the thunderstorms move into the Jacksonville area? (See Figure 9.8.)

1--18Z.   3--20Z.
2--19Z.   4--21Z.

56. (FAA 1877) To aid in making a GO-NO GO decision for the flight, which question should the pilot ask the weather briefer? (See Figure 9.8.)

1--What altitude is the freezing level in the thunderstorms?
2--Where does radar show some possible routes through the area?
3--How can we circumnavigate the front and the thunderstorms?
4--How large is the hail?

57. (FAA 1878) If the flight remains VFR for the entire trip, the greatest hazard described in the briefing in Figure 9.8 is

1--an icing condition.   3--a strong wind.
2--the low ceilings.   4--the wind shear.

58. (FAA 1879) What weather phenomenon can aid the pilot to identify the line of thunderstorms described in the weather briefing in Figure 9.8?

1--Low ceilings.   3--Structural icing.
2--Lightning.   4--Rain.

59. (FAA 1873) When telephoning a weather briefing facility for preflight weather information, pilots should

1--identify themselves as pilots.
2--tell the number of hours they have flown within the preceding
   90 days.
3--state the number of occupants on board and the color of the
   aircraft.
4--state that they possess a current medical certificate.

60. (FAA 1884) What flight turbulence is inferred by the
briefing in Figure 9.9?

1--No turbulence for the entire flight.
2--No turbulence except near thunderstorms in the Boston area and
   the temperature inversion.
3--Light turbulence for the entire flight except moderate
   turbulence near the thunderstorms.
4--Light turbulence from Buffalo to Albany and moderate
   turbulence from Albany to Boston.

61. (FAA 1885 DI) What minimum altitude should provide smooth,
stable air for the flight proposed in Figure 9.9?

1--3,000 ft.   3--11,500 ft.
2--7,500 ft.   4--17,500 ft.

62. (FAA 1886) At what altitude should a pilot expect to
encounter a temperature inversion and possible wind shear during
the flight as briefed in Figure 9.9?

1--Between 3 and 6 thousand, Buffalo to Albany.
2--Between 3 and 6 thousand, Albany to Boston.
3--Between 6 and 9 thousand, Buffalo to Boston.
4--Between 12 and 14 thousand, Buffalo to Boston.

63. (FAA 1887) Which change in conditions at Boston will
eliminate the advection fog reported in the briefing in Figure
9.9?

1--Passage of a thundershower.
2--Increase in wind speed to 15 MPH.
3--Temperature decrease of at least 4°.
4--Wind shift to a land breeze.

64. (FAA 1888) The widespread system affecting the en route
weather described in the weather briefing in Figure 9.9 is

1--a temperature inversion.
2--an occluded front.
3--a warm front.
4--a low-pressure trough.

65. (FAA 1889) Under what minimum flight rules may the flight
depart the airport traffic area at Buffalo at the proposed
departure time according to the weather briefing in Figure 9.9?

1--VFR.
2--IFR.
3--Special VFR.
4--Composite VFR/IFR.

This is a pilot weather briefing from the FSS at Buffalo, New York, for a flight to Boston, Massachusetts. The pilot advises that the departing time will be 18Z and the ETE is approximately 3 hours.

'Conditions are for marginal VFR and occasionally IFR conditions for the entire flight. Airmet PAPA ONE calls for occasional moderate icing in the clouds and precipitation from Buffalo to central Massachusetts from 4 thousand to 6 thousand. Sigmet 12 EASY warns of scattered embedded thundershowers along coastal Massachusetts continuing beyond 18Z.'

'At 12Z, a warm front extended from a low centered near Toledo, Ohio, eastward across central Pennsylvania and into the Atlantic just south of Long Island.'

'Current weather here at Buffalo is a thin obscurement 5 hundred scattered, 9 hundred overcast, 3 miles in rain and light fog, temperature 46, dew point 42, wind 080 at 12, altimeter 29 point 80. At 1435Z, the pilot of a B727 reported mixed icing during descent from 6 thousand to 4 thousand.'

'You should experience marginal VFR and occasionally IFR conditions through western New York to western Massachusetts with marginal to VFR conditions the remainder of the route. At 15Z, Boston reported 12 hundred scattered, measured ceiling 28 hundred broken, 8 thousand overcast, 5 miles in fog, temperature 41, dew point 38, wind 120 at 15, altimeter 29 point 90. Here at Buffalo at departure time, we should have 14 hundred overcast, 5 miles visibility in light rain and fog, wind 090 at 10 gusting to 20.'

'En route, clouds are layered from 1 thousand to above 25 thousand. Visibility will be 5 miles, occasionally 3 miles in the precipitation and fog. Airmet PAPA ONE warns of occasional moderate mixed icing in the clouds and precipitation from 4 to 6 thousand New York to central Massachusetts. Convective Sigmet 12 EASY forecasts a freezing layer, 4 to 6 thousand New York to central Massachusetts, otherwise the freezing level is 12 thousand at Buffalo sloping to 14 thousand at Boston.'

'At 21Z, Boston should have 12 hundred scattered, ceiling 2 thousand overcast, 5 miles in light rain and fog, wind 120 at 10, occasionally ceiling 1 thousand overcast, 3 miles in light rain and fog, and a chance of thundershowers.'

"Winds aloft forecast: From Buffalo to Albany 3 thousand 100 at 18, 6 thousand 120 at 30, 9 thousand and above 240 at 35 to 45. From Albany to Boston 3 thousand 150 at 18, 6 thousand 160 at 35, 9 thousand 270 at 35, 12 thousand and above 270 at 45."

Figure 9.9

66. (FAA 1872) What values are used for Winds Aloft Forecasts?

1--Magnetic direction and knots.
2--Magnetic direction and miles per hour.
3--True direction and knots.
4--True direction and miles per hour.

67. (FAA 1890) What minimum altitude should provide smooth, stable air for the flight proposed in Figure 9.10?

1--3,000 ft.   3--11,500 ft.
2--7,500 ft.   4--17,500 ft.

68. (FAA 1891) Should the pilot expect icing conditions at some level? (See Figure 9.10.)

1--Yes, at 16,000 ft.
2--Yes, at 25,000 ft. in the clouds.
3--No, there is not visible moisture at the freezing level.
4--No, the temperature/dew point spread is too large.

69. (FAA 1892) What hazard to light aircraft is forecast for the flight depicted in Figure 9.10?

1--Icing conditions.
2--Moderate turbulence.
3--Low ceilings over the mountains.
4--Embedded thunderstorms.

70. (FAA 1893) Which altitude offers the best groundspeed advantage for a flight from Reno to Phoenix? (See Figure 9.10.)

1--3,000 ft. AGL.   3--11,500 ft. AGL.
2--7,500 ft. MSL.   4--19,500 ft. MSL.

71. (FAA 1894) The large weather system that dominates the area depicted by the briefing in Figure 9.10 is

1--a high-pressure area.
2--a low-pressure area.
3--quasi-stationary front.
4--cold frontogenesis.

72. (FAA 1902) How should you establish contact with an EFAS (En Route Flight Advisory Service) station, and what service would you normally expect?

1--Call EFAS on 122.2 for routing weather, current reports on
   hazardous weather, and altimeter settings.
2--Call flight assistance on 122.5 for advisory service
   pertaining to severe weather.
3--Call flight watch on 122.0 for information regarding actual
   weather and thunderstorm activity along your route.
4--Call ARTCC on assigned frequency and ask for flight watch
   service.

This is a pilot weather briefing from the Reno FSS on Reno-Cannon International Airport, Nevada, for a flight to Phoenix Sky Harbor in Arizona. The pilot advises that the estimated time of departure is 1 hour later (17Z) and the ETE is approximately 3 hours.

'You can expect moderate turbulence from the surface to 16 thousand MSL in southern Nevada and Arizona due to thermals. There is a large high-pressure system that covers the southwest U.S.'

'Currently here at Reno, we have 4 thousand scattered, 25 thousand scattered with 7 miles visibility. Temperature is 81, dew point 50, wind from 120 at 5 knots, and the altimeter setting is 30 point 02.'

'Along the route, reports indicate scattered clouds 4 to 7 thousand AGL with scattered clouds at or above 20 thousand. Isolated mountain thundershowers are reported.'

'At this time, Phoenix has 6 thousand scattered, 25 thousand thin broken, and 7 miles visibility. Temperature is 91, dew point 45, wind from 160 at 8 knots, and their altimeter setting is 30 point 01. At 1645Z, a B727 reported moderate turbulence from 8 thousand to 3 thousand during descent.'

'If you depart at 18Z, you should have 5 thousand scattered, 25 thousand scattered, and 7 miles visibility. The wind will be from 140 at about 8 and the altimeter setting should be 29 point 99. En route you should expect scattered clouds at 4 thousand AGL, tops to 12 thousand MSL with scattered clouds at or above 20 thousand, occasional moderate turbulence from the surface to 16 thousand MSL. You might see a few isolated mountain thundershowers. The freezing level will be at about 16 thousand MSL for the entire route.'

'At 20Z, Phoenix is forecast to have 25 thousand scattered, 7 miles visibility, wind from 190 at 8 knots, and an altimeter setting of 29 point 95.'

'The winds aloft from Reno to Las Vegas; 3 thousand 160 at 25, 6 thousand 180 at 34, 9 thousand 200 at 37, 12 thousand 220 at 40, 15 thousand 240 at 42, 18 thousand 260 at 45, 21 thousand 270 at 45. From Las Vegas to Phoenix; 3 thousand 220 at 25, 6 thousand 210 at 31, 9 thousand 220 at 38, 12 thousand 230 at 45, 15 thousand 240 at 53, 18 thousand 250 at 62, 21 thousand 250 at 65.'

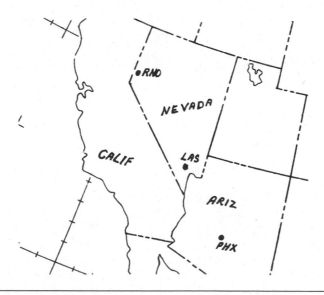

Figure 9.10

73. **(FAA 1895)** What weather phenomenon reported in the TWEB in Figure 9.11 indicates a layer of above-freezing temperatures above 7 or 8 thousand?

1--Snow.
2--Ice pellets.
3--Wind shift aloft.
4--Temperature/dew point spread.

74. **(FAA 1896)** The boundary between the air from the high in Nebraska and the low in New Mexico as described by the TWEB in Figure 9.11 is between

1--the surface and 6,000 ft.
2--6,000 ft. and 9,000 ft.
3--9,000 ft. and 12,000 ft.
4--12,000 ft. and 15,000 ft.

75. **(FAA 1897)** What type of icing may occur in the conditions described in the TWEB in Figure 9.11?

1--Structural and induction icing below 8 thousand and carburetor icing in the warmer air above.
2--Carburetor induction, and structural icing from the surface to 7 or 8 thousand.
3--Structural icing from the surface to 7 or 8 thousand and above 15 thousand.
4--Carburetor, structural, and induction icing from the surface to 7 or 8 thousand and structural icing above 15 thousand.

76. **(FAA 1898)** What type of clouds normally produce the icing mentioned by the PIREP in the TWEB in Figure 9.11?

1--Cumulus.   3--Stratus.
2--Stratocumulus.   4--Cumulonimbus.

77. **(FAA 1899)** What condition described in the TWEB in Figure 9.11 is causing the fog?

1--Precipitation and turbulence.
2--Temperature/dew point spread and light wind.
3--Moist air moving over a colder surface.
4--Temperature below freezing and moisture sublimating directly as ice crystals.

78. **(FAA 1900)** In which direction is the precipitation moving? (See Figure 9.11.)

1--North.
2--Northeast.
3--East.
4--Southeast.

This is a Transcribed Weather Broadcast from Oklahoma City, Oklahoma, with Oklahoma City to Amarillo, Texas route. Valid until 23Z.

"Synopsis: Rapidly developing upper level low-pressure system over eastern New Mexico moving eastward over the area through the period. A strong low-level easterly flow south of an artic high-pressure center in Nebraska continuing over the area."

'Adverse conditions: IFR conditions due to low ceilings and visibilities for the west portion of the route spreading east over the entire route by midafternoon. Occasional moderate to severe icing in the clouds and precipitation from the surface to 8 thousand MSL for the west portion of the route. Occasional moderate turbulence 20 to 30 thousand along the route due to the jet stream. Low-level wind shear along the route due to strong surface winds.'

'Route forecast, Oklahoma City-Amarillo: The west portion of the route, ceilings at or below 1 thousand and visibilities occasionally below 3 miles in snow, ice pellets, and fog. The east portion of the route, ceilings 3 thousand becoming ceilings at or below 1 thousand, and visibilities occasionally below 3 miles in snow and fog by midafternoon. Tops to 17 thousand MSL, also scattered to broken clouds at or above 25 thousand MSL along the entire route.'

'Winds aloft at Oklahoma City: 3 thousand 050 at 30, 6 thousand 070 at 50, 9 thousand 110 at 40, 12 thousand 230 at 40, 18 thousand 230 at 65, 24 thousand 240 at 75, 30 thousand 240 at 87, 34 thousand 240 at 80, 39 thousand 240 at 60. Amarillo: 3 thousand 080 at 40, 6 thousand 090 at 45, 9 thousand 120 at 35, 12 thousand 230 at 40, 18 thousand 240 at 70, 24 thousand 240 at 85, 30 thousand 250 at 96, 34 thousand 250 at 90, 39 thousand 240 at 60.'

'Radar reports: At 1435Z, Amarillo radar shows an area of light snow and ice pellets with 7 tenths coverage extending from Amarillo to Gage to Childress to Amarillo. The area has increased in size from past hour. Maximum tops are uniform at 14 thousand MSL. The area is moving to the northeast at 35 knots.'

'Surface weather reports: Amarillo, 1520Z, measured ceiling 8 hundred overcast, visibility 3 miles in light snow and fog, temperature 25, dew point 23, wind 090 at 25 gusting to 37, altimeter 29 point 67, the ice pellets began at 25 after the hour and ended 40 minutes past the hour; Gage, 1515Z, measured ceiling 3 hundred broken, 10 thousand overcast, visibility 2 miles in light snow and fog, temperature 22, dew point 20, wind 070 at 18 gusting to 30, altimeter 29 point 97. Childress, 5 hundred scattered, measured ceiling 9 hundred overcast, visibility 6 miles in light snow/ice pellets/fog, temperature 30, dew point 28, wind 110 at 25 gusting to 35, altimeter 29 point 88. Hobart, measured ceiling 25 hundred broken, 5 thousand overcast, visibility 7 miles, temperature 27, dew point 24, wind 080 at 20 gusting to 32, altimeter 29 point 94, snow showers of unknown intensity to the west. Oklahoma City, measured ceiling 3 thousand broken, 8 thousand overcast, visibility 7 miles, temperature 21, dew point 16, wind 040 at 15 gusting to 28, altimeter 30 point 02.'

'Pilot reports: 30 miles east of Amarillo, a pilot of a Boeing 727 during a climbout at Amarillo reported moderate rime icing from the surface to 7 thousand, tops of clouds 15 thousand with cirrus above and moderate turbulence from the surface to 5 thousand."

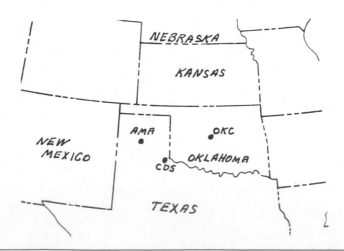

Figure 9.11

79. (FAA 1901) When telephoning a weather briefing facility for preflight weather information, you should state

1--the number of hours you have flown within the preceding 90
    days.
2--that you possess a current medical certificate.
3--whether you intend to fly VFR only.
4--the color of the aircraft and number of occupants on board.

80. (FAA 1903) What service should a pilot normally expect from an En Route Flight Advisory Service station?

1--Actual weather information and thunderstorm activity along the
    route.
2--Preferential routing and provide radar vectoring to
    circumnavigate severe weather.
3--Severe weather information, changes to flight plans, and
    receive routine position reports.
4--Radar vectors for traffic avoidance, routine weather
    advisories, and altimeter settings.

81. (FAA 1904) Transcribed weather broadcasts can be monitored by tuning the appropriate receiver to certain

1--FSS communication frequencies.
2--airport advisory frequencies.
3--VOR and NDB frequencies.
4--NDB frequencies only.

82. Refer to the weather symbols in Figure 9.12. Weather conditions associated with symbol O are

1--an area of thunderstorms.
2--a squall line.
3--rain, drizzle, and fog.
4--stratiform clouds and haze.

83. Refer to the weather symbols in Figure 9.12. Which one properly identifies a stationary front?

1--E   3--G
2--F   4--I

84. Refer to the weather symbols in Figure 9.12. If symbol E is shown on a surface weather map, it indicates that

1--a cold air mass has caught up with a warm air mass and the air
    masses have closed together to form an occluded front.
2--a cold air mass is overtaking and replacing a warm air mass.
3--neither a cold air mass nor a warm air mass is being replaced,
    and the front is stationary.
4--a warm air mass is moving in and replacing colder air.

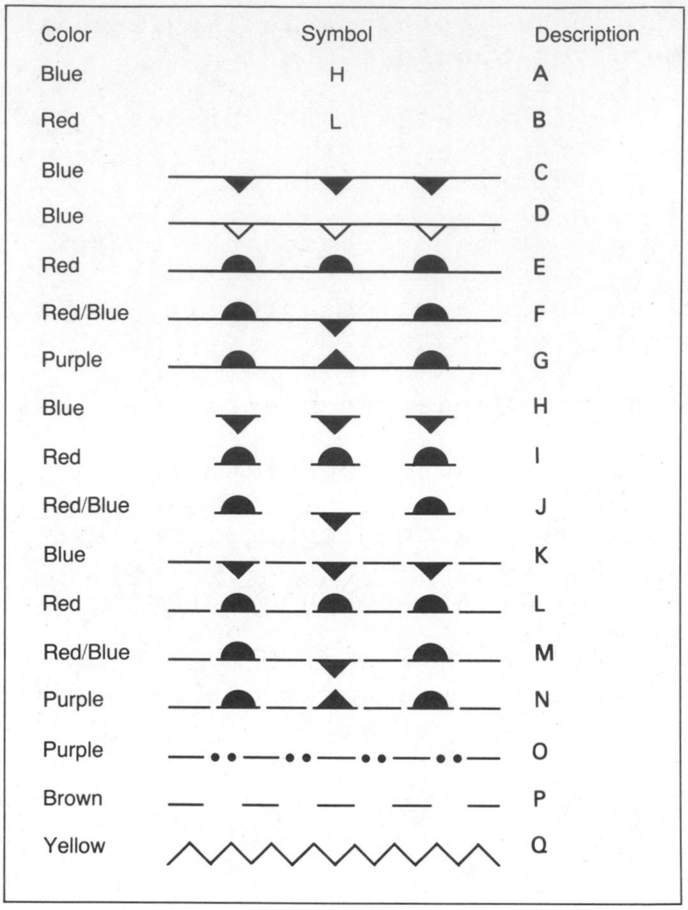

Figure 9.12

ANSWERS

Key Terms and Concepts

1.	L	7.	V	13.	F	19.	S
2.	D	8.	P	14.	K	20.	U
3.	J	9.	T	15.	C	21.	I
4.	A	10.	G	16.	W	22.	M
5.	B	11.	R	17.	O	23.	H
6.	X	12.	E	18.	N	24.	Q

Review Questions

1. 2--The frequency for en route weather advisories (En Route Flight Advisory Service, or EFAS) is 122.0 MHz. They are called "Flight Watch." EFAS normally operates from 6 a.m. to 10 p.m. local time seven days a week.
2. 3--See answer 1.

3.   ?--This question has been designated as unusable by the
     FAA and removed from the FAA Question Selection Sheets.
4.   2--TPA refers to traffic pattern altitude, which at this
     airport is 1,000 ft.
5.   3--Runway lights operate dusk to dawn on Rwy 15/33.
6.   2--The ceiling is the height above the surface of the lowest
     layer of clouds reported as broken or overcast, or an
     obscuration that is not reported as thin or partial.
7.   4--**Embedded** thunderstorms are obscured (contained within)
     cloud layers and cannot be seen.
8.   2--Radar reports report the location of precipitation along
     with type, intensity, and trend.
9.   3--Expected weather at your destination and when you plan to
     arrive (ETA) are contained in Terminal Forecasts.
10. 2--The "-X" means partially obscured, the measured ceiling
     is 700 ft., and the visibility is 1-1/2 with heavy rain
     and fog.
11. 3--See answer 10. JFK has an obscured sky, visibility of
     1/2 mi. and fog. Both of these airports are below the
     required minimums for VFR--1,000 ft. and 3 mi.
12. 3--The key here is "1804," just past temperature and dew
     point numbers. The 18 indicates 180° and 04 indicates 4
     kts. Note that the conditions here are ideal for fog,
     which is also present.
13. 4--The key here is "1112G18," again next to temperature and
     dew point. The first two digits indicate wind direction,
     110° in this case. The next number indicates velocity,
     12 kts. in this example, with gusts to 18 kts.
14. 1--In the second line of OKC, it says "21Z CFP," which means
     that cold front passage is to occur between 21Z and 22Z.
15. 3--Read from the first forecast period for HBC 1500Z-1700Z.
     The winds are forecast for 300° at 10 kts.
16. 4--Between 16Z and 01Z, the ceiling is forecast 10,000 BKN.
17. 4--The outlook (09Z) is for marginal VFR (MVFR), low
     ceilings (CIG) and thunderstorms (TRW).
18. 2--The outlook (09Z) contains the notation WIND, which means
     that winds are expected that exceed 25 kts.
19. 2--At 19Z they are forecast for 180° at 15 kts. with gusts
     to 25 kts. At 23Z they are forecast 320° at 15 kts. with
     gusts to 25 kts. That represents a major shift in wind
     direction from the south to the northwest.
20. 1--Shifting winds, moderate turbulence, and low ceilings are
     expected.
21. 3--Scattered to broken are expected ahead of the front,
     broken to overcast along the front, 10,000 50-100 miles
     beyond the front.
22. 3--"MVFR CIG TRW BCMG VFR 100 MI W OF CDFNT"--VFR in the
     western portion and marginal VFR with thunderstorms in
     the eastern portion.
23. 3--"CSTL WTRS" - Generally 2,500 scattered, widely scattered
     showers developing after 18Z becoming MVFR by midmorning.
24. ?--This question has been designated as unusable by the FAA
     and removed from the FAA Question Selection Sheets.
25. 1--The wind at 3,000 ft. is from the south; it becomes
     more westerly the higher you go.

26.   4--Westerly winds provide the greatest advantage. The winds
      are stronger at 34,000 ft. than they are at 24,000 ft.
27.   1--The first two digits (27) indicate true wind direction,
      or 270°; the next two digits (10) denote wind speed in
      knots, or 10 kts.
28.   1--The winds are mostly southerly at 3,000 ft. They become
      more westerly as you go higher.
29.   3--See answer 27.
30.   2--See answer 27. In addition, note that temperatures are
      always negative above 24,000 ft.
31.   1--Weather depiction charts are used to determine general
      weather conditions. They are useful in flight planning.
      They include directly observed weather and show areas of
      IFR and MVFR.
32.   1--The symbol to the left of the large dot means that there
      were reported thunderstorms.
33.   ?--The two ,, next to the 5 indicate drizzle. The three
      parallel lines next to the circle with the "X" in the
      center at the bottom of the shaded area means fog. In
      both the upper and lower areas, the ceilings are low.
      So, either #4 or #2 could be correct.
34.   1--Symbols on both sides of the line for a front indicate
      that neither air mass has been able to dominate, thus it
      is called a stationary front.
35.   3--The cloud cover indicators along that route of flight
      indicate both broken and scattered clouds at 25,000 ft.
36.   1--The upper panels indicate MVFR to IFR and the bottom
      panels indicate intermittent thundershowers and rain.
37.   3--The chart shows the freezing level moving from 8,000 ft.
      to the surface.
38.   2--Significant Weather Progs are designed to provide
      information about areas and/or altitudes of most
      significant icing and turbulence. They are forecasts.
39.   1--It means moderate turbulence, surface to 18,000 ft.
40.   4--Visual inspection reveals that it will move eastward at
      30 kts. and that it will be accompanied continuous snow
      as indicated by the two **.
41.   1--"FLT PRCTN" contains a summary of hazardous weather.
42.   1--"SIG CLDS AND WX" contains a summary of cloudiness and
      weather significant to flight operations broken down by
      states or other geographical areas.
43.   4--Area Forecasts summarize weather conditions between
      weather reporting stations.
44.   3--Area Forecasts depict freezing levels and areas of
      probable icing aloft.
45.   1--Transcribed weather briefings (TWEBs) are prepared for
      specific routes of flight and can be found on many VOR
      and NDB frequencies.
46.   1--See answer 45.
47.   3--See answer 45.
48.   2--SIGMETs apply to all aircraft. AIRMETs are issued
      primarily for light airplanes.
49.   2--See answer 48.
50.   2--SIGMETs apply to all aircraft, and severe icing would be
      of concern to everyone.

51. 1--A convective SIGMET contains severe weather information
         concerning tornadoes, embedded thunderstorms, and large
         sized hail.
52. 1--Radar Summary Charts show lines and cells of hazardous
         thunderstorms.
53. 2--Preflight weather briefers need to know your intended
         route of flight, destination, and type of aircraft.
54. 2--See Figure 9.8.
55. 4--See Figure 9.8.
56. 3--The best approach is to circumnavigate the hazardous
         weather.
57. 4--See Figure 9.8. Turbulence is the greatest hazard in and
         around thunderstorms.
58. 2--Lightning always accompanies thunderstorms.
59. 1--You should always identify yourself to a weather briefer
         as a pilot. Also, see answer 53.
60. 2--See Figure 9.9.
61. ?--This question has been designated as unusable by the FAA
         and removed from the FAA Question Selection Sheets.
62. 3--See Figure 9.9. Note the Winds Aloft Forecast.
63. 4--See Figure 9.9.
64. 3--See paragraph two in Figure 9.9.
65. 1--The weather at the point of departure is VFR.
66. 3--Winds Aloft Forecasts express wind direction relative to
         true north and wind speed in knots.
67. 4--See Figure 9.10.
68. 2--See Figure 9.10.
69. 2--See Figure 9.10.
70. 4--See the Winds Aloft Forecast in Figure 9.10.
71. 1--See paragraph two of Figure 9.10.
72. 3--EFAS is available for weather advisories on 122.0 MHz.
         See answer 1.
73. 2--See Figure 9.11. Note the wind shift at higher
         altitudes.
74. 3--See Figure 9.11. Note the wind shift.
75. 1--See Figure 9.11.
76. 3--Rime icing is correlated with stratus clouds.
77. 3--See Figure 9.11.
78. 2--See Figure 9.11.
79. 3--The weather briefer needs to know the type of flight.
80. 1--EFAS (122.0) provides actual weather information and
         thunderstorm activity along the route.
81. 3--See answer 1.
82. 2--This defines a squall line.
83. 2--This represents a stationary front.
84. 4--A warm front is replacing cooler surface air.

10/FLIGHT INFORMATION PUBLICATIONS

MAIN POINTS

1.   The FAA expects you to be a knowledgeable and prepared pilot. Familiarity with a number of publications will help to accomplish this task. The critical feature of each publication is its currency.

2.   **Regulatory** and technical information publications include: **Federal Aviation Regulations (FARs)**, which establish and enforce aviation safety and standards and which will be discussed in great detail in Chapter 11; **Airworthiness Directives (ADs)**, which are notices about compulsory maintenance, repair, or inspection; and **FDC Notams**, which contain Flight Data Center regulatory information. **Nonregulatory** and supplemental publications include: **Advisory Circulars (ACs)**, which contain diverse information for pilots and aircraft owners, and **Flight Standards Safety Pamphlets**, which describe operational problems affecting safety and pilot judgment. The **National Transportation Safety Board (NTSB)** publishes aviation statistics and accident and incident reports.

3.   **Aeronautical charts** published by the National Ocean Survey (NOS) include sectional aeronautical charts (primarily for VFR navigation), world aeronautical charts (for high-flying, high-performance aircraft), VFR terminal area charts (large-scale charts that cover the area around a TCA or TRSA), and flight planning charts (large-size, small-scale charts that show the entire United States).

4.   The **Airman's Information Manual (AIM)** provides data on navigational aids (radio navaids, airport marking, and lighting aids), airspace restrictions, air traffic control (for example, ARTCC, FSS, and VFR advisory services; radio techniques; airport operations at controlled and uncontrolled airports; ATC clearance procedures; preflight activities such as weather briefings; pilot and controller responsibility; emergency procedures; and national

security and interception procedures), flight safety procedures, good operating practices, aviation physiology, and terminology. Each AIM issue indicates, on the cover, the next date of issue.

5.   The **Airport/Facility Directory** is an important tool for cross-country flying. It is published every 56 days and covers primarily airports and communication facilities. Refer to Figure 10.1 for a sample entry in the Airport/Facility Directory and its interpretation.

6.   **Notices to Airmen (NOTAMS)** are divided into two classes. Class I NOTAMS are distributed via telecommunications and Class II are published every 14 days. Class II NOTAMs contain current flight safety bulletins. They are available for pilot briefings and are posted in FSS facilities.

7.   **Advisory Circulars (ACs)** are informative and explanatory in nature. They are not regulations; rather, they attempt to supplement the letter of the law with some guidance as to its intent and with techniques for operating safely within its framework. ACs are numbered to correspond with the topics covered by the FARs and are revised when necessary to reflect changing concepts, adoption of new equipment, or the approval of new regulations.

KEY TERMS AND CONCEPTS

Match each term or concept (1-12) with the appropriate description (A-L) below. Each item has only one match.

___	1. NOTAM--Class II	___	7. FARs
___	2. FDC NOTAM	___	8. A/F Directory
___	3. NTSB	___	9. ACs
___	4. WAC	___	10. sectionals
___	5. ADs	___	11. flight planning chart
___	6. terminal area chart	___	12. AIM

A. a guidebook of airports
B. group that publishes aviation safety statistics and accident reports
C. publication that deals with navigation, ATC procedures, flight safety, and medical facts
D. navigational charts used primarily by high-flying, high-performance aircraft
E. large-scale navigational charts that cover the area around a TCA
F. rules establishing and enforcing aviation safety standards
G. large, small-scale, navigational charts that depict the entire United States
H. large-scale navigational charts used primarily by VFR pilots
I. notices of compulsory maintenance, repair, or inspection
J. Flight Data Center regulatory information
K. publications that contain information on flying safety
L. informative and explanatory FAA publications

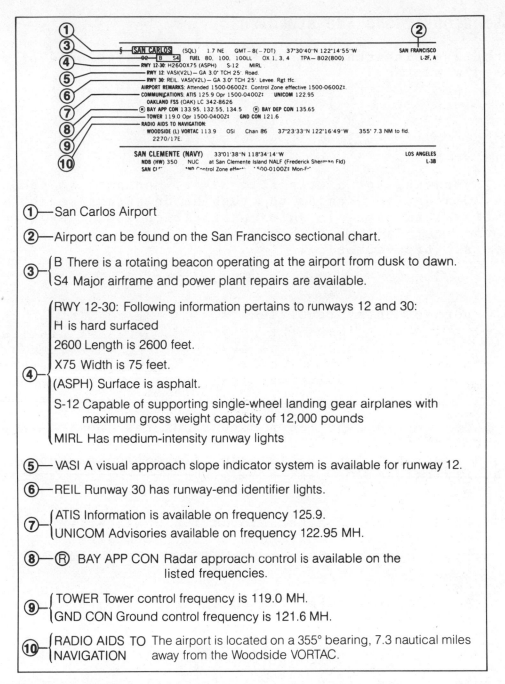

① —San Carlos Airport

② —Airport can be found on the San Francisco sectional chart.

③ —{ B There is a rotating beacon operating at the airport from dusk to dawn.
S4 Major airframe and power plant repairs are available.

④ —{ RWY 12-30: Following information pertains to runways 12 and 30:

H is hard surfaced

2600 Length is 2600 feet.

X75 Width is 75 feet.

(ASPH) Surface is asphalt.

S-12 Capable of supporting single-wheel landing gear airplanes with
maximum gross weight capacity of 12,000 pounds

MIRL Has medium-intensity runway lights

⑤ — VASI A visual approach slope indicator system is available for runway 12.

⑥ — REIL Runway 30 has runway-end identifier lights.

⑦ —{ ATIS Information is available on frequency 125.9.
UNICOM Advisories available on frequency 122.95 MH.

⑧ —Ⓡ BAY APP CON Radar approach control is available on the
listed frequencies.

⑨ —{ TOWER Tower control frequency is 119.0 MH.
GND CON Ground control frequency is 121.6 MH.

⑩ —{ RADIO AIDS TO   The airport is located on a 355° bearing, 7.3 nautical miles
NAVIGATION   away from the Woodside VORTAC.

Figure 10.1

DISCUSSION QUESTIONS AND EXERCISES

1.   What are **Airworthiness Directives** and why should pilots be concerned with them?

2.   In preparing for a trip from Havre, Montana, to Edmonton, Alberta, Canada, you decide to brush up on requirements for international flight. In which publication would you find:

   a.   ATC procedures

   b.   port-of-entry requirements

   c.   operational bulletins

3.   Suppose you want to learn more about the special techniques and hazards of flying at night. What publication would you expect to contain piloting tips and information on the physiology of night flying?

4.   Before using **any** flight information **publication** for flight planning or navigation, what should you check first?

5.   In what flight information document would you find a list of telephone numbers for an FAA Flight Service weather briefing?

6.   Assume you are planning to arrive at Beals Airport at twilight. The following NOTAM is posted for Brule, Nebraska:

   BRULE, BEALS ARPT: for rwy lights rwy 8-26 key freq 121.7 (10/84-2)

What does the NOTAM say and what operational procedures does it contain that are relevant to your arrival?

```
┌─────────────────────────────────────────────────────────────────────────────┐
│ SAN DIEGO                                                                     │
│ § BROWN FIELD MUNI   (SDM)   13 SE   GMT−8(−7DT)   32°34′20″N 116°58′46″W   LOS ANGELES │
│   524   B   S4   FUEL 80, 100, 100LL, JET A   OX 2   TPA— See Remarks   AOE      L-3C │
│     CFR Index Ltd                                                                 IAP │
│   RWY 08L-26R: H7999X200 (ASPH-CONC)    S-80, D-110, DT-175    MIRL               │
│     RWY 26R: REIL. Tree. Rgt tfc.                                                 │
│   RWY 08R-26L: H3032X70 (ASPH)    S-12                                            │
│     RWY 08R: Thld dsplcd 200′. Rgt tfc.           RWY 26L: Thld dsplcd 305′.       │
│   AIRPORT REMARKS: Attended 1430-0630Z‡, thereafter for fee call (714) 426-3410 or 421-0873. Parachute │
│     Jumping. Overngt parking fee. Local Wx observation facility. Rwy 08R-26L dalgt hours only. All asph areas other │
│     than taxiways and Rwy 08L-26R limited to acft under 12,000 pounds gross weight. TPA— 1524(1000) Rwy │
│     08L-26R, 1124(600) Rwy 08R-26L. CFR Index Ltd available 1430-0630Z‡. Flight Notification Service (ADCUS) │
│     available. Control Zone effective 1500-0600Z‡.                                 │
│   COMMUNICATIONS: ATIS 132.35 opr 1500-0600Z‡    UN:COM 122.95                     │
│     SAN DIEGO FSS (SAN) LC 291-6381 NOTAM FILE SDM                                 │
│   ® SAN DIEGO APP CON 119.6    ® SAN DIEGO DEP CON 125.15                          │
│     TOWER 126.9, 128.25 (Rwy 08L-26R) opr 1500-0600Z‡    GND CON 124.4    CLNC DEL 124.4 │
│   RADIO AIDS TO NAVIGATION:                                                        │
│     MISSION BAY (H) ABVORTAC 117.8   ■ MZB   Chan 125   32°46′56″N 117°13′28″W   120° 17.7 NM to │
│     fld. 10/15E                                                                    │
│     TWEB avbl 1400-0500Z‡.                                                         │
│     VORTAC unusable 300°-310° beyond 20 NM below 5000′   310°-330° beyond 10 NM below 3000 │
└─────────────────────────────────────────────────────────────────────────────┘
```

Figure 10.2

8.   Using the entry from the <u>Airport/Facility Directory</u> in Figure 10.2, answer the following questions:

   a.   How would you find out about current conditions and runway use without calling and requesting such information?

   b.   Is 80-octane fuel available?

   c.   During what hours is the control zone in effect?

   d.   How would you obtain a weather briefing at this airport?

   e.   How long is the longest runway?

   f.   What should you know about Rwy 08R?

g. Why do you think Rwy 26R has a right traffic pattern?

h. If you arrive at Brown Field at midnight and the wind is 330° at 15 kts., on what runway would you land and why?

i. What is the tower frequency? On what frequency would you contact ground control?

j. What is the field elevation?

k. How could you receive transcribed weather broadcasts at Brown Field Muni?

8.   T F To determine if UNICOM is available at an airport without a control tower, you should refer to Notices to Airmen (NOTAMs).

9.   T F Sectional charts for the conterminous United States are updated every six months.

10. Briefly describe what you should do when planning a VFR cross-country flight. Be sure to identify what planning aids you would use and what information you would obtain from each.

REVIEW QUESTIONS

       For several of the following questions, you will have to
refer to Appendix A at the end of this book. Appendix A contains
a Sectional Aeronautical Legend and a Directory Legend from the
Airport/Facility Directory.

1.   (FAA 1470) The radar service provided for VFR flights at
Abilene Municipal is (see Figure 10.3)

1--advisory service on a workload basis.
2--radar advisory and sequencing service for participating
   aircraft.
3--radar sequencing and separation service for participating
   aircraft.
4--radar advisory, sequencing, and separation service for all
   aircraft.

2.   (FAA 1471) How can a pilot receive airport advisory service
at Abilene Municipal between the hours of 0500Z-1300Z? (See
Figure 10.3.)

1--FSS on 122.65 MHz.
2--FSS on the tower frequency.
3--Fort Worth APP CON on 127.45 MHz.
4--UNICOM on 122.95 MHz.

3.   (FAA 1472) When approaching Abilene Municipal at 7,500 ft.
from the west, the initial contact with approach control should
be on (see Figure 10.3)

1--121.3 MHz.   3--125.0 MHz.
2--124.1 MHz.   4--126.5 MHz.

4.   (FAA 1473) What special traffic pattern is in effect at
Elmdale Airpark? (See Figure 10.3.)

1--1,000 ft.
2--Right-hand traffic on all runways.
3--As assigned by Abilene RAPCON tower.
4--2,175 (400) ft. or lower.

5.   (FAA 1474) Which runways at Abilene Municipal have right-
hand traffic patterns? (See Figure 10.3.)

1--Rwy 17L-35R.   3--Rwys 17R, 35R, and 22.
2--Rwy 17R-35L.   4--All runways.

6.   (FAA 1475) How can a pilot receive transcribed weather
broadcasts at Abilene Municipal? (See Figure 10.3.)

1--Local telephone call to ABI FSS.
2--Monitor ABI VORTAC on 113.7 MHz.
3--Monitor TQA VOR on 111.6 MHz.
4--Monitor ABI LOM on 353 kHz.

TEXAS

ABERNATHY MUNI (F83) 4.3 E GMT−6(−5DT) 33°50'45"N 101°45'45"W — DALLAS-FT. WORTH
3327 B S4 — L-13A
RWY 17-35: H4000X75 (ASPH) S-8 LIRL
RWY 17: Road. RWY 35: Road. Rgt tfc.
RWY 03-21: H3235X75 (ASPH) S-8
RWY 21: P-line. Rgt tfc.
AIRPORT REMARKS: Attended continuously.
COMMUNICATIONS: UNICOM 122.8
LUBBOCK FSS (LBB) Toll free call dial 0, ask for ENTERPRISE 84044.
RADIO AIDS TO NAVIGATION:
LUBBOCK (L) ABVORTAC 110.8 ■ LBB Chan 45 33°42'18"N 101°54'49"W 031° 11.3 NM to fld.
3310/11E. General outlook only 0400-1100Z‡.

ABILENE

§ **ABILENE MUNI** (ABI) 2.6 SE GMT−6(−5DT) 32°24'41"N 99°40'53"W — DALLAS-FT. WORTH
1790 B S4 FUEL 100LL, JET A, A1 + OX 1 CFR Index A — H-2H, L-13B, 15C
RWY 17L-35R: H7199X150 (ASPH) S-80, D-110, DT-160 HIRL .39% up N — IAP
RWY 17L: VASI(V4L)— GA 3.0° TCH 43'. RWY 35R: SSALR. Pole. Rgt tfc.
RWY 17R-35L: H7201X150 (ASPH) S-75, D-100, DT-160 MIRL .32% up S
RWY 17R: VASI(V4L)— GA 3.0° TCH 58'. Rgt tfc. RWY 35L: VASI(V4L)— GA 3.0° TCH 49'.
RWY 04-22: H3686X100 (ASPH) S-35, D-44, DT-68 MIRL
RWY 22: Rgt tfc.
AIRPORT REMARKS: Attended continuously. Parachute Jumping. Prior Permission Required for certificated air carrier operations 0400-1200Z‡.
COMMUNICATIONS: UNICOM 122.95
ABILENE FSS (ABI) on arpt. 122.65, 122.2, 122.1R, 120.1 (915) 677-4336
Ⓡ APP/DEP CON 126.5 (5500-10,000') 121.3 (5500' & blo) 161-350°, 124.1 (5500' & blo) 351-160°, 125.0 (1300-0500Z‡)
Ⓡ FORT WORTH CENTER APP/DEP CON 127.45 (0500-1300Z‡)
TOWER 120.1 (1500-0100Z‡) GND CON 121.9
STAGE II SVC ctc APP CON within 20 NM
RADIO AIDS TO NAVIGATION:
(H) ABVORTAC 113.7 ■ ABI Chan 84 32°28'53"N 99°51'47"W 105° 10 NM to fld. 1810/10E.
TUSCOLA (L) VOR 111.6 TQA 32°14'08"N 99°48'59"W 020° 13.5 NM to fld.
TOMHI NDB (LOM) 353 AB 32°17'55"N 99°40'26"W 350° 5.4 NM to fld.
ILS 110.3 I-ABI Rwy 35R LOM TOMHI NDB
ASR
COMM/NAVAID REMARKS: ABI FSS will provide AAS 0500-1300Z‡ 120.1.

- -

ELMDALE AIRPARK (6F4) 5.2 E GMT−6(−5DT) 32°27'00"N 99°39'00"W — DALLAS-FT. WORTH
1775 S4 FUEL 100LL TPA— 2175(400) — L-13B, 15C
RWY 17-35: H2950X25 (ASPH) S-4 LIRL
RWY 17: Thld dsplcd 180'. Trees. RWY 35: Thld dsplcd 50'. Road. Rgt tfc.
AIRPORT REMARKS: Attended dalgt hours. Parachute Jumping. TPA 400' or lower unless authorized by Abilene RAPCON/tower. Ngt lndgs not recommended for pilots unfamiliar with arpt. Rwy condition poor on South half of runway.
COMMUNICATIONS:
ABILENE FSS (ABI)
RADIO AIDS TO NAVIGATION:
ABILENE (H) ABVORTAC 113.7 ■ ABI Chan 84 32°28'53"N 99°51'47"W 090° 11.0 NM to fld. 1810/10E.

- -

ZIMMERLE (6F2) 13.9 SE GMT−6(−5DT) 32°16'13"N 99°35'51"W — DALLAS-FT. WORTH
2057 Not insp.
RWY 15-33: 1200X100 (TURF)
RWY 15: Rgt tfc. RWY 33: Rgt tfc
AIRPORT REMARKS: Unattended
COMMUNICATIONS:
ABILENE FSS (ABI)

ACTON 32°26'04"N 97°39'49"W — DALLAS-FT. WORTH
(L) VORTAC 110.6 AQN Chan 43 173° 12.5 NM to Cleburne Muni. 848/09E — L-13C, 15D, 17A
LRCO 122.1R 110.6T (FORT WORTH FSS)

ADDISON (See DALLAS)

Figure 10.3

7.   (FAA 1476) How can a pilot receive the general outlook as indicated at Abernathy Municipal in Figure 10.3?

1--Monitor LBB VORTAC.
2--Telephone LBB FSS.
3--Contact Lubbock radio.
4--Contact Abernathy UNICOM.

8.   (FAA 1502) The letters VHF/DF appearing in the <u>Airport/Facility Directory</u> for a certain airport, indicate that

1--this airport is designed as an airport of entry.
2--the flight service station has equipment with which to
   determine your direction from the station.
3--this airport has a direct-line phone to the flight service
   station.
4--this airport is a defense facility.

9.   (FAA 1507) What special procedures are in effect for landings on Rwy 19R at John Wayne Airport after 0700 local standard time? (See Figure 10.4.)

1--Right traffic, tower frequency (126.8), and no local training
   or touch-and-go operations.
2--Left traffic, closed to turbojet traffic, and no local
   training or touch-and-go operations.
3--Left traffic, key transmitter for MALSR, and no local training
   or touch-and-go operations.
4--Right traffic and tower frequency 126.8.

10.   (FAA 1508) Which traffic patterns are in effect at South County Airport of Santa Clara Co.? (See Figure 10.4.)

1--All runways 1,000 ft. AGL, right-hand traffic pattern.
2--All runways, 1,281 ft. AGL, left-hand traffic pattern except
   at night.
3--Rwy 14, 1,000 ft. AGL, left-hand traffic; Rwy 32, 1,000 ft.
   AGL, right-hand traffic.
4--Rwy 14, 1,281 ft. AGL, left-hand traffic; Rwy 32, 1,281 ft.
   AGL, right-hand traffic.

11.   (FAA 1509) What is the procedure for setting the altimeter for a takeoff when the Orange County Control Tower is shut down at night? (See Figure 10.4.)

1--Make a local telephone call to the Ontario FSS for the
   altimeter setting.
2--Contact Coast Departure Control on 128.1 for the altimeter
   setting.
3--Set the altimeter to 54 ft., the elevation of the airport.
4--Call Santa Ana RCO on the field for the altimeter setting.

AIRPORT/FACILITY DIRECTORY
CALIFORNIA

SAN MARTIN

SOUTH COUNTY AIRPORT OF SANTA CLARA CO (Q99) .9 E GMT−8(−7DT)   SAN FRANCISCO
   37°04′55″N 121°35′45″W   L-2F
   281   S4   FUEL 100   TPA— 1281(1000)
   RWY 14-32: 3100X75 (ASPH)   S-3   .32% up NW
     RWY 14: Trees.   RWY 32: Rgt tfc.
   AIRPORT REMARKS: Attended 1500Z‡-dusk. Arpt restricted to day.
   COMMUNICATIONS: UNICOM 122.7
     SALINAS FSS (SNS) LC 683-4660
   RADIO AIDS TO NAVIGATION:
     SAN JOSE (L) VOR/DME 120°/23.3 NM

SAN NICHOLAS   33°14′10″N 119°26′56″W   LOS ANGELES
   NDB (HW) 203   NSI   at San Nicholas Island, O.L.F.   L-3B
   SAN NICHOLAS O.L.F. Control Zone effective 1600-0030Z‡ Mon-Fri

---

SANTA ANA

§   **JOHN WAYNE AIRPORT/ORANGE CO**   (SNA)   4.3 S   GMT−8(−7DT)   LOS ANGELES
    33°40′32″N 117°52′02″W   H-2F, L-3C, A
    54   B   S4   FUEL 80, 100, 100LL, JET A   OX 1, 2   CFR Index B   IAP
    RWY 01L-19R: H5700X150 (ASPH)   S-70, D-95, DT-150   HIRL
     RWY 01L: VASI— GA 3.0° TCH 55′. P-line.
     RWY 19R: MALSR. VASI— GA 3.0° TCH 46.5′. Rgt tfc. (left tfc when tower closed).
    RWY 01R-19L: H2887X75 (ASPH)   S-25, D-38   MIRL   .41% up S
     RWY 01R: Rgt tfc.   RWY 19L: REIL.
    AIRPORT REMARKS: Attended continuously. Rwy 19R MALSR unattended when twr closed. Noise sensitive area all
      quadrants; pilots use recommended noise abatement procedures avbl on request. Closed to turbo-jet acft
      0700-1500Z‡ except approved emergency or mercy flights and Cessna Citation-500, Learjet 35/36,
      Hawker-Siddeley 125-700; Falcon-10, Westwind 1124, and Sabreline-65 not on training or demonstration flights.
      Other turbo-jet acft of equally low noise level excepted with prior approval of arpt manager. Migratory bird flyways
      over and in vicinity of arpt. For MALSR 19R when twr closed, key 126.8 7 times in 5 seconds for high, 5 times
      in 5 seconds for medium 3 times in 5 seconds, for low intensity. No local training or touch/go operations
      0700-1400Z‡. Control Zone effective 1415-0645Z‡.
    COMMUNICATIONS: ATIS 126.0 (714) 546-2279   UNICOM 122.95
     ONTARIO FSS (ONT) LC 547-9592 NOTAM FILE SNA
     SANTA ANA RCO 122.45 (ONTARIO FSS)
    Ⓡ COAST APP CON 121.3   Ⓡ COAST DEP CON 128.1
     ORANGE COUNTY TOWER 119.9 (Rwy 01R-19L) 126.8 (Rwy 01L-19R), 128.35 opr 1415-0645Z‡.   GND CON
      120.8, 121.85   CLNC DEL 118.0
    RADIO AIDS TO NAVIGATION:
     SEAL BEACH (L) VORTAC 110°/11.3 NM
     SANTA ANA (T) VORW 109.4   SNA   33°41′02″N 117°51′43″W   at fld.
      VOR and ILS unmonitored when tower closed
     ILS/DME 108.3 I-SNA Chan 52 Rwy 19R

SANTA ANA   33°41′02″N 117°51′43″W   LOS ANGELES
   (T) VORW 109.4   SNA   at John Wayne Airport/Orange Co.   L-3C, A
   RCO 122.45 (ONTARIO FSS)

Figure 10.4

12.   (FAA 1510)   How should a pilot secure local weather and
airport conditions at John Wayne Airport?   (See Figure 10.4.)

1--Orange County Ground Control on 120.8.
2--ATIS on 126.0 or phone 546-2279.
3--Local telephone call to Ontario FSS.
4--Santa Ana radio on 109.4.

13.   (FAA 1511)   Through which radio facility can a pilot contact
the Ontario FSS while at John Wayne Airport?   (See Figure 10.4.)

1--Santa Ana RCO (frequency 122.45) on the airport.
2--Clearance Delivery (118.0) on the airport.
3--Santa Ana VOR (122.1T-109.4R) on the airport.
4--Seal Beach VORTAC (122.1T-115.7R) 11.3 NM east of the airport.

14.   (FAA 1512)   What initial contact should be made to ATC for a
landing at John Wayne Airport?   (See Figure 10.4.)

1--Monitor ATIS, then contact Orange County Tower upon entering
   the traffic pattern.
2--Contact Orange County Tower about 15 mi. from the airport,
   then monitor ATIS.
3--Contact Clearance Delivery for a clearance to enter the
   airport traffic area.
4--Monitor ATIS, then contact Coast Approach Control about 15 mi.
   from the airport.

15.   (FAA 1513)   Information concerning parachute jumping sites
may be found in the

1--Graphic Notices and Supplemental Data.
2--legend of sectional aeronautical charts only.
3--Airport/Facility Directory.
4--NOTAMS.

16.   (FAA 1526)   FAA Advisory Circulars (some free, others at
cost) are available to all pilots and are obtained by

1--distribution from the nearest FAA district office.
2--ordering those desired.
3--subscribing to the Federal Register.
4--subscribing to FARs.

17.   Sectional charts for the conterminous United States are
updated every

1--3 months.
2--6 months.
3--12 months.
4--24 months.

AIRPORT/FACILITY DIRECTORY
ARKANSAS

§ **JONESBORO**   (JBR) 2.6 E GMT-6(-5DT) 35°49'51''N 90°38'47''W   **MEMPHIS**
   262   B   S4   **FUEL** 80, 100, JET A+   CFR Index A   **H-4F, L-14F**
   RWY 05-23: H5599X150 (ASPH) S-80, D-90, DT-140 MIRL   **IAP**
    RWY 23: VASI
   RWY 14-32: H4101X150 (ASPH) S-30 MIRL
    RWY 14: Thld dsplcd 130'   RWY 32: Railway 700' thld dsplcd 160'
   RWY 18-36: H3943X60 (ASPH) S-30
    RWY 18: Trees 1800'. Thld dsplcd 160'.   RWY 36: Bldg 1400' from thld. Thld dsplcd 347'.
   AIRPORT REMARKS: Attended 1200-0100Z‡. Control Zone effective 1200-0400Z‡.
   COMMUNICATIONS: UNICOM 123.0
    JONESBORO FSS (JBR) on fld 123.6 122.3 122.2 122.1R 108.6T (501) 935-3471
     Opr 1200-0400Z‡, DL- dial 0, ask for ENTERPRISE 0246 O/T ctc Memphis FSS
   RADIO AIDS TO NAVIGATION:
    (T) **BVOR** 108.6 JBR 35°52'30''N 90°35'18''W 222° 3.1 NM to fld. Unmonitored 0400-1200Z‡

PINE BLUFF   34°14'48''N 91°55'34''W   **MEMPHIS**
   (L) **BVORTAC** 116.0 PBF Chan 107 181° 3.9 NM to Grider Fld   **L-14F**
    VOR unusable 054°-075° beyond 35 NM below 5000'
          170°-185° beyond 30 NM below 2000'
          236°-249° beyond 20 NM below 6000', or beyond 26 NM below 13000'.
    TACAN az unusable 091°-129° beyond 20 NM below 3500'

PINE BLUFF
GRIDER FLD   (PBF) 4.3 SE GMT-6(-5DT) 34°10'32''N 91°56'07''W   **MEMPHIS**
   206   B   S4   **FUEL** 80, 100, JET A   CFR Index A   **H-4F, L-14F**
   RWY 17-35: H6000X150 (ASPH) S-50, D-70, DT 110 HIRL   **IAP**
    RWY 17: MALSR, VASI. Key 118.4 7 times in 5 sec for high, 5 times in 5 sec for med, 3 times in 5 sec
     for low intensity.
    RWY 35: VASI
   AIRPORT REMARKS: Attended 1300-0500Z‡. On call other hrs. Control Zone effective 1200-0400Z‡.
   COMMUNICATIONS:
    LITTLE ROCK FSS (LIT) DL-536-8466
    PINE BLUFF FSS (PBr) 123.6 on arpt (501) 536-8466 Opr 1400-2200Z‡
     Flight planning/briefing svc only
    PINE BLUFF RCO 122.6 122.2 122.1R 116.0T (LITTLE ROCK FSS)
    PINE BLUFF APP/DEP CON 118.4 Opr 1200-0400Z‡
    LITTLE ROCK APP/DEP CON 124.2 0400-1200Z‡
    PINE BLUFF TOWER: 118.4 Opr 1200-0400Z‡   **GND CON:** 122.7
   RADIO AIDS TO NAVIGATION:
    PINE BLUFF (L) **BVORTAC** 116.0 PBF Chan 107 34°14'48''N 91°55'34''W 181° 3.9 NM to fld
     VOR unusable 054°-075° beyond 35 NM below 5000'
           170°-185° beyond 30 NM below 2000'
           236° 249° beyond 20 NM below 6000', or beyond 26 NM below 13000'
     TACAN az unusable 091°-129° beyond 20 NM below 3500'
    ILS 111.7 I-PBF RWY 17 LOC only

LITTLE ROCK   34°40'39''N 92°10'49''W   **MEMPHIS**
   (H) **BVORTAC** 113.9 (LIT) Chan 86 315° 3.8 NM to Adams Fld   **H-4F, L-14E**

LITTLE ROCK FSS (LIT) on Adams Fld   **MEMPHIS**
   122.55, 122.35, 122.2, 122.1R, 113.9T (501) 376-0721   **H-4B, L-14E**

LITTLE ROCK
§ **ADAMS FIELD**   (LIT) 1.7 E GMT-6(-5DT) 34°43'48''N 92°13'59''W   **MEMPHIS**
   257   B   S4   **FUEL** 80, 100 JET A   OX 1, 3   LRA   CFR Index C   **H-4B, L-14E**
   RWY 04-22: H7010X150 (ASPH) S-70, D-90, DT-140 HIRL   **IAP**
    RWY 04: SSALR Thld dsplcd 127'   RWY 22: MALSR, VASI
   RWY 17-35: H5125X150 (ASPH) S-30, D-45, DT-70 MIRL
    RWY 17: Road 260'. Thld dsplcd 270'   RWY 35: Road 33' AL'SF1
   RWY 14-32: H4032X150 (ASPH) S-26 MIRL
    RWY 14: Road 220'. Thld dsplcd 365'   RWY 32: Trees 3000'
   AIRPORT REMARKS: Landing fee. Rwy 14-32 closed to air carriers.
    Transient acft parking at airline terminal ramp ctc arpt police at airline concourse for reentry to locked
    operations area.
   COMMUNICATIONS: ATIS 125.6 1200-0600Z‡   **UNICOM** 123.0
    LITTLE ROCK FSS (LIT) on fld. 122.55 122.35 122.2 122.1R 113.9T (501) 376-0721
   Ⓡ **LITTLE ROCK APP CON:** 124.2 042°-221° 119.5 222°-041° 118.1
    TOWER: 118.7 123.85   **GND CON:** 121.9
   Ⓡ **LITTLE ROCK DEP CON:** 124.2 041°-220° 119.5 221°-040° 118.1
    STAGE III SVC ctc **APP CON** 20 NM, check ATIS
   RADIO AIDS TO NAVIGATION:
    LITTLE ROCK (H) **BVORTAC** 113.9 LIT Chan 86 34°40'39''N 92°10'49''W 315° 3.8 NM to fld.
    LASKY NDB (H-SAB) 353 LI 34°57'09''N 92°01'09''W 041° 4.6 NM to fld
    ILS 110.3 J-LIT Rwy 04 LOM LASKY NDB
    ASR 110.3 I-AAY Rwy 22

Figure 10.5

18. Which of the following publications is regulatory and/or technical in nature?

1--Advisory Circular (AC).
2--Airworthiness Directive (AD).
3--National Transportation Safety Board publications.
4--VFR terminal area charts.

19. Refer to the <u>Airport/Facility Directory</u> in Figure 10.5. Which statement is true about Jonesboro Airport?

1--Aircraft and power plant maintenance are not available.
2--For airport advisory service, contact UNICOM on 122.8 MHz.
3--Rwy 36 threshold is displaced 347 ft.
4--The airport elevation is 2,620 ft. MSL.

20. Refer to the <u>Airport/Facility Directory</u> data in Figure 10.5 for Adams Field at Little Rock and select the true statement.

1--Grade 115 gasoline is available.
2--Low-pressure oxygen replacement bottles are available.
3--Rwy 35 threshold is displaced 356 ft.
4--The longest hard surface runway available for takeoffs is Rwy 22.

21. Refer to the <u>Airport/Facility Directory</u> data in Figure 10.5 for Adams Field at Little Rock. The proper sequence of radio frequencies for departing this airport southbound using ATIS, ground control, tower, departure control, and the Flight Service Station is

1--123.0, 121.7, 118.7, 124.2, and 113.9 MHz.
2--124.2, 123.85, 121.9, 119.5, and 122.55 MHz.
3--125.6, 121.9, 118.7, 124.2, and 122.2 MHz.
4--125.6, 124.2, 121.9, 118.1, and 122.35 MHz.

ANSWERS

Key Terms and Concepts

1.	K	4.	D	7.	F	10.	H
2.	J	5.	I	8.	A	11.	G
3.	B	6.	E	9.	L	12.	C

Discussion Questions and Exercises

2.   a. AIM.
     b. <u>Airport/Facility Directory</u>; AIM.
     c. NOTAMs.
3.   AIM.
4.   Its currency.
5.   <u>Airport/Facility Directory.</u>

6.   For Beals Airport at Brule, Nebraska. Runway lights are
        available for Rwy 8-26 by keying the radio on 121.7 MHz.
7. a. Listen to ATIS on 132.35.
    b. Yes, 80-octane is available.
    c. 1500-0600 zulu.
    d. Call San Diego FSS on the phone (291-6381), local call.
    e. 7,999 ft.
    f. It has a 200-ft. displaced threshold and uses a right-
       hand traffic pattern. It also is for use only during
       daylight hours and has a 600-ft. TPA.
    g. The right-hand pattern helps avoid aircraft arriving and
       departing on Rwy 26L.
    h. You would use 26R since it is lighted and it has the best
       angle with respect to the wind.
    i. 126.9; 124.4.
    j. 524 ft. MSL.
    k. TWEB is available on ABVORTAC 117.8 as evidenced by the
       enclosed square next to MZB.
9. F--You should refer to the appropriate Airport/Facility
       Directory.
10. T--See the sectional chart legend.

Review Questions

1.   2--Refer to Figure 10.3 and Appendix A. Stage II radar
        service is available to all participating VFR traffic and
        provides sequencing and advisories.
2.   2--Refer to Figure 10.3 and Appendix A. Use the tower
        frequency to receive airport advisories between 0500Z and
        1300Z.
3.   4--Refer to Figures 10.3 and Appendix A. Contact approach
        control on 126.5 for flights approaching from the west
        between 5,500 and 10,000 ft.
4.   4--Refer to Figure 10.3 and Appendix A. The TPA is 2,175
        (400) ft. or lower.
5.   3--Refer to Figure 10.3 and Appendix A.
6.   2--Refer to Figure 10.3 and Appendix A. The enclosed square
        next to the ABI indicates TWEB.
7.   1--Refer to Figure 10.3. The general outlook is available
        between 0400Z and 1100Z on the LBB VORTAC.
8.   2--The VHF direction finder indicates your magnetic
        direction from the station.
9.   4--Refer to Figure 10.4 and Appendix A. The key to this one
        is to determine the time in zulu. 0700 PST is 1500Z, as
        indicated by the GMT conversion numbers (GMT -8), so the
        tower is open.
10. 3--Refer to Figure 10.4 and Appendix A.
11. 3--Refer to Figure 10.4 and Appendix A.
12. 2--Refer to Figure 10.4 and Appendix A.
13. 1--Refer to Figure 10.4 and Appendix A.
14. 4--Refer to Figure 10.4 and Appendix A.

15. 3--Information about parachute jumping sites and other such
     events will be found in the <u>Airport/Facility Directory.</u>
16. 2--FAA ACs are available by ordering them.
17. 2--Sectional charts are updated every 6 months.
18. 2--ADs are notices about compulsory maintenance, repair, or
     inspection.
19. 3--Refer to Figure 10.5 and Appendix A. "Rwy 36 ... thld
     dsplcd 347" means Rwy 36 has a displaced threshold of 347
     ft.
20. 4--Rwy 04-22 has the longest hard-surfaced runway.
21. 3--This is the proper sequence: ATIS, ground control,
     tower, departure control, and FSS.

11/FEDERAL AVIATION REGULATIONS

MAIN POINTS

1. FAR PART 1: DEFINITIONS AND ABBREVIATIONS.

Airport: area of land or water used for takeoffs and landings.

Airport traffic area (ATA): airspace from the ground up to but not including 3,000 ft. AGL; 5-mi. radius from the center of an airport with an operating control tower.

Air traffic clearance: authorization to proceed within controlled airspace.

Calibrated airspeed: airspeed corrected for position and installation errors.

Ceiling: height AGL of the lowest layer of broken or overcast clouds.

Controlled airspace: ATAs, control zones, transition areas, navigation routes, and the continental control area are examples.

Flight crewmember: pilot, navigator, or engineer with assigned duty during flight time.

Flight visibility: forward horizontal distance.

IFR: Instrument Flight Rules, in effect when weather conditions are below VFR minimums.

Indicated airspeed: pitot-static airspeed, pressure, uncorrected for airspeed system errors.

Positive control: control of all air traffic within designated airspace by air traffic control.

Prohibited area: no flight allowed.

Restricted area: restrictions to flight apply.

True airspeed: airspeed relative to undisturbed air.

2. FAR PART 61: CERTIFICATION: Pilots and Flight Instructors; Subpart A, General.

61.3   (a) You must have your pilot's certificate in your possession to act as pilot-in-command.

61.15   Convicted drug dealers and users can have certificates or ratings revoked and are ineligible for certification for one year following conviction.

61.17   Temporary certificates are good for 120 days.

61.19   Student pilot certificates expire at the end of the 24th month after issue. Other pilot certificates are issued without a specific expiration date.

61.23   Third Class Medical Certificates, common for student and private pilots, expire at the end of the same calendar month in which they were issued, 24 mo. after the exam.

61.31   In general, one must receive instruction and hold a category and class rating for an airplane to act as pilot-in-command.

61.39   To qualify for the flight test, one must have passed the written test within two years, hold a current medical certificate, and have a CFI's written statement, issued within 60 days, stating that the candidate is prepared to take the flight test.

61.51   Reliable records of required flight time and experience must be kept. Student pilots must also carry logbooks on solo cross-country flights.

61.57   (a, b) Biennial Flight Reviews (BFR) by a CFI or other person designated by the Administrator are mandatory at least every 24 mo. to act as pilot-in-command. They expire exactly 24 mo. to the day after they are issued. The BFR includes questions on general operating procedures, flight rules, and flight maneuvers. (c) To act as pilot-in-command carrying passengers, one must have made at least three takeoffs and landings in the same category and class of aircraft within the prior 90 days. (d) To act as pilot-in-command during the period from one hour after sunset to one hour before sunrise, a person must have made at least three takeoffs and three full-stop landings during that period within the preceding 90 days.

   3.   FAR 61: CERTIFICATION: Subpart B, Aircraft Ratings and Special Certificates.

61.63   To receive additional ratings, the pilot must present a logbook endorsed by an authorized instructor and pass a flight test.

   4.   FAR PART 61: CERTIFICATION: Subpart C, Student Pilots.

61.83   The minimum age is 16, English language competency is required, and you must hold at least a Third Class Medical Certificate.

61.87   Solo flight requires an endorsement by your CFI that appropriate training has been given; it is good for 90 days.

61.89   Student pilots cannot carry passengers nor can they act as required crewmembers.

61.93   Cross-country flights (more than 25 NM) require a CFI's
          endorsement on the student pilot certificate and in the
          logbook for **each** cross-country flight.

    5.   FAR PART 61: CERTIFICATION: Subpart D, Private Pilots.

61.103   The minimum age is 17; you must have English language
          competence; you must hold at least a Third Class Medical
          Certificate; and you must pass the written test and the
          oral and flight tests.
61.109   To take the oral and flight tests, the applicant must
          have at least 20 hr. of flight instruction and 20 hr. of
          solo time, as outlined in Chapter 1.
61.118   Private pilots may not act as a pilot-in-command of an
          aircraft that is carrying passengers for compensation or
          hire. A private pilot may share operating expenses with
          passengers. A private pilot may also act as pilot-in-
          command to demonstrate an aircraft if he/she is an
          aircraft salesperson with at least 200 hr. of logged
          flight time.

    6.   FAR PART 67: MEDICAL STANDARDS AND CERTIFICATION. If
a situation warrants it, the Administrator may request medical
records.

    7.   FAR Part 91: GENERAL OPERATING AND FLIGHT RULES:
          Subpart A, General.

91.3   The pilot-in-command is responsible for the aircraft.
          Deviations from a rule to meet an emergency must be
          reported to the Administrator in writing only upon
          request.
91.5   The pilot must consider weather and runway lengths
          before any flight not in the vicinity of an airport.
91.7   Seatbelts and, when available, shoulder harnesses are
          required of crewmembers when at stations.
91.11   Alcohol cannot be consumed by the pilot-in-command
          within 8 hr. of a flight, nor can the pilot carry
          passengers who are obviously under the influence of
          alcohol or drugs, unless under medical supervision.
91.14   The pilot-in-command is responsible for ensuring that
          each person knows how to fasten and unfasten his/her
          safety belt and for instructing each person to fasten
          that belt prior to takeoff and landing.
91.21   Simulated instrument flights require the presence of an
          appropriately rated safety pilot.
91.22   Minimum fuel reserves are 30 min. for a day VFR flight
          and 45 min. for a night VFR flight.
91.24   Group I TCAs require a Mode 3/A 4096 code transponder as
          well as a Mode C encoding altimeter. Group II TCAs
          require only a Mode 3/A 4096 code transponder.
91.27   An airworthiness certificate must be openly displayed
          inside the aircraft.
91.31   The pilot must comply with the aircraft's operating
          limitations and have the following required documents on

board: airworthiness certificate, registration, radio
station operator's permit, operating limitations, and
weight and balance data. A convenient way to remember
these is to use the acronym ARROW.

91.32   Supplemental oxygen is required for the flight crew for
flights of more than 30 min. between 12,500 and 14,000
ft. MSL. For **any** flights above 14,000 ft. MSL, the crew
must use oxygen, and passengers must be provided with
oxygen for flights above 15,000 ft. MSL.

91.52   An operable emergency locator transmitter (ELT) is
required for flights outside the local area.

8.   FAR PART 91: GENERAL OPERATING AND FLIGHT RULES:
     Subpart B; Flight Rules.

91.67   (a) Vigilance shall be maintained (VFR, and IFR when
weather permits) to see and avoid other aircraft. (b)
An aircraft in distress has the right-of-way over all
other aircraft. (c) The aircraft to the right has the
right-of-way when two aircraft of the same category are
converging. Balloons have the right-of-way over other
categories. A glider has the right-of-way over
airships, airplanes, and rotorcraft. An airship has the
right-of-way over airplanes and rotorcraft. (d) When
approaching head-on, alter course to the right. (e) An
aircraft that is overtaking another aircraft shall give
right-of-way and pass on the right. (f) Aircraft on
final approach have the right-of-way over aircraft on
the ground; also, aircraft at lower altitudes have the
right-of-way on landing.

91.70   (a) The speed limit below 10,000 ft. MSL and in TCAs is
250 kts. (288 MPH). (b) The ATA speed limit is 156 kts.
(180 MPH) for reciprocating engine aircraft and 200 kts.
(230 MPH) for turbine-powered aircraft. (c) The limit
under overhanging TCA airspace is 200 kts. (230 MPH).

91.71   Acrobatic flight is prohibited over congested areas or
open-air assemblies, within control zones, on Federal
airways, below 1,500 ft. AGL, or when visibility is less
than 3 mi.

91.73   Position lights are required for night operation
(on the ground and in the air); anticollision lights are
required for night flight.

91.75   Except in an emergency, the pilot must comply with ATC
instructions. ATC should be notified of any deviations
as soon as possible. ATC may request a written report
of any emergency within 48 hr.

91.77   ATC light signals were covered in Chapter 7. If you
cannot remember them, go back for review.

91.79   (a) Aircraft must always be operated at an altitude that
would allow an emergency landing without undue hazard to
people and property on the surface. (b) Over congested
areas, aircraft must be at least 1,000 ft. above the
highest obstacle within 2,000 ft. of the aircraft. (c)
Over sparsely populated areas, aircraft must stay at
least 500 ft. away from people, structures, or vehicles.

91.81   For flights below 18,000 ft. MSL, altimeters must be set
        to a station within 100 NM along the route or, if no
        station is within the prescribed area, to an appropriate
        current reported altimeter setting. If no radio is
        available, set in the field elevation of the departure
        airport.

91.83   Flight plans shall include the aircraft ID number; type
        of aircraft; pilot's name and address; time and point of
        departure; proposed route, altitude, and TAS; point of
        first landing and ETE; fuel on board; alternate airport
        (IFR only); and number of persons on board. (Note: VFR
        flight plans are optional but **highly** recommended,
        particularly for cross-country flights.

91.87   Pilots going to or from or operating on an airport with
        a control tower are to follow instructions issued by the
        controller. Two-way radio communication is mandatory
        unless other arrangements have been made in advance or
        the radio fails in flight. (Note: A clearance to "taxi
        to" the takeoff runway is a clearance to cross other
        runways, but not to taxi onto or cross any portion of
        the active runway.)

91.90   Operating within a Group I TCA requires an operable VOR
        or TACAN receiver, an operable two-way radio with
        appropriate frequencies, ATC clearance, and a Mode 3/A
        4096 code transponder with a Mode C encoding altimeter.
        A private pilot's certificate is the minimum certificate
        required to land at or take off from the primary airport
        within a Group I TCA.

91.91   Special flight restrictions are issued as **Notices to
        Airmen (NOTAMs)**.

    9.   FAR PART 91: GENERAL OPERATING AND FLIGHT RULES:
         Subpart C, Visual Flight Rules.

91.105 Cloud clearance and flight visibilities:

Surface to 1,200 ft. AGL:

        Controlled airspace: 3 mi. visibility and 500 ft.
        below, 1,000 ft. above, and 2,000 ft. horizontal
        cloud clearance.

        Uncontrolled airspace: 1 mi. visibility and clear
        of clouds.

1,200 ft. AGL to 10,000 ft. MSL:

        Controlled airspace: 3 mi. visibility and 500 ft.
        below, 1,000 ft. above, and 2,000 ft. horizontal
        cloud clearance.

        Uncontrolled airspace: 1 statute mile visibility
        and the same cloud clearance as for controlled
        airspace.

More <u>than 1,200 ft. AGL and at or above 10,000 ft. MSL</u>:

> Visibility of 5 statute miles and 1,000 ft. above and below clouds with a 1 mi. horizontal cloud clearance.

91.107   Special VFR weather minimums that apply in a control zone are clear of clouds and 1 statute mile visibility. Special VFR at night requires both the airplane and the pilot to be capable of IFR flight.

91.109   VFR cruising altitudes for flights above 3,000 ft. AGL: Below FL 18: on **magnetic** courses of 0° to 179°: odd thousand foot MSL atitudes plus 500 ft.; on **magnetic** courses of 180° to 359°: even thousand foot MSL altitudes plus 500 ft.
FL 18 to FL 29: on **magnetic** courses of 0° to 179°: any odd flight level plus 500 ft.; on **magnetic** courses of 180° to 359°: any even flight level plus 500 ft.

10.   FAR PART 91:   GENERAL OPERATING AND FLIGHT RULES: Subpart C, Maintenance, Preventive Maintenance, and Alterations.

91.163   The owner/operator is responsible for maintaining the aircraft in an airworthy condition. Only authorized persons can perform certain maintenance and alterations.

91.165   Maintenance and inspections are the owner's/operator's responsibility.

91.167   Passengers may not be carried until an alteration or repair has been checked and entered in the aircraft log. There are two logs, one for the airframe and one for the engine.

91.169   (a) Annual inspections are required for all aircraft. (b) 100-hr. inspections are required for all aircraft that are operated for hire, including those used for flight instruction for hire. (c) Progressive inspections may be authorized to replace (a) and (b).

91.173   Records of inspections, alterations, total airframe time, and overhauls must be maintained by the owner or operator.

11.   PART 830:   NATIONAL TRANSPORTATION SAFETY BOARD: Accidents.

830.1   Accidents must be reported.

830.5   The operator must report an accident to the NTSB at once. Other flight incidents that must be reported include flight control system failure or malfunction, inability of a required crewmember to perform normal duties due to injury or illness, in-flight fire, an airborne collision, or an aircraft that is overdue or is believed to have been involved in an accident.

830.10   Wreckage, cargo, and all records must be preserved.

830.15   Written reports are due within 10 days of an accident. A report of an incident in 830.5 may also be requested.

DISCUSSION QUESTIONS AND EXERCISES

The following questions emphasize what the FAA tests on the Private Pilot Written Exam.

1. T F An airport traffic area (ATA) extends from the surface to 12,000 ft. MSL and for a radius of 5 mi. from the control tower.

2. T F Calibrated airspeed (CAS) is indicated airspeed corrected for position and installation errors.

3. T F The ceiling is defined as the height above the ground of the lowest layer of clouds that are classified as broken or overcast.

4. T F Transition areas are examples of controlled airspace.

5. T F Indicated airspeed (IAS) refers to pitot-static airspeed corrected for airspeed system errors.

6. T F Flights in prohibited areas must be cleared by ATC in a manner identical to that for flights in other control zones.

7. What does each of the following acronyms mean?

   a. AGL

   b. ATC

   c. CAS

   d. DME

   e. IAS

   f. IFR

   g. MSL

   h. NDB

   i. TAS

   j. Vx

   k. Vy

   l. VFR

   m. VHF

8. T   F   To act as pilot-in-command of an airplane, you must have your pilot certificate and medical certificate in your possession.

9. T   F   Temporary certificates are typically good for six months from the day on which they were issued.

10. T   F   Student pilot certificates are issued without a specific expiration date.

11. T   F   FARs recommend but do not require student pilots to log their flight time.

12. T   F   FARs recommend but do not require private pilots to have a Biennial Flight Review once every two years.

13. T   F   To act as a pilot-in-command of a single-engine airplane carrying passengers, you must have made three takeoffs and landings in the past 30 days in the same category and class of aircraft.

14. T   F   Student pilots can carry passengers on local flights but not on cross-country flights, unless authorized by a CFI.

15. T   F   Obtaining weather information is mandatory only for cross-country flights that may encounter weather below IFR or marginal VFR minimums.

16. T   F   Alcohol cannot be consumed less than 8 hr. before you perform duties as a crewmember of a civil aircraft operated under FAR Part 91.

17. T   F   Each passenger is responsible for knowing how to fasten and unfasten safety belts.

18. T   F   Minimum fuel reserves for a daytime VFR flight of more than 2-hr. duration is 45 min.

19. T   F   Simulated instrument flights require the presence of an appropriately rated safety pilot.

20. What aircraft documents must you carry on board the airplane? They are represented by the letters ARROW.

21. Who has the right-of-way in the following circumstances?

    a. aircraft of the same category converging on your right

   b. two aircraft approaching head-on

   c. one aircraft overtaking another

   d. aircraft on final approach

22. Supplemental oxygen is required for the minimum flight crew
of a civil aircraft if the flight exceeds 30 min. above a certain
minimum cabin pressure altitude. What is that altitude? What is
the minimum cabin pressure altitude at or above which the minimum
flight crew must use oxygen continuously? At what minimum cabin
pressure altitude must **every** occupant of the aircraft be provided
with supplemental oxygen?

23. T F If two aircraft are on final approach, the aircraft at
           the lower altitude has the right-of-way.

24. T F The speed limit in a TCA below 10,000 ft. MSL is 150
           kts. (180 MPH).

25. T F Supplemental oxygen is required for all crewmembers
           for any flight over 30 min. at 10,000 ft. MSL or
           higher.

26. T F Acrobatic flight is prohibited over congested areas,
           below 3,000 ft., and with visibility less than 1 mi.

27. T F Position lights are required for night flight but not
           for taxiing an aircraft from one place to another on
           the airfield at night.

28. T F Except for an emergency or if in violation of FARs,
           the pilot must comply with ATC instructions when
           flying in an ATA unless he/she obtains an amended
           clearance.

29. T F Except for takeoffs and landings, aircraft must always
           be operated at an altitude that would allow an
           emergency landing without undue hazard to people and
           property on the surface.

30. T F Over sparsely populated areas, you must stay at least
           1,000 ft. away from people or structures.

31. T F Flight plans are required for VFR flights over 200 mi.

32.   What are the minimum safe altitudes (except for takeoffs and landing) and horizontal clearance from obstacles required for operation over a congested area? What are the minimums over a sparsely populated area?

33.   T F   For flights below 18,000 ft. MSL, the altimeter must be set to a station within 50 mi. of the route of flight.

34.   T F   Flight plans are mandatory on any cross-country flight in which the pilot-in-command is a student pilot.

35.   A controller clears you to "taxi to" the active runway. Between you and the active runway are several other intersecting runways. How will you proceed?

36.   T F   At airports without control towers, turns are typically made to the left unless markings (such as a segmented circle) indicate otherwise.

37.   T F   To land at or take off from the primary airport within a Group I TCA, the pilot must posses at least a private pilot's certificate and have at least 20 hr. of instrument flight instruction.

38. List the basic VFR flight visibilities and cloud clearances both within and outside of controlled airspace for flights:

     a. at 1,200 ft. or less AGL

     b. at more than 1,200 ft. AGL but less than 10,000 ft. MSL

     c. at more than 1,200 ft. AGL and at or above 10,000 ft. MSL

39.   T F   Special VFR minimums in a control zone include visibility of at least 3 mi. and clear of clouds.

40.   T F   A private pilot may not act as pilot-in-command for hire or compensation, but he/she may share operating expenses of a flight with passengers.

REVIEW QUESTIONS

1.   (FAA 1519)   If an aircraft is involved in an accident which
resulted in substantial damage to the aircraft, the nearest NTSB
field office should be notified

1--immediately.   3--within 7 days.
2--within 5 days.   4--within 10 days.

2.   (FAA 1520)   Of the following incidents, which would
necessitate an immediate notification to the nearest NTSB field
office?

1--An in-flight generator/alternator failure.
2--An in-flight fire.
3--An in-flight loss of VOR receiver capability.
4--Ground damage to the propeller blade.

3.   (FAA 1521)   Which of these incidents would require that an
immediate notification be made to the nearest NTSB field office?

1--An overdue aircraft that is believed to be involved in an
   accident.
2--An in-flight radio (communication) failure.
3--An in-flight generator or alternator failure.
4--In-flight hail damage.

4.   (FAA 1522)   Which incident requires immediate notification to
the nearest NTSB field office?

1--A forced landing due to engine failure.
2--Landing gear damage, due to a hard landing.
3--Inability of any required crewmember to perform normal flight
   duties due to in-flight injury or illness.
4--Substantial aircraft ground fire with no intention of flight.

5.   (FAA 1523)   An aircraft is involved in an accident that
results in substantial damage to the aircraft, but no injuries to
the occupants. When must the pilot or operator of the aircraft
notify the nearest NTSB field office of the occurrence?

1--Within 1 week.   3--Within 48 hr.
2--Immediately.   4--Within 10 days.

6.   (FAA 1524)   The operator of an aircraft that has been
involved in an accident is required to file an accident report
within how many days?

1--3.   3--7.
2--5.   4-10.

7. (FAA 1525) The operator of an aircraft that has been in an incident is required to submit a report to the nearest field office of the NTSB

1--within 3 days.
2--within 10 days.
3--within 7 days.
4--only if requested to do so.

8. (FAA 1540) To act as pilot-in-command of an airplane that has more than 200 hp., a person is required to do which of the following, if no pilot-in-command time in a high-performance airplane was logged prior to November 1, 1973?

1--Make three solo takeoffs and landings in such an airplane.
2--Receive flight instruction in an airplane that has more than 200 hp.
3--Pass a flight test in such an airplane.
4--Hold a 200-hp. class rating.

9. (FAA 1541) To act as pilot-in-command of an airplane with passengers aboard, the pilot must have made at least three takeoffs and three landings in an aircraft of the same category and class within the preceding

1--120 days.   3--24 mo.
2--90 days.   4--12 mo.

10. (FAA 1547) If recency of experience requirements for night flight are not met and official sunset is 1830, the latest time passengers may be carried is

1--1730.   .   3--1900.
2--1830.   4--1930.

11. (FAA 1548 DI) To act as pilot-in-command, a flight review or proficiency check must have been satisfactorily completed within the preceding

1--12 calendar months.   3--24 calendar months.
2--18 calendar months.   4--36 calendar months.

12. (FAA 1561) According to regulations pertaining to general privileges and limitations, a private pilot may

1--not be paid in any manner for the operating expenses of a flight.
2--be paid for the operating expenses of a flight if at least three takeoffs and three landings were made by the pilot within the preceding 90 days.
3--share the operating expenses of a flight with the passengers.
4--charge a reasonable fee for acting as pilot-in-command.

13.   (FAA 1554) To act as pilot-in-command of an aircraft, one must show by logbook endorsement the satisfactory (1) accomplishment of a flight review, or (2) completion of a pilot proficiency check within the preceding

1--6 mo.   3--24 mo.
2--12 mo.   4--36 mo.

14.   (FAA 1555) The takeoffs and landings required to meet recency of experience requirements for carrying passengers in a tailwheel airplane

1--may be touch-and-go in the same class airplane.
2--must be touch-and-go in any class airplane.
3--may be full stop in any class airplane.
4--must be full stop in the same class airplane.

15.   (FAA 1557) What experience requirements are needed to carry passengers?

1--Checkout by a flight instructor.
2--Three takeoffs and landings.
3--Proficiency check by a pilot examiner or an FAA inspector.
4--Application for a waiver.

16.   (FAA 1558) When a certified pilot changes permanent mailing address and fails to notify the FAA Airmen Certification Branch of the new address, the pilot is entitled to exercise the privileges of the pilot certificate for a period of only

1--30 days after the date of the move.
2--60 days after the date of the move.
3--90 days after the date of the move.
4--120 days after the date of the move.

17.   (FAA 1549) To meet recent flight experience requirements for acting as pilot-in-command carrying passengers at night, a pilot must have made, within the preceding 90 days and at night, at least

1--three takeoffs and three landings to a full stop in the same
   category and class of aircraft to be used.
2--three touch-and-go landings in the same category and class of
   aircraft to be used.
3--three takeoffs and three landings, either full stop or touch-
   and-go, but must be accompanied by a certified flight
   instructor who meets the recent experience for night flight.
4--three takeoffs and three landings to a full stop in the same
   category but not necessarily in the same class of aircraft to
   be used.

18. (FAA 1562) What exception, if any, permits a private pilot to act as pilot-in-command of an aircraft carrying passengers who pay for the flight?

1--There is no exception.
2--If the passengers pay all the operating expenses only.
3--If the flight is sponsored by a charitable organization and a donation is made by the passengers.
4--If the pilot acts as second-in-command of an aircraft that requires more than one pilot.

19. (FAA 1563) Which is one privilege of a private pilot?

1--Act as pilot-in-command of any powered aircraft.
2--Act as pilot-in-command of an aircraft carrying passengers.
3--Act as pilot-in-command of an aircraft carrying property for hire if in connection with a business or employment.
4--Act as second-in-command, for compensation or hire, of an aircraft requiring more than one pilot.

20. (FAA 1569) If an in-flight emergency requires immediate action, a pilot-in-command may

1--deviate from FARs to the extent required to meet the emergency, but must submit a written report to the Administrator within 24 hr.
2--not deviate from FARs unless prior to the deviation approval is granted by the Administrator.
3--deviate from FARs to the extent required to meet that emergency.
4--not deviate from FARs unless permission is obtained from air traffic control.

21. (FAA 1571 DI) The final authority as to the operation of an aircraft is the

1--FAA.
2--aircraft flight manual data.
3--owner or operator of the aircraft.
4--inspector who performs the periodic check.

22. (FAA 1572) Which preflight action is required for every flight?

1--Check weather reports and forecasts.
2--Determine runway lengths at airports of intended use.
3--Determine alternatives if the flight cannot be completed.
4--Check for any known traffic delays.

23. (FAA 1573) Preflight action, as required by regulations for
all flights away from the vicinity of an airport, shall include a
study of the weather, taking into consideration fuel requirements
and

1--an operational check of the navigation radios.
2--the designation of an alternate airport.
3--the filing of a flight plan.
4--an alternate course of action if the flight cannot be
   completed as planned.

24. (FAA 1574) In addition to other preflight actions for a VFR
flight away from the vicinity of the departure airport,
regulations require the pilot-in-command to

1--file a flight plan.
2--check each fuel tank visually to ensure that it is full.
3--check the accuracy of the omninavigation equipment and the
   emergency locator transmitter.
4--determine runway lengths of airports of intended use and the
   airplane's takeoff and landing distance data.

25. (FAA 1575) Prior to each flight, the pilot-in-command must

1--check the personal logbook for appropriate recent experience.
2--become familiar with all available information concerning that
   flight.
3--calculate the weight and balance of the aircraft to determine
   if the CG is within limits.
4--check with ATC for the latest traffic advisories and any
   possible delays.

26. (FAA 1576) Under what condition, if any, may a pilot allow
a person who is obviously under the influence of intoxicating
liquors or drugs to be carried aboard an aircraft?

1--Under no condition.
2--Only if the person is a medical patient under proper care or
   in an emergency.
3--Only if the person does not have access to the cockpit or
   pilot's compartment.
4--Only if a second pilot is aboard.

27. (FAA 1577) A person may not act as a crewmember of a civil
aircraft if alcoholic beverages have been consumed by that person
within the preceding

1--8 hr.   3--24 hr.
2--12 hr.   4--48 hr.

28. (FAA 1586) A chair-type parachute must have been packed by
a certified and appropriately rated parachute rigger within the

1--preceding 30 days.   3--preceding 90 days.
2--preceding 60 days.   4--preceding 120 days.

29. (FAA 1582) Regulations require that seatbelts in an airplane be properly secured about the

1--occupants during takeoffs and landings.
2--crewmembers only, during takeoffs and landings.
3--occupants during flight in moderate or severe turbulence only.
4--passengers and crewmembers during the entire flight.

30. (FAA 1584) What obligation does a pilot-in-command have concerning the use of seatbelts?

1--The pilot-in-command's seatbelt must be fastened during the entire flight.
2--The pilot-in-command must brief the passengers on the use of seatbelts and notify them to fasten the seatbelts during takeoff and landing.
3--The pilot-in-command must instruct the passengers to keep seatbelts fastened for the entire flight except for brief rest periods.
4--The pilot-in-command must instruct the passengers on the use of seatbelts, but does not have the authority to require their use, except during an emergency.

31. (FAA 1585) When must each occupant of an aircraft wear an approved parachute?

1--When an aircraft is being tested after major repair.
2--When flying over water beyond gliding distance to the shore.
3--When a door is removed from the aircraft to facilitate parachute jumpers.
4--When an intentional bank that exceeds 60° is to be made.

32. (FAA 1578) Under what condition may a person act as pilot-in-command of an aircraft after consuming alcohol which may affect that person's faculties?

1--Passengers may not be carried.
2--A waiver must be obtained.
3--Only after expiration of 8 hr.
4--Only after release from a doctor.

33. (FAA 1594) Considering the wind and forecast weather conditions, no person may begin a flight in an airplane at night unless there is enough fuel to fly to the first point of intended landing at normal cruising speed and after that to fly at least

1--30 min.   3--1 hr.
2--45 min.   4--1 hr., 15 min.

34. (FAA 1595) What is the fuel requirement for flight under
VFR at night?

1--Full fuel tanks.
2--Enough to complete the flight at normal cruising flight with
    adverse wind conditions.
3--Enough to fly to the first point of intended landing and to
    fly after that for 30 min. at normal cruising speed.
4--Enough to fly to the first point of intended landing and to
    fly after that for 45 min. at normal cruising speed.

35. (FAA 1596) What is the fuel requirement for flight under
VFR during daylight hours?

1--Full fuel tanks.
2--Enough to complete the flight at normal cruising speed with
    adverse wind conditions.
3--Enough to fly to the first point of intended landing and to
    fly after that for 30 min. at normal cruising speed.
4--Enough to fly to the first point of intended landing and to
    fly after that for 45 min. at normal cruising speed.

36. (FAA 1605) Regarding certificates and documents, no person
may operate an aircraft unless it has within it an

1--airworthiness certificate, aircraft and engine logbooks, and
    owner's handbook.
2--airworthiness certificate and owner's handbook.
3--airworthiness certificate, registration certificate, and
    operating limitations.
4--airworthiness certificate, and aircraft and engine logbooks.

37. (FAA 1606) What documents or records must be aboard an
aircraft during flight?

1--Operating limitations and an Aircraft Use and Inspection
    Record.
2--Operating limitations; a registration certificate; and an
    appropriate, current, and properly displayed airworthiness
    certificate.
3--Repair and alteration forms and a registration certificate.
4--Aircraft and engine logbooks and a registration certificate.

38. (FAA 1607) No person may operate a civil aircraft unless
the airworthiness certificate, or special flight permit or
authorization required by regulations, is

1--on file in the owner's operation office where the aircraft is
    based.
2--filed with the other required certificates or documents within
    the aircraft to be flown.
3--displayed at the cabin or cockpit entrance so that it is
    legible to passengers or crewmembers.
4--included in the approved logbooks for the aircraft to be
    flown.

39. (FAA 1608) Who is responsible for determining if an aircraft is in condition for safe flight?

1--A certified aircraft mechanic.
2--A certified aircraft maintenance inspector.
3--The pilot-in-command.
4--The owner or operator.

40. (FAA 1612) An aircraft's operating limitations may be found

1--on the airworthiness certificate.
2--in the airplane flight manual, approved manual material,
     markings, and placards, or any combination thereof.
3--only in the FAA approved airplane flight manual.
4--only in the owner's handbook published by the aircraft
     manufacturer.

41. (FAA 1613) Unless each occupant is provided with supplemental oxygen, no person may operate a civil aircraft of U.S. registry above a cabin pressure altitude of

1--15,000 ft. MSL.   3--12,500 ft. MSL.
2--14,000 ft. MSL.   4--10,000 ft. MSL.

42. (FAA 1616) Regulations require that supplemental oxygen be provided to each occupant when the aircraft is operated above a cabin pressure altitude of

1--12,500 ft. MSL.   3--14,500 ft. MSL.
2--14,000 ft. MSL.   4--15,000 ft. MSL.

43. (FAA 1620) Which is normally prohibited when operating a restricted category civil aircraft?

1--Flight under instrument flight rules.
2--Flight over a densely populated area.
3--Flight within a control zone.
4--Flight within the continental control area.

44. (FAA 1624) An airplane and an airship are converging. If the airship is left of the airplane's position, which aircraft has the right-of-way?

1--The pilot of the airplane should give way; the airship is to
     the left.
2--The airship has the right-of-way.
3--Each pilot should alter course to the right.
4--The airplane has the right-of-way; it is more maneuverable.

45. (FAA 1630) Which aircraft has the right-of-way over all other aircraft?

1--Balloon.   3--Experimental aircraft.
2--Airship.   4--Restricted aircraft.

46.   (FAA 1625)   When two or more airplanes are approaching an
airport for the purpose of landing, the right-of-way belongs to
the airplane

1--that has the other to its right.
2--that is the least maneuverable.
3--that is either ahead of or to the other's right regardless of
    altitude.
4--at the lower altitude, but it shall not take advantage of this
    rule to cut in front of or to overtake another.

47.   (FAA 1626)   What action is required when two airplanes
converge at the same altitude, but not head-on?

1--The more maneuverable airplane shall give way.
2--The faster airplane shall give way.
3--The airplane on the left shall give way.
4--Each airplane shall give way to the right.

48.   (FAA 1627)   What action should be taken, if during a night
flight, a pilot observes a steady green light and a flashing red
light on another aircraft at the same altitude?

1--Exercise extreme caution, the other aircraft is in distress.
2--Give way to the right, the other aircraft is approaching head-
    on.
3--Be alert, the other aircraft should give way.
4--Give way to the left, the other aircraft is passing to the
    right.

49.   (FAA 1628)   What action should be taken if a glider and an
airplane approach each other at the same altitude and on a head-
on collision course?

1--The airplane should give way because the glider has the right-
    of-way.
2--The airplane should give way because it is more maneuverable.
3--Both should give way to the right.
4--The airplane should climb and the glider should descend so as
    to pass each other by at least 500 ft.

50.   (FAA 1629)   Which aircraft has the right-of-way over all
other aircraft?

1--A balloon.
2--An aircraft in distress.
3--An aircraft on final approach to land.
4--An aircraft towing or refueling another aircraft.

51.   (FAA 1631)   When flying beneath the lateral limits of a
terminal control area, the maximum speed authorized is

1--250 kts.   3--180 kts.
2--200 kts.   4--156 kts.

52. (FAA 1632) Unless otherwise authorized or required by air
traffic control, what is the maximum indicated airspeed at which
a person may operate an aircraft below 10,000 ft. MSL?

1--156 kts.   3--200 kts.
2--180 kts.   4--250 kts.

53. (FAA 1633) Unless otherwise authorized or required by air
traffic control, the maximum indicated airspeed at which a
reciprocating engine equipped aircraft should be flown within an
airport traffic area is

1--156 kts.   3--200 kts.
2--180 kts.   4--288 kts.

54. (FAA 1634) In which controlled airspace is acrobatic flight
prohibited?

1--All controlled airspace.
2--Control zones and Federal airways.
3--Control zones, Federal airways, and control areas.
4--Control zones, Federal airways, control areas, and traffic
   control areas.

55. (FAA 1635) According to regulations, no person may operate
an aircraft in acrobatic flight

1--over any congested area of a city, town, or settlement.
2--within 5 mi. of a Federal airway.
3--below an altitude of 2,000 ft. above the surface.
4--when flight visibility is less than 5 mi.

56. (FAA 1636) No person may operate an aircraft in acrobatic
flight when flight visibility is less than

1--3 mi.   3--7 mi.
2--5 mi.   4--10 mi.

57. (FAA 1637) What is the lowest altitude permitted for
acrobatic flight?

1--1,000 ft. AGL.
2--1,500 ft. AGL.
3--1,000 ft. above the highest obstacle within 5 mi.
4--1,500 ft. above the highest obstacle within 5 mi.

58. (FAA 1638) When an aircraft is being operated at night, it
must display lighted position lights during the period from

1--1 hr. before sunset to 1 hr. after sunrise.
2--30 min. after sunset to 30 min. after sunrise.
3--30 min. before sunset to 30 min. after sunrise.
4--sunset to sunrise.

59. (FAA 1640) During what time period should lighted position lights be displayed on an aircraft?

1--30 min. before sunset to 30 min. after sunrise.
2--1 hr. before sunset to 1 hr. after sunrise.
3--1 hr. before sunset to 1 hr. before sunrise.
4--sunset to sunrise.

60. (FAA 1641) What action should be taken if a pilot receives a clearance that will cause a deviation from a rule?

1--Accept the clearance, because the pilot is not responsible for the deviation.
2--Accept the clearance and advise air traffic control when deviation occurs.
3--Refuse the clearance as stated and request that it be amended.
4--Accept the clearance and advise air traffic control that you believe a rule deviation will occur.

61. (FAA 1642) When may air traffic control request a detailed report of an emergency even though a rule has not been violated?

1--When priority has been given.
2--Anytime an emergency occurs.
3--When the emergency occurs in controlled airspace.
4--Only when an accident results from an emergency.

62. (FAA 1643) What action, if any, is appropriate if the pilot deviates from an air traffic control instruction during an emergency and you are given priority?

1--Take no special action since you are pilot-in-command.
2--If requested, file a detailed report within 48 hr. to the chief of the air traffic control facility.
3--File a report to the FAA Administrator within 48 hr.
4--File a report to the chief of that air traffic control facility within 24 hr.

63. (FAA 1644) An air traffic control clearance provides

1--authorization for flight in uncontrolled airspace.
2--priority over all other traffic.
3--adequate separation from all traffic.
4--authorization to proceed under specified traffic conditions in controlled airspace.

64. (FAA 1652) Except when necessary for takeoff or landing, an aircraft may not be operated closer than what distance from any person, vehicle, or structure?

1--2,000 ft.   3--500 ft.
2--1,000 ft.   4--100 ft.

65. (FAA 1654) To operate an aircraft over any congested area,
a pilot should maintain an altitude of at least

1--500 ft. above the highest obstacle within a horizontal radius
     of 500 ft.
2--500 ft. above the highest obstacle within a horizontal radius
     of 1,000 ft.
3--1,000 ft. above the highest obstacle within a horizontal
     radius of 2,000 ft.
4--2,000 ft. above the highest obstacle within a horizontal
     radius of 1,000 ft.

66. (FAA 1657) Except when necessary for takeoff or landing,
what is the minimum safe altitude for a pilot to operate an
aircraft anywhere?

1--An altitude allowing, if a power unit fails, an emergency
     landing without undue hazard to persons or property on the
     surface.
2--An altitude of 500 ft. above the surface and no closer than
     500 ft. to any person, vessel, vehicle, or structure.
3--An altitude of 500 ft. above the highest obstacle within a
     horizontal radius of 1,000 ft.
4--An altitude of 1,000 ft. above the highest obstacle within a
     horizontal radius of 2,000 ft.

67. (FAA 1658) Except when necessary for takeoff or landing,
what is the minimum safe altitude required for a pilot to operate
an airplane over congested areas?

1--An altitude allowing, if a power unit fails, an emergency
     landing without undue hazard to persons or property on the
     surface.
2--An altitude of 500 ft. above the surface and no closer than
     500 ft. to any person, vessel, vehicle, or structure.
3--An altitude of 500 ft. above the highest obstacle within a
     horizontal radius of 1,000 ft.
4--An altitude of 1,000 ft. above the highest obstacle within a
     horizontal radius of 2,000 ft.

68. (FAA 1659) Except when necessary for takeoff or landing,
what is the minimum safe altitude for a pilot to operate an
airplane over other than a congested area?

1--An altitude allowing, if a power unit fails, an emergency
     landing without undue hazard to persons and property on the
     surface.
2--An altitude of 500 ft. AGL except over open water or sparsely
     populated area which requires 500 ft. from any person, vessel,
     vehicle, or structure.
3--An altitude of 500 ft. above the highest obstacle within a
     horizontal radius of 1,000 ft.
4--An altitude of 1,000 ft. above the highest obstacle within a
     horizontal radius of 2,000 ft.

69. (FAA 1707) The responsibility for ensuring that an aircraft
is maintained in an airworthy condition is primarily that of the

1--pilot-in-command of the aircraft.
2--owner or operator of the aircraft.
3--maintenance shop.
4--certified mechanic who signs the aircraft maintenance records.

70. (FAA 1708) An airworthiness certificate remains valid

1--until ownership is transferred.
2--provided the aircraft has not had major damage.
3--until surrendered, suspended, or revoked.
4--provided the aircraft is maintained and operated according to
   FARs.

71. (FAA 1710) Before passengers can be carried in an aircraft
that has been altered in a manner that may have appreciably
changed its flight characteristics, it must be flight tested by
an appropriately-rated pilot with at least a

1--commercial pilot certificate and an instrument rating.
2--private pilot certificate.
3--commercial pilot certificate and a mechanic's certificate.
4--commercial pilot certificate.

72. (FAA 1714) Completion of an annual inspection and the
return of the aircraft to service should always be indicated by

1--the relicensing date on the registration certificate.
2--an appropriate notation in the aircraft maintenance records.
3--an inspection sticker placed on the instrument panel that
   lists the annual inspection completion date.
4--the issuance date of the airworthiness certificate.

73. (FAA 1715) An aircraft's last annual inspection was
performed on July 12, this year. The next annual inspection will
be due no later than

1--July 31, next year.
2--July 13, next year.
3--100 flight hours following the last annual inspection.
4--12 calendar months after the date shown on the airworthiness
   certificate.

74. (FAA 1716) To determine the expiration date of the last
annual aircraft inspection, a person should refer to the

1--airworthiness certificate.
2--registration certificate.
3--aircraft maintenance records.
4--owner/operator manual.

75. (FAA 1717) How long does the airworthiness certificate of
an airplane remain valid?

1--As long as the aircraft has a current registration
   certificate.
2--Indefinitely, unless the aircraft suffers major damage.
3--As long as the airplane is maintained and operated as required
   by FARs.
4--Indefinitely, unless the prescribed operating limitations are
   exceeded.

76. (FAA 1722) Which record or documents shall the owner or
operator of an aircraft keep to show compliance with an
applicable Airworthiness Directive?

1--The aircraft maintenance records.
2--Airworthiness certificate and owner's handbook.
3--Airworthiness and registration certificates.
4--Aircraft flight manual and owner's manual.

ANSWERS

Discussion Questions and Exercises

   1.   F--It extends up to but does not include 3,000 ft. AGL and
            for a 5-mi. radius. See FAR 1.1.
   2.   T--See FAR 1.1.
   3.   T--See FAR 1.1.
   4.   T--See FAR 1.1.
   5.   F--This is called calibrated airspeed. See FAR 1.1.
   6.   F--Flights are prohibited in prohibited areas. See FAR 1.1.
   7.   a. Above ground level.
        b. Air traffic control.
        c. Calibrated airspeed.
        d. Distance measuring equipment.
        e. Indicated airspeed.
        f. Instrument Flight Rules.
        g. Mean sea level.
        h. Nondirectional beacon.
        i. True airspeed.
        j. Best angle-of-climb airspeed.
        k. Best rate-of-climb airspeed.
        l. Visual Flight Rules.
        m. Very high frequency.
   8.   T--See FAR 61.3.
   9.   F--They are good for 120 days. See FAR 61.17.
  10.   F--Student pilot certificates expire at the end of 24 mo.
            after they were issued. See FAR 61.17.
  11.   F--Student pilots must log flight time. See FAR 61.51.
  12.   F--BFRs are required every two years. See FAR 61.57.
  13.   F--You must have made three takeoffs and landings in the
            past 90 days. See FAR 61.57.
  14.   F--Student pilots cannot carry passengers. See FAR 61.89.

15.   F--Obtaining weather information is mandatory on all IFR and
      cross-country flights. See FAR 91.5.
16.   T--See FAR 91.11.
17.   F--It is the responsibility of the pilot-in-command to
      inform passengers how to fasten and unfasten their seat
      belts. See FAR 91.14.
18.   F--The minimum fuel reserve for a daytime VFR flight is 30
      min. The minimum for a night flight is 45 min. See FAR
      91.22.
19.   T--See FAR 91.21.
23.   T--See FAR 91.67.
24.   F--The speed limit is 250 kts. (288 MPH). See FAR 91.70.
25.   F--Supplemental oxygen is required of all crewmembers for
      any flight over 30 min. from 12,500 ft. to 14,000 ft.
      MSL; above 14,000 ft. MSL, it is required for crewmembers
      at all times; and above 15,000 ft. MSL, it is required
      for all occupants. See FAR 91.32.
26.   F--Acrobatic flight is prohibited over congested areas,
      below 1,500 ft. AGL, and when visibility is less than 3
      mi. See FAR 91.71.
27.   F--They are required for any night operation, on the ground
      or in the air. See FAR 91.73.
28.   T--See FAR 91.75.
29.   T--See FAR 91.79.
30.   F--Over sparsely populated areas, you must stay at least 500
      ft. away from people or structures. See FAR 91.79.
31.   F--VFR flight plans are optional, but highly recommended.
      See FAR 91.83.
33.   F--The station needs only be within 100 mi., if one exists.
      See FAR 91.81.
34.   F--VFR flight plans are optional, but highly recommended.
      See FAR 91.83.
36.   T--See FAR 91.89.
37.   F--A private pilot's certificate is the minimum certificate
      required to land at or take off from an airport within
      the horizontal and vertical limits of a Group I TCA. A
      two-way radio, operable VOR or TACAN, Mode 3/A 4096 code
      transponder, and a Mode C encoding altimeter are required
      to operate within a Group I TCA. See FAR 91.90.
39.   F--Special VFR minimums are clear of clouds and visibility
      of 1 mi. They apply to daytime operations **only**. See FAR
      91.107.
40.   T--See FAR 61.118

Review Questions

 1.   1--Accidents must be reported to the NTSB by the operator
      immediately. See NTSB 830.5.
 2.   2--In-flight fires must be reported to the NTSB immediately.
      See NTSB 830.5.
 3.   1--See NTSB 830.5.
 4.   3--See NTSB 830.5
 5.   2--See answer 1 and NTSB 830.5.
 6.   4--A report must be filed within 10 days. See NTSB 830.15.

7.   4--The NTSB will request a report of an incident. See NTSB
         830.15.
8.   2--See FAR 61.31.
9.   2--See FAR 61.57.
10. 4--The pilot must meet recent experience requirements for
         night flight. See FAR 61.57.
11. ?--This question has been designated as unusable by the FAA
         and removed from the FAA Question Selection Sheets.
12. 3--A private pilot may share operating expenses with his/her
         passengers. See FAR 61.118.
13. 3--See FAR 61.57.
14. 4--Tailwheel airplanes require full stop landings. See FAR
         61.57.
15. 2--See FAR 61.57.
16. 1--You must notify the FAA Airman Certification Branch within
         30 days of any change in address. Your author has been
         delinquent. See FAR 61.60.
17. 1--You are required to make three takeoffs and landings to a
         full stop at night to satisfy night flight requirements.
         See FAR 61.57.
18. 3--See FAR 61.118.
19. 2--As a private pilot, you are able to carry passengers. See
         FAR 61.118.
20. 3--You may deviate from Part 91 operating rules to meet an
         emergency. A report is necessary if the FAA requests it.
         See FAR 91.3.
21. ?--This question has been designated as unusable by the FAA
         and removed from the FAA Question Selection Sheets.
22. 2--You must determine runway lengths and your airplane's
         ability to use that runway prior to every flight. See FAR
         91.5.
23. 4--See FAR 91.5.
24. 4--See FAR 91.5.
25. 2--See FAR 91.5.
26. 2--Only in an emergency or if a person is under medical
         supervision may a pilot allow someone under the influence
         of alcohol or drugs to be carried aboard an aircraft. See
         FAR 91.11.
27. 1--Eight hours is the minimum between the bottle and the
         throttle. See FAR 91.11.
28. 4--It must have been packed within the previous 120 days.
         See FAR 91.15.
29. 1--Seatbelts are required during takeoffs and landings. See
         FAR 91.14.
30. 2--The pilot-in-command is responsible for explaining the use
         of seatbelts to his/her passengers. See FAR 91.14.
31. 4--Only if you intend to exceed banks of 60° are parachutes
         required. See FAR 91.15.
32. 3--See answer 27 and FAR 91.11.
33. 2--Night VFR flights require a 45-min. fuel reserve. See FAR
         91.22.
34. 4--See answer 33 and FAR 91.22.
35. 3--Daytime VFR requires a 30-min. fuel reserve. See FAR
         91.22.

36. 3--ARROW: airworthiness certificate, registration, radio
         station operator's permit, operating limitation, and
         weight and balance data. See FARs 91.27 and 91.31.
37. 2--See answer 36 and FARs 91.27 and 91.31.
38. 3--See answer 36 and FARs 91.27 and 91.31.
39. 3--See FAR 91.3.
40. 2--See answer 36 and FARs 91.27 and 91.31.
41. 1--See FAR 91.32.
42. 4--See FAR 91.32.
43. 2--See FAR 91.39.
44. 2--Airships have the right-of-way over airplanes. See FAR
         91.67.
45. 1--A balloon has the right-of-way over all other aircraft,
         except an aircraft in distress. See FAR 91.67.
46. 4--The lower aircraft has the right-of-way. See FAR 91.67.
47. 3--The one on the right has the right-of-way. See FAR 91.67.
48. 3--Since the aircraft is passing from the left to the right,
         you should be on the alert as the other aircraft should
         give way. See FAR 91.67.
49. 3--When approaching head-on at the same altitude, both
         aircraft should turn to the right. See FAR 91.67.
50. 2--An aircraft in distress has the right-of-way over all
         other aircraft. See FAR 91.67.
51. 2--See FAR 91.70.
52. 4--See FAR 91.70.
53. 1--See FAR 91.70.
54. 2--Acrobatic flight is not allowed over congested areas, over
         an open assembly of people, within a control zone or
         Federal airway, below 1,500 ft. AGL, or when flight
         visibility is less than 3 mi. See FAR 91.71.
55. 1--See answer 54 and FAR 91.71.
56. 1--See answer 54 and FAR 91.71.
57. 2--See answer 54 and FAR 91.71.
58. 4--See FAR 91.73.
59. 4--See FAR 91.73.
60. 3--See FAR 91.3.
61. 1--ATC may request a detailed report. See FAR 91.75.
62. 2--See FAR 91.75.
63. 4--See FAR 91.3.
64. 3--The pilot should **always** operate so as to be able to land
         anywhere without undue hazard to persons or property
         should an emergency arise. Over congested areas, the
         pilot should maintain an altitude of 1,000 ft. above the
         highest obstacle within a horizontal radius of 2,000 ft.
         Over areas that are not congested, the pilot should not
         operate closer than 500 ft. from any person, vessel,
         vehicle, or structure. See FAR 91.79.
65. 3--See answer 64 and FAR 91.79.
66. 1--See answer 64 and FAR 91.79.
67. ?--There are two correct answers. First, alternative (1) is
         absolutely correct. See answer 64 and FAR 91.79. You
         should always fly so as to be able to land should an
         emergency arise. But, the intent of the question is to
         ask for limits over a congested area, which makes
         alternative (4) the best choice.

68. ?--As with question 67, there are two correct answers.
     Again, the intent of the question is to ask for limits
     over areas that are not congested, which makes alternative
     (2) the best choice.
69. 2--See FAR 91.161.
70. 4--See FAR 91.161.
71. 2--See FAR 91.167.
72. 2--See FAR 91.173.
73. 1--See FAR 91.169.
74. 3--See FAR 91.173.
75. 3--See FAR 91.161.
76. 1--See FAR 91.173.

12/BASICS OF AIR NAVIGATION

MAIN POINTS

1.   All navigation problems share certain things in common. Before you go somewhere, you need to know exactly where you are. Determining your position on the Earth's surface is called a fix. A fix is made when two **lines of position (LOPs)** intersect. There are five basic methods of navigation: pilotage, reference to visual landmarks; **dead reckoning (DR)**, figuring one's position from time-distance computations using one or more LOPs; **radio navigation**, observing one's bearings from a radio source; **celestial**, navigation by reference to the sun and stars; and **inertial**, navigation based on continuous computations performed by the airplane's instruments. Pilotage, dead reckoning, and radio navigation are the most common methods used by private pilots. Pilotage and dead reckoning are covered in this chapter. Radio navigation is covered in Chapter 13.

2.   Lines that run between the poles are called **lines of longitude**, or **meridians**. Lines that run parallel to the equator are called **lines of latitude**, or **parallels**. Taken together, lines of latitude and lines of longitude provide a geographic coordinate system or grid. The **prime meridian** runs north and south through Greenwich, England, and lines to the east and west are numbered in degrees. The **equator** divides Northern and Southern Hemispheres and is 0° latitude. Latitude degrees increase as you approach the poles. For finer discriminations, latitude and longitude are further subdivided into 60 arc minutes per degree and 60 arc seconds per minute. A **nautical mile (NM)** is one minute of longitude marked off vertically on a meridian. The shortest distance between two points on the earth's surface is along a **great circle**, one whose plane runs through the center of the Earth.

3.   The Earth is divided into 24 time zones of approximately 15° longitude each. Most aviation-related time is reported relative to a standard: Greenwich mean time (GMT) or zulu time.

To convert from local time to GMT in the Western Hemisphere, add
hours to local time depending on which time zone you are in and
whether the area is observing standard or daylight savings time.
Times are typically reported on the 24-hr. clock. Thus, 6 p.m.
becomes (1200 + 600) or 1800. An a.m. time such as 4 a.m.
becomes 0400. All GMT or zulu times are given on the 24-hr.
clock.

   4.   The three most common types of VFR navigational charts
are **sectional charts, terminal area charts, and world
aeronautical charts.** Sectional charts are used primarily for
low-altitude, low-airspeed flight by reference to visible
landmarks. They contain three general types of information:
   (a) topographical features: cities, roads, obstructions (MSL
with AGL given in parentheses), contour lines, lakes, and so on
   (b) aeronautical data: airports, communication information,
airspace restrictions, and so forth
   (c) legend and notes: airport directory, VFR rules, and so
on.
Terminal area charts are available for selected TCAs and are
larger in scale than sectional charts. **World aeronautical charts
(WACs)** are for higher altitude flights where less detail is
needed.

   5.   **Pilotage** requires little more than a clear day, a
sectional chart, and a straightedge such as a **navigational
plotter.** The plotter consists of a straightedge, mileage scales
for sectional charts and WACs, and a protractor. To determine a
course, select the appropriate chart, make a general inspection
of the terrain, draw a true course line, and determine the **true
course,** or TC. To determine the true course, use your plotter
and a meridian line near the center of your course. Next,
measure the distance you intend to travel. Divide the course
into intervals every 10-20 mi. and select prominent landmarks
(plainly visible ones that suggest direction as well as position)
along the intended route of flight. Also select landmarks to
either side of your intended route so as to **bracket** the course.

   6.   **Dead reckoning (DR)** begins with determining a true
course line and then making corrections for wind and magnetic
variation. The angle between the wind and your intended route of
flight is called the **wind correction angle (WCA).** When the value
of the WCA is added to or subtracted from the true course, the
result is called the **true heading (TH).** Angular **variation (VAR)**
between magnetic and geographic north is shown on the chart and
is used to obtain the **magnetic heading (MH).** Magnetic heading
must be corrected for **compass deviation (CD)** to obtain the
compass heading (CH). Finally, determine **true airspeed (TAS)** by
referring to you flight manual. After adjusting for the effects
of the wind, you will know what your **groundspeed** will be, which
is crucial in determining how fast you will travel from fix to
fix.

7.   Once you have completed all steps normally involved in pilotage, you perform numerous calculations for dead reckoning, all of which are facilitated by using a **flight computer** or **electronic flight calculator**. With a flight computer, you use the wind face to calculate both the wind correction angle and groundspeed for any given true course, true airspeed, and wind condition (obtained from the Winds Aloft Forecast). Any crosswind component will affect your true course, and the effect is magnified as the crosswind component increases. Second, to determine the magnetic heading, you need to correct for the magnetic variation shown on the chart. **Isogonic lines** are lines of equal variation. (The line of no variation is referred to as the **agonic line**.) If the variation is to the west, you **add** it to the true heading; if it is to the east, you **subtract** it. To remember this, use the rhyme: "East is least (-) and west is best (+)." **Magnetic course (MC)** is the true course corrected for variation **without** taking wind into account; it is used by the FAA to regulate cruising altitudes. **Magnetic heading (MH)** is the true heading corrected for **both** wind (WCA) and variation (VAR). Once magnetic heading is corrected for compass deviation errors, you will have determined the compass heading you will need to hold in order to fly a desired track.

8.   Your flight computer or electronic calculator will greatly simplify your ability to solve dead reckoning problems, but you should also understand the concepts behind each computation. Most computations are based on the relationship between distance (D), time (T), and rate (R) such that $R = D/T$. Endurance can be calculated by dividing fuel on board by fuel flow (measured in gallons per hour). The computer can be used to solve these and similar proportional problems. You will need to spend time with your computer to learn to read the scales accurately.

9.   In addition to groundspeed, endurance, fuel flow, and time, the computer has a scale to convert calibrated airspeed into true airspeed (airspeed corrected for the effects of altitude and temperature on air density). TAS must be determined before the wind correction angle or groundspeed can be computed. During preflight planning, you will use an estimated TAS from the performance charts in the POH. To compute TAS in-flight, you need to know the pressure altitude, temperature, and indicated airspeed.

10.   The flight computer also allows you to compute true altitudes for non-standard temperatures, density altitude for a given temperature and pressure altitude, and to make conversions such as statute miles to nautical miles, pounds of fuel to gallons, and Fahrenheit to Celsius.

11.   The wind correction angle (WCA) is the number of degrees you must add or subtract from your true course to obtain a true heading that compensates for drift. Groundspeed is affected by both the speed and the direction of the wind. Both groundspeed and true heading can be derived from the computer by

following specific instructions written on most computers. After you have done several practice exercises, these computations will become routine.

12.   The choice of a flight computer or electronic calculator is an individual one. In either case, it is wise to understand the concepts and relationships so that you can intuitively arrive at an estimate of what the answer should be. The computer or calculator can then give you a precise answer.

13.   If you use an electronic flight calculator during the FAA written test, any information printed on the case relating to regulations, ATC signals, and so forth must be obscured. Memory circuits must be cleared before and after the test, and a tape printout, if produced, must be surrendered at the end of the test. Finally, you cannot use the operations manual during the test, nor are you allowed to use prewritten programs.

KEY TERMS AND CONCEPTS, PART 1

Match each term or concept (1-20) with the appropriate description (A-T) below. Each item has only one match.

___ 1. zulu
___ 2. inertial
___ 3. pilotage
___ 4. protractor
___ 5. equator
___ 6. latitude
___ 7. bracketing
___ 8. sectional chart
___ 9. longitude
___ 10. fix

___ 11. compass heading
___ 12. magnetic heading
___ 13. prime meridian
___ 14. celestial
___ 15. true heading
___ 16. plotter
___ 17. dead reckoning
___ 18. easterly
___ 19. agonic line
___ 20. terminal control area
         chart

A.   TH plus or minus VAR
B.   navigation by reference to visual landmarks
C.   a typical one has a straightedge, two mileage scales, and
     protractor
D.   line that runs through Greenwich, England
E.   lines that run between the poles
F.   selecting landmarks on both sides of a course line
G.   MH corrected for compass deviation
H.   navigation by reference to the sun and stars
I.   large-scale aeronautical chart that depicts a TCA (scale of 1
     in. = 4 mi.)
J.   parallel lines that circle the earth
K.   point where two lines of position (LOPs) intersect
L.   instrument used to measure angles
M.   line that divides North and South Hemispheres
N.   TC plus or minus the WCA
O.   type of time used to make most weather reports and for filing
     flight plans

P. navigation by making time-distance computations along a line
    of position
Q. navigation based on computations performed by the airplane's
    instruments
R. chart used primarily for low-altitude, low-airspeed flight by
    reference to visual landmarks (scale of 1 in. = 8 mi.)
S. type of magnetic variation subtracted from TH to obtain MH
T. line of no variation

KEY TERMS AND CONCEPTS, PART 2

Match each term or concept (1-12) with the appropriate
description below (A-L). Each item has only one match.

___ 1. groundspeed   ___ 7. true altitude
___ 2. indicated airspeed   ___ 8. wind correction angle
___ 3. pressure altitude   ___ 9. magnetic course
___ 4. compass heading   ___ 10. westerly
___ 5. true airspeed   ___ 11. calibrated airspeed
___ 6. magnetic variation   ___ 12. isogonic line

A. IAS corrected for pitot-static errors
B. true course corrected for magnetic variation but not wind
    effects
C. number of degrees added to or subtracted from TC to yield TH
D. lines of equal variation
E. airspeed corrected for density altitude
F. difference between true north and magnetic north
G. type of magnetic variation added to TH to obtain MH
H. altitude read from the altimeter when 29.92" Hg is set in the
    Kollsman window.
I. airspeed read directly from the airspeed indicator
J. actual height above mean sea level
K. TAS corrected for wind effects
L. magnetic heading corrected for compass deviation

DISCUSSION QUESTIONS AND EXERCISES

1.   What is the relationship between a line of position (LOP)and
a fix?

2.   Name the three basic methods of navigation commonly used by
private pilots.

3.   What does the expression 55°40'N, 104°20'W represent and how would it be interpreted by a pilot or navigator?

4.   T F The prime meridian runs east and west while the equator runs north and south.

5.   T F Longitude lines are of the same length, whereas lines of latitude decrease in length as the distance from the equator increases.

6.   T F Longitude lines converge at the poles, whereas lines of latitude converge at the equator.

7.   How are surface obstructions and terrain relief depicted on sectional charts? What information about them would you expect to find in a sectional chart?

8.   Refer to your San Francisco sectional chart. What is the highest obstruction (above ground level) within a 5-NM radius of Hanford Airport (approximately 36°19'N, 119°38'W)? What kind of obstruction is it? How high is it above the ground?

9.   Locate the quadrangle on your San Francisco sectional chart in which the city of Stockton is located (approximately 37°54'N, 121°15'W). At a quick glance, can you tell how high above sea level the highest obstruction is? How is this reported on the map?

10.   Find Fresno Air Terminal on your San Francisco sectional
chart (approximately 36°46'N, 119°43'W). Explain what each of
the various numbers and symbols above and below the name mean.

11.   Name two major limitations to pilotage as a primary means of
air navigation.

12.   What is bracketing? Explain why it is an important
procedure to follow in pilotage.

13.   Why is dead reckoning considered a means of coordinating
other methods of navigation?

14.   Given:

              True course   280°
              Variation   15° east
              Cruising altitude   8,500 ft. MSL
              Winds at 9,000 ft.   240° at 35 kts.
              True airspeed   110 kts.

Find the magnetic heading, the correct wind angle, and the
groundspeed.

15. Suppose you plan a flight of 225 statute miles at an
anticipated groundspeed of 123 MPH. The airplane has 36 gal. of
usable fuel on board and a fuel consumption rate of 9 gal./hr.
How much fuel will you have left when you land at your
destination? How much flying time does this represent?

16. Assume you depart Hays, Kansas, at 1330 CDT for a 2-hr.
flight to Colorado Springs, Colorado. What would be your landing
time, expressed in GMT, or zulu time?

17. Suppose the OAT is 68° F and you are flying at a CAS of 120
kts. at a pressure altitude of 6,500 ft. MSL. What is your TAS
in kts.? What is your TAS in MPH?

18. For this problem, refer to your San Francisco sectional
chart. You are going to fly a cross-country trip from Salinas
Airport (approximately 36°40'N, 121°36'W) to Fresno Chandler
Airport. Given:

Usable fuel	29 gal.
Fuel consumption	5.3 gal./hr.
Cruise altitude	7,500 ft.
TAS	91 kts.
Winds at 6,000 ft.	210° at 25 kts.
Winds at 9,000 ft.	230° at 35 kts.

   a. How far is it from Salinas to Fresno?

   b. What is your true course?

   c. What is your magnetic course?

d.   What is the wind correction angle and magnetic heading?

e.   What will your groundspeed be? How long will it take you to fly the trip?

19.   Given the same conditions as in question 18, answer the following questions for a return trip to Salinas. Use a cruise altitude of 6,500 ft. and assume the winds at 6,000 ft. are representative of your proposed altitude.

a.   What is your true course?

b.   What is your magnetic course?

c.   What is the WCA? What is the magnetic heading?

d.   What will your groundspeed be? How long will it take you to complete the trip?

e.   How much fuel will you burn?

REVIEW QUESTIONS

(NOTE: Because the FAA combines questions on pilotage, dead reckoning, and radio navigation in one section, many of the questions that would fall in this chapter appear at the end of the next chapter on radio navigation.)

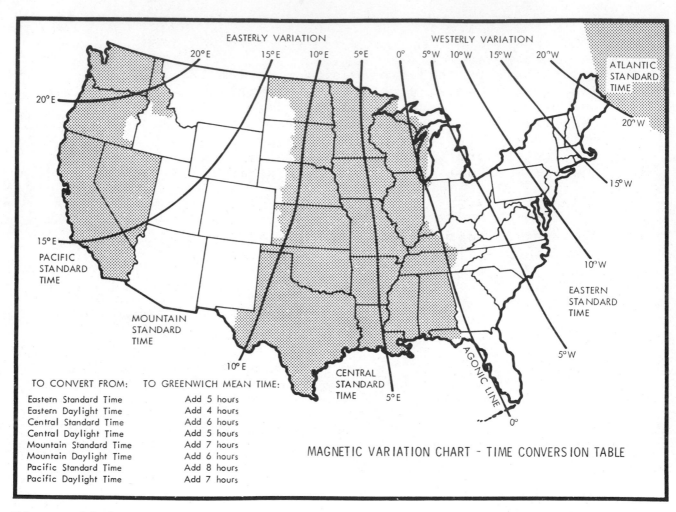

Figure 12.1

1. (FAA 1244) Refer to Figure 12.1. An aircraft departs an airport in the eastern daylight time zone at 0945 EDT for a 2-hr. flight to an airport located in the central daylight time zone. At what Greenwich mean time should the landing be?

1--1145Z.   2--1345Z.   3--1445Z.   4--1545Z.

2. (FAA 1245) Refer to Figure 12.1. An aircraft departs an airport in the central standard time zone at 0930 CST for a 2-hr. flight to an airport located in the mountain standard time zone. What should the landing time be?

1--0930 MST.   2 1030 MST.   3--1130 MST.   4--1230 MST.

3.   (FAA 1246 DI) Refer to Figure 12.1. An aircraft departs an airport in the central daylight time zone at 0845 CST for a 2-hr. flight to an airport located in the mountain daylight time zone. At what Greenwich mean time should the landing be?

1--1345Z.   2--1445Z.   3--1545Z.   4--1645Z.

4.   (FAA 1247) Refer to Figure 12.1. An aircraft departs an airport in the central standard time zone at 0730 CST for a 3-hr. flight to an airport located in the mountain standard time zone. What should the landing time be?

1--0830 MST.   2--0930 MST.   3--1000 MST.   4--1030 MST.

5.   (FAA 1248) Refer to Figure 12.1. An aircraft departs an airport in the Pacific standard time zone at 1230 PST for a 3-hr. flight to an airport located in the central standard time zone. At what Greenwich mean time should the landing be?

1--1630Z.   2--2130Z.   3--2330Z.   4--0030Z.

6.   (FAA 1249) Refer to Figure 12.1. An aircraft departs an airport in the mountain standard time zone at 1615 MST for a 2-hr., 15 min., flight to an airport located in the Pacific standard time zone. What should be estimated time of arrival at the destination airport?

1--1630 PST.   2--1730 PST.   3--1830 PST.   4--1930 PST.

7. · (FAA 1250) Refer to Figure 12.1. An aircraft departs an airport in the Pacific standard time zone at 1030 PST for a 4-hr. flight to an airport located in the central standard time zone. At what Greenwich mean time should the landing be?

1--2030Z.   2--2130Z.   3--2230Z.   4--2330Z.

8.   (FAA 1251) Refer to Figure 12.1. An aircraft departs an airport in the mountain standard time zone at 1515 MST for a 2-hr., 30-min., flight to an airport located in the Pacific standard time zone. What is the estimated time of arrival at the destination airport?

1--1645 PST.   2--1745 PST.   3--1845 PST.   4--2345 PST.

9.   (FAA 1252) In the conterminous United States, sectional charts are updated each

1--6 mo.   2--8 mo.   3-12 mo.   4--24 mo.

10. (FAA 1253) What is the highest cruising altitude that may be selected from Parker Airport (f) to Bullhead City Airport (b)? See Figure 12.2.

1--9,500 ft.   3--13,500 ft.
2--10,500 ft.   4--14,500 ft.

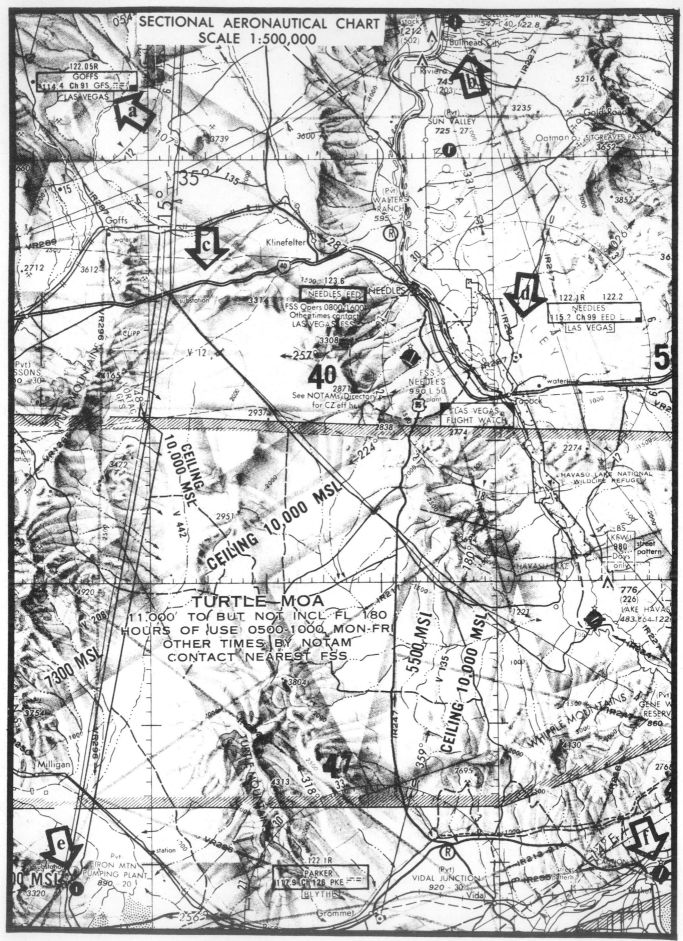

SECTIONAL AERONAUTICAL CHART
SCALE 1:500,000

Figure 12.2

11. (FAA 1254) What VFR cruising altitude(s) is (are) available
on V208 southwest bound from Needles VORTAC (d) in Figure 12.2?

1--8,500 ft. MSL only.
2--7,500, 9,500 ft. MSL only.
3--8,500, 10,500, 12,500, 14,500, and 16,500.
4--7,500, 9,500, 11,500, 13,500, 15,500, and 17,500.

12. (FAA 1255) What is the magnetic course from Goffs VORTAC
(a) to Parker Airport (f)? (See Figure 12.2.)

1--115°.   2--129°.   3--143°.   4--157°.

13. (FAA 1256) Determine the groundspeed for a flight from
Parker Airport (f) to Bullhead City Airport (b) in Figure 12.2.
The wind is from 030° at 30 kts. and the TAS is 165 kts.

1--138 kts.   2--142 kts.   3--146 kts.   4--187 kts.

14. (FAA 1257) If an aircraft crosses over Bullhead City
Airport (b) at 1557 and Highway 40 at 1606, at what time should
it arrive over Iron Mountain Airport (e)? (See Figure 12.2.)

1--1623.   2--1627.   3--1631.   4--1640.

15. (FAA 1258) Determine the magnetic heading from Bullhead
City Airport (b) to Iron Mountain Airport (e)? (See Figure
12.2.) The wind is from 300° at 25 kts. and the average TAS is
145 kts.

1--180°.   2--190°.   3--200°.   4--205°.

16. (FAA 1259 DI) What is the magnetic heading for a flight
from Parker Airport (f) to Bullhead City (b), Figure 12.2, if the
wind is from 210° at 25 kts. and the TAS is 170 kts.

1--227°.   2--237°.   3--241°.   4--255°.

17. (FAA 1260 DI) If an aircraft crosses over Goffs VORTAC (a)
at 0946 and then over Highway 40 (c) at 0955, what time should it
arrive over Parker Airport (f)? (See Figure 12.2.)

1--1016.   2--1023.   3--1030.   4--1037.

18. (FAA 1261 DI) What lighting service is indicated for
Fairfield Airport (e) in Figure 12.3?

1--Pilot controlled lighting.
2--Lighting on prior request.
3--Beacon sunset to sunrise and runway lights on request.
4--Beacon and runway lights limited to certain hours during the
    night.

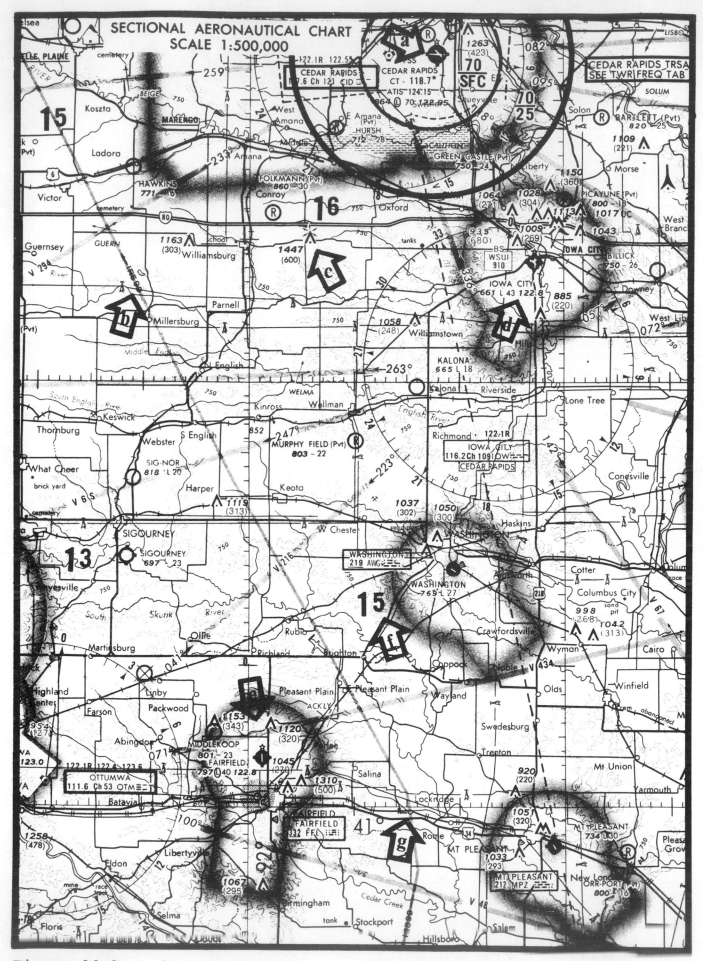

Figure 13.3

19. (FAA 1262) The symbol depicted at (c) in Figure 12.3 represents

1--a TV or radio tower.
2--a rotating beacon light.
3--an obstruction with high-intensity lights.
4--a visual checkpoint used in conjunction with the TRSA.

20. (FAA 1264) What does the symbol 1^5 (f) in Figure 12.3 represent?

1--Minimum obstacle clearance altitude for VFR flight.
2--Maximum elevation figure for the quadrangle bounded by the ticked lines of latitude and longitude.
3--The degree of change in variation over an average of 10 years.
4--The floor of the control area where not designated at 1,200 ft.

21. (FAA 1265 DI) What is the height of the obstruction depicted approximately 2 mi. northwest of the Iowa City Airport (d) in Figure 12.3?

1--269 ft.   2--680 ft.   3--935 ft.   4--1,009 ft.

22. (FAA 1269) What military flight operations are conducted on IR 599 (b)? (See Figure 12.3.)

1--RNAV instrument training flights in IFR weather conditions.
2--Instrument rotorcraft training flights above 1,500 ft. AGL.
3--Training flights below 1,500 ft. AGL in VFR conditions at speeds in excess of 150 kts.
4--Training flights above 1,500 ft. AGL under IFR regardless of the weather and at speeds in excess of 250 kts.

23. (FAA 1270) Which service is provided for aircraft operating within the Cedar Rapids TRSA (a)? (See Figure 12.3.)

1--Priority for participating aircraft for the purpose of vectors, sequencing, landing, and takeoff.
2--Separation between all participating VFR and IFR aircraft.
3--Radar vectoring and separation between all aircraft.
4--Radar vectoring and sequencing of all traffic to all of the airports within the TRSA.

24. (FAA 1271 DI) Which is a proper procedure for entry into the Cedar Rapids TRSA (Figure 12.3) for landing at (a)?

1--Contact the tower on 118.7 MHz at least 10 mi. from the airport and request landing instructions.
2--Monitor the FSS broadcast at 15 min. past the hour on 117.6 MHz for proper frequencies and instructions.
3--Monitor the ATIS broadcast on 118.7 MHz for terminal information and frequencies for contacting approach control.
4--Contact Cedar Rapids Approach Control on 122.95 MHz and request a vector to the traffic pattern.

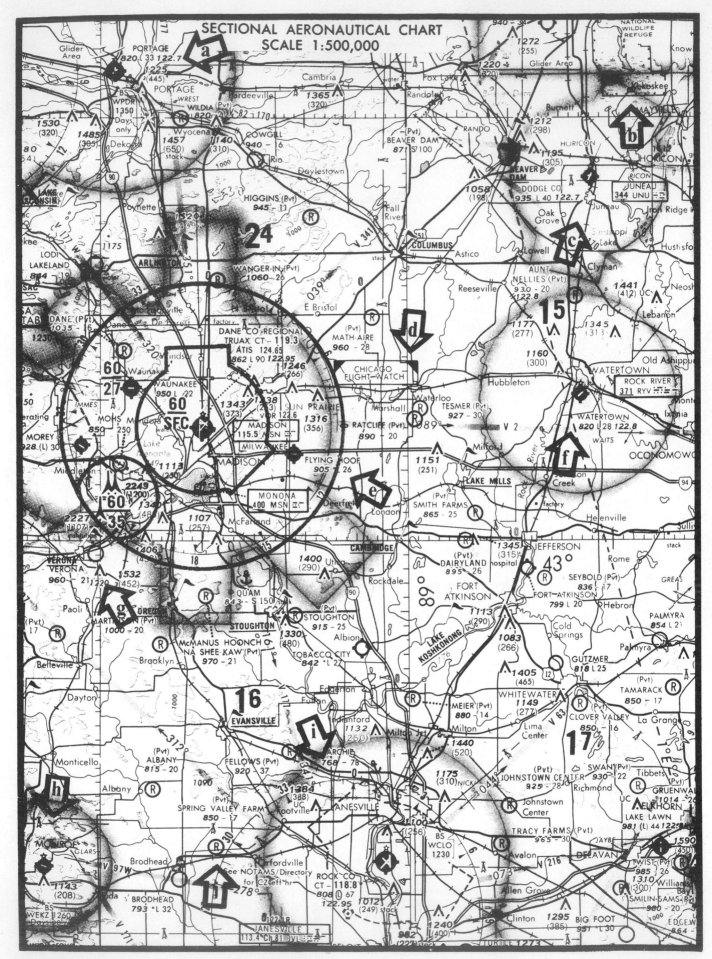

Figure 12.4

25.   (FAA 1273)   The top of the obstruction (g) in Figure 12.4 is

1--1,220 ft. MSL.   3--1,532 ft. MSL.
2--1.220 ft. AGL.   4--1,532 ft. AGL.

26.   (FAA 1274)   What minimum altitude is requested for aircraft
over National Wildlife Refuges such as the one depicted at (b)?
(See Figure 12.4.)

1--500 ft. AGL.   3--2,000 ft. AGL.
2--1,000 ft. AGL.   4--3,000 ft. AGL.

27.   (FAA 1275)   What is the base of the TRSA at Madison (e) in
Figure 12.4 over Flying Hood Airport?

1--Surface.   3--2,105 ft. MSL.
2--1,605 ft. MSL.   4--2,700 ft. MSL.

28.   (FAA 1276 DI)   The lighting at Rock County Airport (i) in
Figure 12.4 is

1--pilot controlled.
2--operated part-time.
3--available on request.
4--limited to the beacon light.

29.   (FAA 1277)   The only control zone(s) on the chart excerpt
is(are) at (see Figure 12.4)

1--Madison (e).
2--Madison (e) and Anesville (i).
3--Madison (e), Anesville (i), and Horicon (c).
4--Madison (e), Anesville (i), Horicon (c), Watertown (f),
   Portage (a), and Monroe (h).

30.   (FAA 1278)   What UNICOM frequency, if any, is indicated for
Rock County Airport (i) in Figure 12.4?

1--None is listed.   3--122.8 MHz.
2--118.8 MHz.   4--122.95 MHz.

31.   (FAA 1279)   According to the contours on the excerpt of the
chart (Figure 12.4), the highest terrain elevation is

1--1,000 ft.
2--2,400 ft.
3--between 750 and 1,000 ft.
4--between 1,000 and 1,250 ft.

32.   Navigation by figuring one's position through reference to
lines of position and time-distance computation is called

1--celestial navigation.   3--inertial navigation.
2--dead reckoning.   4--pilotage.

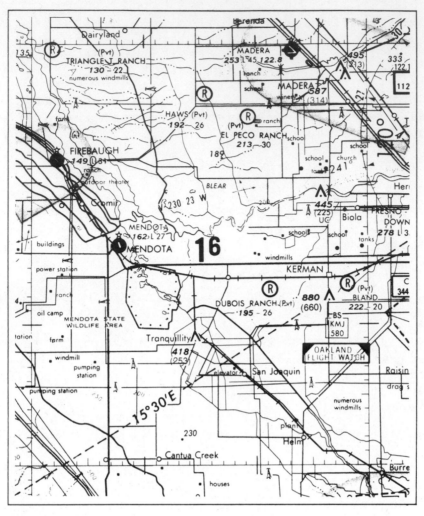

Figure 12.5

33. Refer to Figure 12.5. The maximum elevation of the terrain
and obstructions such as towers within the quadrangle bounded by
the ticked lines of latitude and longitude is

1--495 ft. AGL.   3--1,600 ft. AGL.
2--880 ft. MSL.   4--1,600 ft. MSL.

34. The large number (1^6) in the center of the quadrangle in
Figure 12.5

1--does not include the maximum elevation of vertical
    obstructions within the area.
2--indicates the base of the controlled airspace over the area.
3--is a maximum elevation figure, including terrain and
    obstructions.
4--is the latitude and longitude coordinate of the area bounded
    by the ticked lines.

35.   Refer to the obstruction near the town of Tranquillity in Figure 12.5.   The top of the obstruction is

1--253 ft. AGL.   3--418 ft. AGL.
2--253 ft. MSL.   4--1,600 ft. MSL.

36.   Refer to the obstruction south of Kerman in Figure 12.5. Select the true statement regarding this obstruction.

1--It is a free-standing tower with no guy wires, the base of
    which is 880 ft. MSL.
2--The top of the obstruction is 660 ft. AGL.
3--This is a group obstruction, the tops of which are 600 ft.
    AGL.
4--This is a single obstruction with guy wires, the top of which
    is 880 AGL.

37.   Refer to the obstructions near Madera in Figure 12.5.   Which of the following statements is correct?

1--Each of the these is a single obstruction, neither of which is
    more than 1,000 ft. MSL.
2--This is a group obstruction; the base of one is 213 ft. AGL
    and the base of the other is 314 ft. AGL.
3--This is a group obstruction; the tops are less than 1,000 ft.
    AGL.
4--This is a group obstruction; the maximum top is 587 ft. MSL.

38.   Refer to Figure 12.5.   Assume you are flying over the city of Madera from east to west.   In accordance with regulations, which of the following altitudes would be the minimum safe altitude to fly over the highest obstruction shown?

1--600 ft. AGL.   3--1,400 ft. AGL.
2--1,000 ft. MSL.   4--3,000 ft. AGL.

39.   Refer to Figure 12.5.   What statement, if any, about Madera Airport is correct?

1--FSS at this airport operates on 122.8.
2--Landing lights are available on request.
3--The airport is 253 ft. AGL.
4--None of the statements is correct.

40.   Refer to Figure 12.5.   Suppose the following surface aviation weather report has been issued for Madera.   At what indicated altitude above Madera would you expect to find the base of the clouds?

MAD SA 1351 80 BKN 25 090/29/20/0000/975

1--800 ft. MSL.   3--8,000 ft. MSL.
2--1,053 ft. MSL.   4--8,253 ft. MSL.

41. Contour lines placed on a sectional aeronautical chart are
to show points of the same

1--MSL elevation.   3--longitude.
2--latitude.   4--variation.

42. A star (*) at the top of an airport symbol on a sectional
chart indicates that

1--high-performance airplanes are permitted to land.
2--it has a rotating beacon in operation from sunset to sunrise.
3--services and fuel are available.
4--these are military airports.

43. Assume an airplane is serviced with 38 gal. of usable fuel,
and an average groundspeed of 138 MPH is anticipated on a flight
of 260 statute miles. At a fuel consumption rate of 12 gal./hr.,
what would be the maximum flying time available with the fuel
remaining after reaching your destination?

1--1 hr., 2 min.   3--2 hr., 5 min.
2--1 hr., 17 min.   4--2 hr., 30 min.

44. Refer to Figure 12.6. Given variation of 17° east in this
area, what is the true course from Olivehurst (A) to Yolo Co.
(B)?

1--180°.   2--186°.   3--197°.   4--203°.

45. Refer to Figure 12.6. Given variation of 17° east in this
area, what is the magnetic course from Yolo Co. (B) to Auburn
(C)?

1--41°.   2--44°.   3--58°.   4--61°.

46. Refer to Figure 12.6. What is the approximate distance from
Olivehurst (A) to Auburn (C)?

1--25 NM.   3--50 NM.
2--28 NM.   4--57 NM.

47. Refer to Figure 12.6. Suppose you fly the traffic pattern
at 1,000 ft. AGL at Auburn Airport (C). If the altimeter is
properly adjusted to the latest altimeter setting, it would
indicate a pattern altitude of

1--1,000 ft.   3--2,520 ft.
2--2,390 ft.   4--2,710 ft.

48. Refer to Figure 12.6. Given a TAS of 130 MPH, 17° east
variation, forecasted winds from 110° at 15 kts., what is the
magnetic heading and GS from Auburn (C) to Yolo Co. (B)?

1--215°, 140 MPH.   3--238°, 147 kts.
2--218°, 144 MPH.   4--252°, 138 kts.

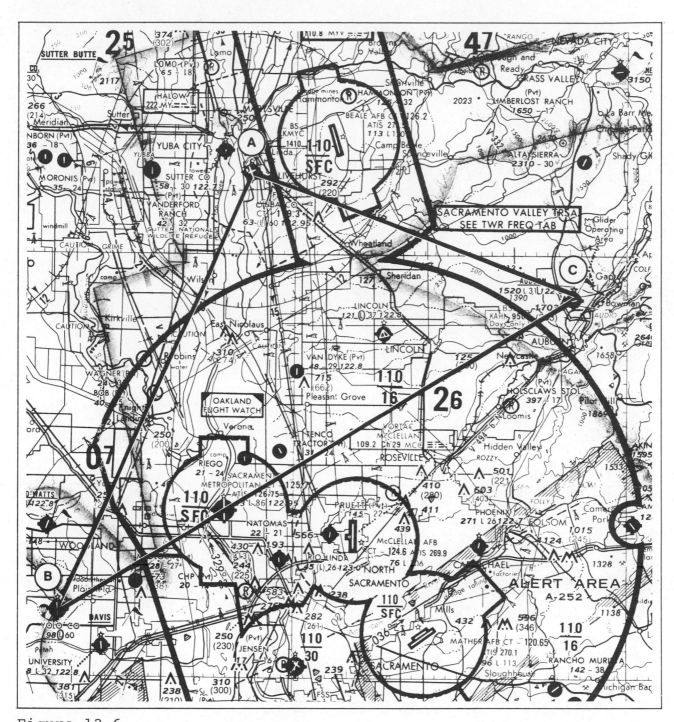

Figure 12.6

49. Refer to Figure 12.6. Given a true airspeed of 109 MPH, variation of 17° east, a compass deviation of +1°, and winds from 210° at 20 kts., calculate the compass heading and time en route from Olivehurst (A) to Auburn (C).

1--83°, 18 min.   3--111°, 25 min.
2--107°, 15 min.   4--141°, 18 min.

50. Refer to Figure 12.6. What is the true course from Olivehurst (A) to Yolo Co (B), from Yolo Co (B) to Auburn (C), and from Auburn (C) to Olivehurst (A)? Variation in this area is 17° east.

1--185°; 45°; 274°.   3--202°; 59°; 291°.
2--195°; 55°; 284°.   4--219°; 79°; 308°.

ANSWERS

Key Terms and Concepts, Part 1

1. O	6. J	11. G	16. C
2. Q	7. F	12. A	17. P
3. B	8. R	13. D	18. S
4. L	9. E	14. H	19. T
5. M	10. K	15. N	20. I

Key Terms and Concepts, Part 2

1. K	4. L	7. J	10. G
2. I	5. E	8. C	11. A
3. H	6. F	9. B	12. D

Discussion Questions and Exercises

4.   F--It is just the opposite.
5.   T--Longitude lines are all the same length and converge at the poles, while lines of latitude are parallel to one another and become shorter as they apporach the poles.
6.   F--Lines of latitude do not converge; they run parallel to one another.
9.   The highest obstruction, a tower (KNGS radio), is 305 ft. AGL and 553 ft. MSL.
10. 3,700 ft. MSL, as indicated by the maximum elevation figures (MEF). MEFs use large numbers to represent thousands of feet and small numbers to represent hundreds of feet above MSL of the highest obstruction in the quadrangle bounded by the ticked lines of longitude and latitude.
11. The control tower frequency is 118.2. Automatic Terminal Information Service (ATIS) is broadcast on 110.3 and 119.3. The field is 332 ft. above sea level (MSL), the

longest runway is 9,200 ft., and UNICOM is available on
122.95. The FSS above the name indicates that there is a
Flight Service Station located on the field.

14.   The true heading is 280°. From that you subtract the
        wind correction angle (WCA) of -11° and the easterly
        variation (VAR) of -15° to obtain a magnetic heading (MH)
        of approximately 254°.

15.   On your trip, you will travel 1.83 hr. (225/123), or
        about 110 min. (1 hr., 50 min.), and you will burn 16.5
        gal. (1.83 x 9) of fuel. The fuel remaining is 19.5 gal.
        (36 - 16.5), which represents about 2.17 hr. (19.5/9), or
        2 hr., 10 min. To convert from a decimal representation
        of hours, such as 2.17, to hours and minutes, you need to
        multiply the decimal portion (.17) times 60 to determine
        the number of minutes.

16.   Your zulu departure from Hays would be 1330 + 5 hr., or
        1830Z. You would land at Colorado Springs 2 hr. later,
        or 2030Z. To look at it another way, when you land at
        Colorado Springs, the local time would be 1330 + 2 hr.
        (flying time) - 1 hr. (time zone change), or 1430 local
        time. Adding 6 hr. gives you 2030Z.

17.   Converting 68° F to Celsius yields 20° C. TAS is 136
        kts., which converts to 156 MPH.

18. a. 87 NM; use your plotter (NM on the sectional chart side).
     b. 87°; use your plotter and plot a true course.
     c. 87° (TC)-15° (VAR) = 72°, the magnetic course.
     d. To determine the winds at 7,500 ft., interpolate between
        6,000 and 9,000 ft. In this example interpolation is
        easy since 7,500 ft. is exactly one-half way. Use 220°
        at 30 kts. The wind correction (WCA) will be about 14°
        to correct for the effects of the southwesterly winds.
        So, to obtain your magnetic heading (MH), you must
        compute 87° (TC) + 14° (WCA) - 15° (VAR) to obtain 86°
        (MH).
     e. Use your navigation computer or calculator to determine
        your groundspeed, 109 kts. Divide 85 kts. by 109 to find
        your estimated time en route, .78 hr. or about 47 min.
     f. Multiply .78 hr. x 5.3 gal./hr. to determine your fuel
        consumption, 4.1 gal.

19. a. 267°.
     b. 267° (TC) - 15° (VAR) = 252°, your magnetic course.
     c. 267° (TC) - 14° (WCA) - 15° (VAR) = 238° (MH).
     d. 75 kts.; 1.16 hr., or 1 hr., 10 min.
     e. 6.1 gal.

Review Questions

1. 4--0945 EDT + 4 hr. = 1345Z departure. 1345 + 2 hr. = 1545Z
        arrival.
2. 2--0930 CST = 0830 MST. 0830 MST + 2 hr. = 1030 MST.
3. ?--This question has been designated as unusable by the FAA
        and removed from the FAA Question Selection Sheets.
4. 2--0730 CST = 0630 MST. 0630 MST + 3 hr. = 0930 MST.
5. 3--1230 PST + 8 hr. = 2030Z + 3 hr. = 2330Z.

6. 2--1615 MST = 1515 PST + 2 hr., 15 min. = 1730 PST.
7. 3--1030 PST + 8 hr. = 1830Z + 4 = 2230Z.
8. 1--1515 MST = 1415 PST + 2 hr., 30 min. = 1645 PST.
9. 1--Sectional charts are updated every 6 mo.
10. 4--Your true course (TC) is 347°. The variation numbers on
       the chart indicate that VAR is 14°E, which you must
       subtract from your TC, to give you a MC of 333°. Since
       you are flying a westerly heading, use even thousand feet
       plus 500 ft. The highest alternative is 14,500 ft. MSL.
11. 1--The airway ceiling is 10,000 ft. MSL, and you are on a
       westerly heading on V208, so use even thousand feet plus
       500 ft. The only altitude available is 8,500 ft. MSL.
12. 2--The magnetic course is read from the compass rose on the
       Goffs VORTAC. We will cover VORTACs and radio navigation
       in Chapter 13. The value I read is 128°. You can also
       use your plotter to determine the TC (142°), from which
       you subtract easterly variation of 14° to arrive at a
       magnetic course of 128°.
13. 2--The true course (TC) is 347°. Using your flight computer
       or electronic calculator, determine the groundspeed at
       142 kts.
14. 3--It takes the aircraft 9 min. to travel 18 NM, so it is
       going 200 kts. It will take it 25 min. to travel the
       next 50 NM. You can also use the same logic using MPH.
15. 3--The true course is 204°. The wind correction angle (WCA)
       is +10°. The easterly variation is -14°. MH is 200°.
16. ?--This question has been designated as unusable by the FAA
       and removed from the FAA Question Selection Sheets.
17. ?--This question has been designated as unusable by the FAA
       and removed from the FAA Question Selection Sheets.
18. ?--This question has been designated as unusable by the FAA
       and removed from the FAA Question Selection Sheets.
19. 3--This symbol indicates an obstruction with high-intensity
       lights.
20. 2--The maximum elevation in the quadrangle is 1,500 MSL.
       The large number represents thousands of feet and the
       small number indicates hundreds of feet.
21. ?--This question has been designated as unusable by the FAA
       and removed from the FAA Question Selection Sheets.
22. 4--The "IR" means that above 1,500 ft. AGL, there are IFR
       training flights regardless of weather conditions.
23. 2--Separation in the TRSA is provided for all **participating**
       IFR and VFR traffic. VFR participation is voluntary.
24. ?--This question has been designated as unusable by the FAA
       and removed from the FAA Question Selection Sheets.
25. 3--The elevation of the obstruction given in bold numbers is
       1,532 MSL. The height of the obstruction, indicated by
       the number in parentheses, is 452 ft. AGL.
26. 3--Aircraft are requested to stay at least 2,000 ft. AGL
       when flying over a National Wildlife Refuge.
27. 4--Flying Hoff is in the second tier of the Madison TRSA,
       which is denoted by the elevation figures 60/27. The 27
       refers to 2,700 ft. MSL and defines the base of the TRSA.
       The 60 refers to 6,000 ft. MSL and defines the top.

28. ?--This question has been designated as unusable by the FAA
     and removed from the FAA Question Selection Sheets.

29. 2--Control zones are shown with dashed "-----" lines, which
     are almost impossible to see around Madison Airport.
     Also, Wisconsin residents please note that "Anesville"
     should be "Janesville."

30. 4--The UNICOM frequency shown on the chart is 122.95 MHz.

31. 4--Good luck reading the terrain contours. 2,400 ft. MSL
     would make sense given the maximum elevation figure in
     the northwest quadrangle of the excerpt. However, the
     contour lines indicate that the highest terrain elevation
     is between 1,000 and 1,250 ft.

32. 2--LOPs and time-distance computations are the basis of dead
     reckoning.

33. 4--Large numbers in the center of a quadrangle (referred to
     as the maximum elevation figure, MEL) indicate that the
     highest elevation is 1,600 ft. MSL.

34. 3--See answer 33.

35. 1--MSL is indicated in bold figures; AGL is indicated in
     parentheses.

36. 2--MSL is indicated in bold figures; AGL is indicated in
     parentheses.

37. 1--These are separate obstructions. One is 495 MSL and
     (213) AGL; the other is 587 MSL and (314) AGL.

38. 3--According to FARs, the minimum required altitude over an
     obstruction is 1,000 ft. The tallest obstruction is 314
     ft. AGL, so 1,400 ft. AGL would satisfy the requirement.

39. 4--No FSS facility is available at this airport (it would be
     indicated by the letters FSS); landing lights are in
     operation from sunset to sunrise at Madera and do not
     need to be requested; and the airport is 253 ft. MSL.

40. 4--The report indicates that the clouds are at 8,000 ft. and
     broken. This measure of the cloud base is referenced to
     ground level, so you must add field elevation (253 ft.)
     to determine the indicated altitude.

41. 1--Contour lines indicate points of equal elevation above
     sea level (MSL).

42. 2--The star (*) means that the airport has a rotating beacon
     in operation from sunset to sunrise.

43. 2--Divide distance (260 mi.) by speed (138 MPH) to determine
     time (1.88 hr.). Multiply time (1.88 hr.) by the fuel
     flow (12 gal./hr.) to determine total consumption (22.6
     gal.). This leaves 15.4 gal, which divided by 12
     gal./hr. leaves 1.28 hr. of unburned fuel. Converting
     .28 hr. to 17 min. (.28 x 60) yields an answer of 1 hr.,
     17 min. Another way to solve the problem is to figure
     total endurance (38 hr./12 gal./hr.), or 3.16 hr. (3 hr.,
     10 min.). Subtract 1.88 from 3.16 to get 1.28 hr.

44. 2--203° (TC) - 17° (VAR) = 186° (MC).

45. 3--58° (TC).

46. 1--Use your plotter to measure 25 NM or 29 statute miles.
     Be sure to use the sectional side of the plotter.

47. 3--The field elevation is 1,520 ft., to which you add 1,000
     ft. to determine the indicated pattern altitude.

48.   1--Your TC is 238°, wind speed is 17.25 MPH; groundspeed is
        140 MPH, and the WCA is -6°. 238° (TC) -6° (WCA) - 17°
        (VAR) = 215° (MH).
49.   2--TC is 111°, the wind is 20 kts. (23 MPH), the WCA is 12°,
        and the groundspeed is 110 MPH. 111° (TC) + 12° (WCA) -
        17° (VAR) = 106° (MH). 106° (MH) + 1° (DEV) = 107° (CH).
        Groundspeed (110 MPH) divided into distance (28 mi.)
        yields roughly .25 hr., or 15 min.
50.   3--Use your plotter to determine the TC for each segment.

13/RADIO NAVIGATION AIDS

MAIN POINTS

    1.   Radio navigation allows you to track predetermined LOPs and to fly to and from a radio source.

    2.   **Very-high-frequency (VHF) omnidirectional ranges** or **VORs** broadcast bearings called **radials** for all 360° of the compass. Stations using ultra-high frequencies (UHF) are called **tactical air navigation** stations, or **TACAN**, and are used by the military. They also have **distance measuring equipment** or **DME**. Most VOR and TACAN stations have been combined and are called **VORTACs**. Stations similar to commercial radio stations use low and medium frequencies to broadcast signals for direction-finding purposes. They are called **nondirectional beacons** or **NDBs** and are part of the **automatic direction finding (ADF)** system.

    3.   VORs transmit on frequencies between 108.0 and 117.95 MHz. Each radial is named for its magnetic course **from** the station. Magnetic course and the VOR radial are the same only when flying outbound from the station; when flying inbound to the station, the magnetic course is the reciprocal of the VOR radial. VHF airways, called **Victor airways**, connect VOR stations.

    4.   VOR cockpit controls include a tuning knob to obtain the desired frequency, an **omnibearing selector (OBS)** to obtain the desired radial, a **course deviation indicator (CDI)** to measure deviations from the radial, a **TO/FROM indicator** to determine whether the course selected takes you to or from the VOR, and an **Off flag** to indicate when the station level is too weak. The Off flag will indicate that the signal is too weak when the station is off, when you are too far away, when you are passing directly overhead (the zone of confusion), or when you are 90° off the radial selected in the OBS. VOR receivers also have an audio channel with Morse code or voice identifiers. You should always verify that the station is operating and is tuned correctly by listening for its identifier.

5.   VOR navigation provides a great deal of flexibility with its omnidirectional characteristics; it is relatively free from atmospheric interference, and it is relatively accurate (within 1° of the magnetic course), as long as the receiver is functioning properly. The main disadvantage of VOR navigation is that its signals operate on a line-of-sight basis, thus giving them a fairly limited range and necessitating a large network of stations.

6.   There are a number of ways to use a VOR.
Always tune and identify the station by selecting the desired frequency and listening to the Morse code or voice identifier. Monitor the station as long as you are using it. Select the desired OBS position and check the position of the Off flag.
If you simply want to fly to the station, turn the OBS until the CDI is centered and you have a TO indication. Then fly the magnetic heading indicated. When the needle moves off center, recenter it and select a new course with the OBS. This is not very efficient since it involves a curved flight path unless you have a direct headwind or tailwind.
If you wish to **proceed direct** to the station, center the CDI with a TO indication, fly the indicated heading, and then keep the CDI indicator centered. When the CDI moves off center (due to crosswind), turn the airplane into the wind (toward the CDI needle). Make minor changes until have you have determined the amount of crosswind correction needed to keep the needle centered.
To **intercept a radial** you need to establish an **intercept heading**. The angle the aircraft makes with the radial is called the **angle of interception**, and how fast you approach it is called the **rate of interception**. After setting the inbound course on the OBS with a TO indication, turn to the same heading as the inbound course. Note if the CDI points to the left or right and turn toward the needle to establish an intercept angle. If the CDI moves slowly, proceed slowly until you intercept it; if it moves rapidly, cut the intercept angle to slow the needle's movement.
To **intercept an outbound radial**, set the course and check for a FROM indication. Fly the heading of the outbound course to an intercept heading and follow the intercept procedures outlined above.
Station passage occurs when you pass through the **cone of confusion** above the station. Station passage is confirmed when the TO/FROM indicator makes its first positive change from TO to FROM.
Time-distance checks can also be made using the VOR by timing how long it takes to fly a 10° bearing change while heading approximately 90° to the radial. The time in seconds between bearings divided by the degree of bearing change (in this procedure, 10) tells you how many minutes you are from the station. Multiply this figure by your speed and you can calculate how far away the station is.
To **establish a VOR fix** once you have established a course with the CDI centered, tune and identify a second station. Turn

the OBS until the CDI is centered and the TO/FROM indicator
states FROM. Read the radial from the OBS and use it to
establish your fix. You can use any two stations to obtain a
fix, but the fix may not be totally accurate unless you can use
two VORs simultaneously. If you know the distance from the
station, you can use a single VOR to obtain a single-station fix.
Since you know the distance to the station, you can also
calculate groundspeed by dividing the distance by how long it
takes you to fly there. Finally, to preflight your VOR
equipment, many airports have VOR **test** (VOT) facilities. VOT
facilities are published in the <u>Airport/Facility Directory</u>. Turn
the OBS until the CDI is centered; it should read 000° with a
FROM indication or 180° with a TO indication (plus or minus 4°).
If a VOT is not available, refer to the <u>Airport/Facility
Directory</u> for airborne or ground checkpoints.

7.   **Distance measuring equipment (DME)** uses distance-fixing
information from the VORTAC system. It measures the slant range,
not horizontal distance, from the source to the aircraft and is
subject to line-of-sight restrictions. DME is valuable not only
for distance readings but also as a source of groundspeed
calculations.

8.   **Automatic direction finding (ADF)** radio compasses use
low frequencies (190 to 1750 kHz) to provide bearing information.
Unlike VHF and UHF signals, they are not limited to line-of-sight
transmission. A serious drawback to ADF navigation, however, is
that the compasses are susceptible to interference.
        A station that broadcasts a low-frequency navigation signal
is called a **nondirectional beacon** or **NDB**. ADF radios have a
bearing indicator, an On-Off and volume control knob, and a
selector knob.
        The first step in ADF navigation is to tune and identify the
station. Magnetic bearing to the station is equal to the
relative bearing indicated on the ADF plus the magnetic heading.
The **magnetic heading** is the heading indicated on the compass.
The **relative bearing** is the direction of the station relative to
the nose of the airplane. The **magnetic bearing** is the magnetic
heading that would point the airplane directly toward the
station. Magnetic heading (MH) + relative bearing (RB) =
magnetic bearing to the station (MB). ADF homing is similar to
VOR homing; it also involves a curved track to the station if
there is any crosswind.
        ADF is usually used as a supplement to VOR navigation--for
example, to establish fixes.

9.   Airway navigation does not always represent the shortest
distance between destinations. An alternative is **area navigation**
(RNAV), which involves creating phantom stations (waypoints)
using VOR and/or DME information with special receiving
equipment.

10.   Radar allows air traffic controllers to monitor traffic
on a radar screen. The system provides vectors to a destination,
but the VFR pilot still has navigational responsibility (vital if

a transponder fails or if service is terminated due to heavy IFR traffic). Radar navigation, rather than being a passive system on the pilot's part, involves several procedures, including VFR cloud and visibility minimums. The transponder has 4096 four-digit codes available for broadcast and several switches. **Standby** is used to warm up the system. **On** activates Mode A operation. **Alt** activates the Mode C encoding altimeter, if installed. **Reply** responds when the unit receives a signal from the ground. **Ident** broadcasts a special signal to the controller. Controllers use the word **squawk** to refer to airborne transponder transmissions. For example, **squawk** means to select the frequency requested by the controller, **squawk ident** means to push the Ident button, **squawk standby** means to turn the selector to Standby, and so on. General transmission codes are 1200 for VFR, 7500 for hijacking, 7600 for loss of two-way radio communication, and 7700 for airborne emergency. Use of precision approach radar (PAR) or airport surveillance radar (ASR) is reserved for IFR pilots or emergencies.

11. Some safety factors to consider in flight planning when using radio navigation include: checking minimum altitudes, since VORs transmit line-of-sight; checking the Airport/Facility Directory and NOTAMs; and checking equipment. Using one method of navigation to complement another (**composite navigation**) is both practical and, for many VFR flights, almost a necessity.

KEY TERMS AND CONCEPTS, PART 1

Match each term or concept (1-16) with the appropriate description (A-P) below. Each item has only one match.

___	1. VOR	___	9. Off flag
___	2. homing	___	10. TACAN
___	3. rate of intercept	___	11. airways
___	4. Victor airway	___	12. radial
___	5. NDB	___	13. relative bearing
___	6. CDI	___	14. TO/FROM
___	7. radio navigation	___	15. omnibearing selector
___	8. magnetic bearing	___	16. VORTAC

A. magnetic heading that would point the airplane directly toward an NDB
B. station that uses low or medium radio frequencies
C. part of the VOR cockpit display that shows the aircraft's deviation from a radial or course
D. how fast you encounter a desired course
E. direction of an NDB station relative to the airplane's nose
F. VHF omnidirectional range station
G. establishing courses and fixes by reference to radio signals broadcast from ground stations
H. part of the VOR cockpit display that designates where the selected radial will take you relative to the VOR
I. a VHF airway connecting two VOR stations
J. bearings that run from one ground station to another

K.   flying toward a radio source
L.   UHF tactical air navigation station
M.   part of the VOR cockpit display that indicates a weak signal
N.   a LOP sent out by a VOR station
O.   combination of "F" and "L" above
P.   part of the VOR cockpit display that allows you to select a
      radial

KEY TERMS AND CONCEPTS, PART 2

       Match each term or concept (1-12) with the appropriate
description (A-L) below. Each item has only one match.

___ 1.   1200   ___ 7.   station passage
___ 2.   cone of confusion   ___ 8.   nondirectional beacon
___ 3.   ADF   ___ 9.   DME
___ 4.   7600   ___ 10.   7700
___ 5.   squawk ident   ___ 11.   composite navigation
___ 6.   intercept heading   ___ 12.   area navigation (RNAV)

A.   navigational system not limited by line-of-sight transmission
B.   using one method of navigation to complement another
C.   this is confirmed when the TO/FROM indicator makes its first
      positive change from TO to FROM
D.   homing beacon that offers low-frequency navigation signal
E.   activate the transponder to broadcast a special signal to air
      traffic control
F.   VFR transponder frequency
G.   heading used to get to a desired radial or course
H.   area above a VOR station where radial signals cannot be
      interpreted correctly
I.   uses UHF to fix distance from a VORTAC system
J.   transponder frequency to indicate an airborne emergency
K.   navigation system in which phantom stations are created using
      VOR and/or DME information
L.   transponder frequency used to indicate a loss of two-way
      radio communication capability

DISCUSSION QUESTIONS AND EXERCISES

1.   What are two major differences between VOR and TACAN radio
transmissions?

2.   How are signals from a VOR station radiated relative to
actual headings? What is the difference between TO and FROM
bearings?

3.   Briefly describe the VOR cockpit controls and their functions.

4.   Name two advantages and two disadvantages of VOR as a navigation aid.

5.   Outline the VOR procedure for each of the following:

   a.   flying direct to the station

   b.   intercepting a course inbound to the station

   c.   intercepting a course outbound from the station

6.   What is the cockpit indication that you have passed over a VOR station?

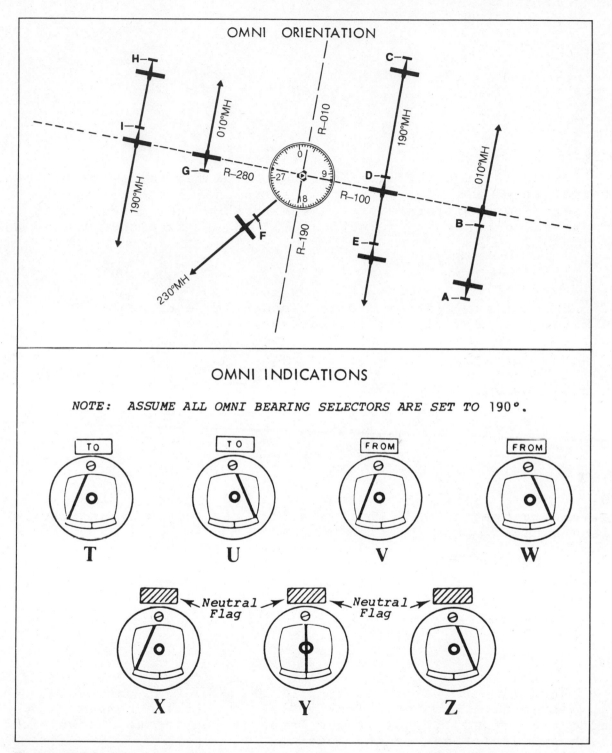

Figure 13.1

7.   Outline how you can use a single VOR to obtain a time and distance check.

8.   Refer to Figure 13.1. Match each VOR indicator in the bottom of the figure with the appropriate airplane in the top of the figure. Some indicators may have more than one correct match. Assume the omnibearing selector (OBS) in all airplanes is set to read 190°.

9.   Briefly explain how distance measuring equipment (DME) works, where it derives radio information, and how station passage is indicated.

10. What is one major advantage of a low-frequency (ADF) navigation over a VOR? What is a low-frequency station's biggest disadvantage?

11. Given an ADF pointing to 190° and a magnetic compass indication of 135°, what is the magnetic bearing to the station? Assume 0° deviation for the heading.

12. Explain each of these as they relate to radar navigation:

   a. 4096

   b. Ident

   c. Standby

   d. squawk code 0413

   e. code 1200

   f. code 7500

   g. code 7600

   h. code 7700

   i. PAR and ASR for VFR pilots

13. Refer to Figure 13.2. Match the correct ADF (fixed compass card) indication in the bottom of the figure with the airplanes shown in flight in the top of the figure for flight in the vicinity of a typical nondirectional radio beacon (NDB). Indicators may have no match, one match, or two matches.

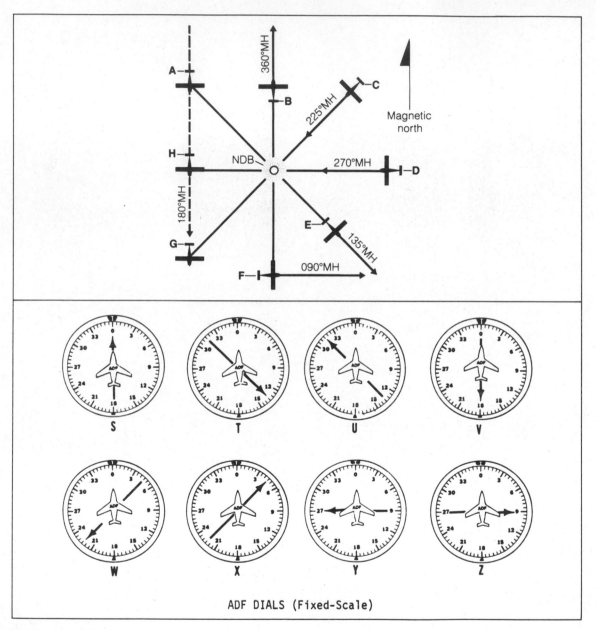

Figure 13.2

REVIEW QUESTIONS

1.   (FAA 1127 DI) When the CDI (course deviation indicator)
needle is centered during an omnireceiver check using a VOT, the
omnibearing selector and the TO/FROM indicator should read

1--180° FROM, only if the pilot is due north of the VOT.
2--0° TO or 180° FROM, regardless of the pilot's position from
    the VOT.
3--0° FROM or 180° TO, regardless of the pilot's position from
    the VOT.
4--0° TO, only if the pilot is due south of the VOT.

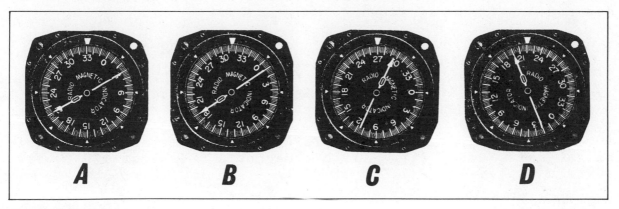

Figure 13.3.

2.   (FAA 1342) Determine the magnetic bearing TO the station as
indicated by RMI A in Figure 13.3.

1--030°.   2--150°.   3--210°.   4--330°.

3.   (FAA 1343) What magnetic heading should the pilot use to fly
to the station as indicated by RMI B in Figure 13.3?

1--010°.   2--134°.   3--190°.   4--316°.

4.   (FAA 1344) Determine the approximate heading to intercept
the 180° bearing TO the station using the information depicted by
RMI B in Figure 13.3.

1--040°.   2--160°.   3--220°.   4--340°.

5.   (FAA 1345) What is the magnetic bearing FROM the station as
indicated by RMI C in Figure 13.3?

1--090°.   2--115°.   3--270°.   4--295°.

6.   (FAA 1346) What RMI indication in Figure 13.3 represents the
aircraft tracking to the station with a right crosswind?

1--A.   2--B.   3--C.   4--D.

7.   (FAA 1347) What outbound bearing is being crossed by the
aircraft represented by RMI A in Figure 13.3?

1--030°.   2--150°.   3--210°.   4--330°.

8.   (FAA 1503) To use VHF/DF facilities for assistance in
locating an aircraft's position, the aircraft must have

1--a VHF transmitter and receiver.
2--an IFF transponder.
3--a VOR receiver and DME.
4--an ELT.

9.   (FAA 1516) What procedure is recommended when climbing or
descending VFR on an airway?

1--Offset 4 mi. or more from center line on the airway before
   changing altitude.
2--Squawk 1400 on the aircraft transponder with Mode C selected.
3--Advise the nearest FSS of the desired altitude change.
4--Climb or descend on the right side of the airway center line.

10. (FAA 1703) Which VFR cruising altitude is acceptable for a
flight on a Victor airway with a magnetic course of 175°? The
terrain is less than 1,000 ft.

1--4,000 ft.   2--4,500 ft.   3--5,000 ft.   4--5,500 ft.

11. (FAA 1704) Which VFR cruising altitude is acceptable for a
flight on a Victor airway with a magnetic course of 185°? The
terrain is less than 1,000 ft.

1--4,000 ft.   2--4,500 ft.   3--5,000 ft.   4--5,500 ft.

12. (FAA 1705) Each person operating an aircraft under VFR in
level cruising flight at an altitude of more than 3,000 ft. above
the surface, and below 18,000 ft. MSL, shall maintain an odd-
thousand plus 500-foot altitude while on a

1--magnetic heading of 180° through 359°.
2--magnetic course of 0° through 179°.
3--true course of 180° through 359°.
4--true heading of 0° through 179°.

13. (FAA 1296) Large numbers, such as 3^2 or 1^6 depicted on
Sectional Aeronautical Charts

1--do not include maximum elevation of vertical obstructions
   within the areas.
2--indicate the base of the controlled airspace over the areas.
3--are maximum elevation figures (including terrain and
   obstructions) shown in quadrangles bounded by ticked lines of
   latitude and longitude.
4--are latitude and longitude coordinates of the areas bounded by
   ticked lines.

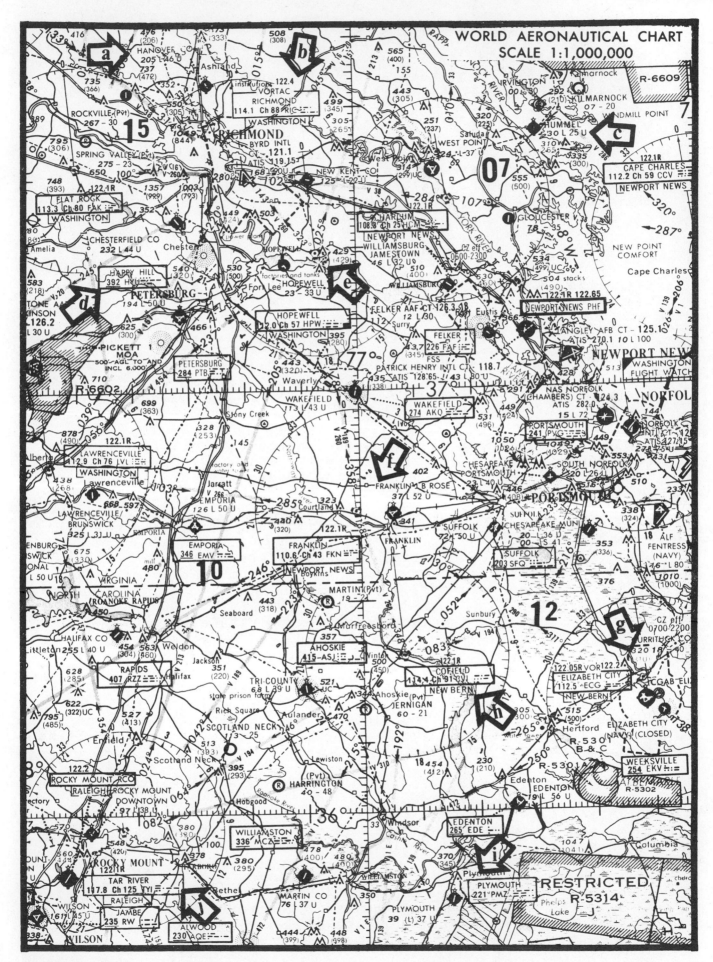

Figure 13.4

14. (FAA 1285) The true course from Hopewell VORTAC (e) to
Franklin VORTAC (f) (Figure 13.4) is

1--171°.   2--178°.   3--183°.   4--189°.

15. (FAA 1286) How much fuel would be saved if the cruising
speed was reduced from 185 kts. to 165 kts., reducing the fuel
consumption from 26.3 gal./hr. to 21.5 gal./hr. on a flight from
Plymouth Airport (i) to Hanover County Airport (a)? (See Figure
13.4.) The wind is calm.

1--1.4 gal.   2--3.5 gal.   3--4.8 gal.   6.4 gal.

16. (FAA 1287) What is the true course from Tar River VORTAC
(j) to Hopewell VORTAC (e)? (See Figure 13.4.)

1--017°.   2--024°.   3--205°.   4--212°.

17. (FAA 1288) Determine the magnetic course from Happy Hill
NDB (d) to Plymouth NDB (i). (See Figure 13.4.)

1--153°.   2--160°.   3--167°.   4--347°.

18. (FAA 1289) Calculate the fuel needed for a flight direct
from Hummel Airport (c) to Tar River VORTAC (j) (Figure 13.4)
under no-wind conditions if the fuel consumption is 14.5 gal./hr.
and the airspeed is 145 kts.

1--5.8 gal.   2--6.6 gal.   3--11.5 gal.   4--13.2 gal.

19. (FAA 1290) What action is appropriate with regard to
Pickett MOA for a southbound VFR flight from Flat Rock VORTAC on
V155 (d)? (See Figure 13.4.)

1--Avoid the MOA and select a cruising altitude above 6,000 ft.
   AGL.
2--Exercise extreme caution when passing through the MOA.
3--Report entering and departing to the controlling agency.
4--File a DVFR flight plan prior to entering the MOA.

20. (FAA 1291) Determine the magnetic heading for a flight from
Hopewell VORTAC (e) to Tar River VORTAC (j) (Figure 13.4) if the
wind is from 080° at 35 kts. and the TAS is 185 kts.

1--195°.   2--201°.   3--208°.   4--215°.

21. (FAA 1294) What is the aircraft position if the OMNI
receivers indicate the 143 radial of Richmond VORTAC (b) and the
323 radial of Hopewell VORTAC (e)? (See Figure 13.4.)

1--Northwest of Byrd International Airport.
2--Between Richmond and Hopewell VORTACs.
3--Over the James River southeast of Hopewell VORTAC.
4--Approaching Hopewell VORTAC on 323° magnetic course.

22. (FAA 1292) To approach Cofield VORTAC (h) (Figure 13.4)
from the southwest on V194, the MONI bearing selector should be
set to

1--066°.   2--083°.   3--194°.   4--246°.

23. (FAA 1293) What is the approximate aircraft position on
V213 between Tar River VORTAC (j) and Hopewell VORTAC (e) if the
OMNI receiver indicates crossing the 297 radial of Cofield VORTAC
(h)? (See Figure 13.4.)

1--Crossing the Virginia, North Carolina border.
2--Crossing the 246 radial of Franklin VORTAC.
3--North of the railroad between Jarrett and Suffolk.
4--Crossing the road between Wildon and Jackson.

24. (FAA 1295) What is the aircraft position on V310 en route
to Tar River VORTAC (j) from Elizabeth City VOR (g)? (See Figure
13.4.) The No. 2 OMNI is tuned to Cofield VORTAC (h) with the
OBS set on 192°, the course deviation needle deflected full scale
to the right, and the ambiguity meter indicating FROM.

1--Approaching the 192 radial.
2--On the 192 radial.
3--Past the 192 radial.
4--Impossible to identify unless on a 192° heading.

25. (FAA 1297) What is a frequency for transcribed weather
broadcasts in the southern half of the chart excerpt in Figure
13.5?

1--122.0 MHz.   3--112.2 MHz.
2--122.4 MHz.   4--329 kHz.

26. (FAA 1298) What is a frequency for transcribed weather
broadcasts in the northern half of the chart excerpt in Figure
13.5?

1--115.1 MHz.   3--126.2 MHz.
2--122.6 MHz.   4--126.6 MHz.

27. (FAA 1299) According to the WAC chart excerpt (Figure
13.5), what pilot services are available at Great Falls FSS (b)?

1--Hourly weather broadcasts.
2--Hourly weather broadcasts and EFAS.
3--TWEB and EFAS.
4--TWEB, EFAS, and ATIS.

28. (FAA 1300) Estimate the time en route from Whitehall VORTAC
(e) to Augusta Airport (a) via V21 to Helena VORTAC (d) then
direct to Augusta. The wind is from 040° at 20 kts. and the
expected TAS is 135 kts. (See Figure 13.5.)

1--40 min.   2--45 min.   3--50 min.   4--54 min.

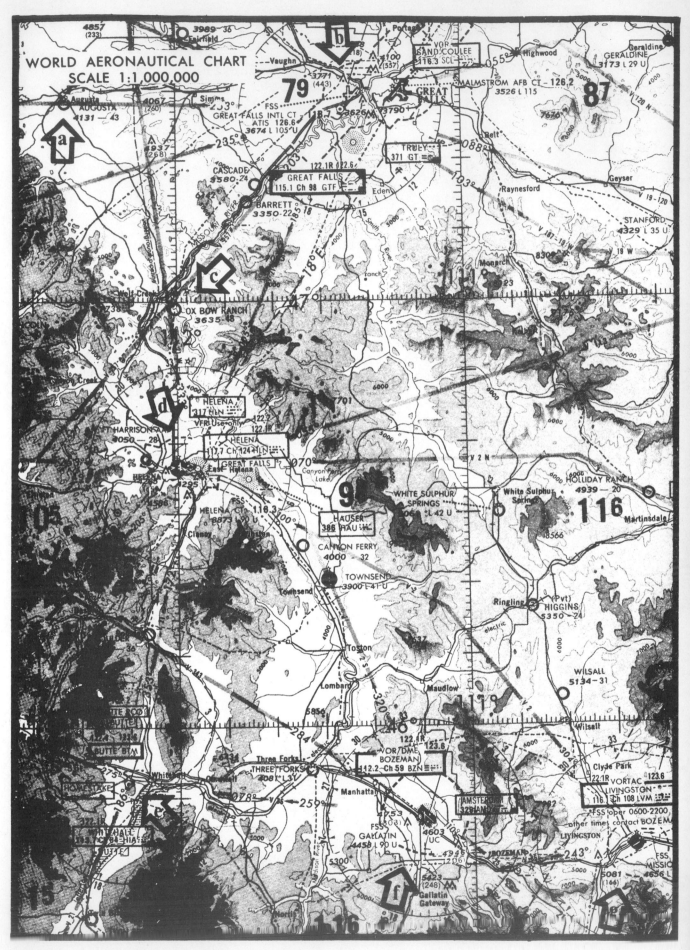

Figure 13.5

29. (FAA 1301) Estimate the time en route from Gallatin Airport
(f) to Great Falls VORTAC (b) via Ox Bow Ranch (c). The wind is
from 350° at 25 kts. and the expected TAS is 110 kts. (See
Figure 13.5.)

1--1 hr., 17 min.   3--1 hr., 26 min.
2--1 hr., 21 min.   4--1 hr., 30 min.

30. (FAA 1302) While en route on V21, the aircraft crosses over
Whitehall VORTAC (e) at 1557 and then over the intersection of
V21 and V343 at 1607. At approximately what time should the
aircraft arrive over the Great Falls VORTAC (b)? (See Figure
13.5.)

1--1641.   2--1648.   3--1658.   4--1704.

31. (FAA 1303) A cross-country flight crosses over Boseman
VOR/DME (f) at 1535 and over the railroad on the east side of
Lombard at 1547. At approximately what time should the flight
cross over Augusta Airport (a)? (See Figure 13.5.)

1--1626.   2--1630.   3--1634.   4--1638.

32. (FAA 1304) Estimate the time en route from Mission Airport
(g) to Great Falls VORTAC (b). The wind is from 195° at 20 kts.
and the expected TAS is 114 kts. (See Figure 13.5.)

1--45 min.   2--48 min.   3--52 min.   4--56 min.

33. (FAA 1305) Determine the magnetic heading for a flight from
Great Falls VORTAC (b) to Mission Airport (g). The wind is from
070° at 22 kts. and the expected TAS is 125 kts. (See Figure
13.5.)

1--131°.   2--141°.   3--151°.   4--171°.

34. (FAA 1306) Select a VFR route from Great Falls
International Airport (b) to Gallatin Airport (f) that assures
1,000 ft. terrain clearance and 500 ft. cloud clearance if the
base of the clouds is at 6,500 ft. MSL. (See Figure 13.5.)

1--Follow Smith River to White Sulphur Springs and then follow
   the railroad to Gallatin.
2--Follow Smith River, then the road that goes through the pass
   to Canyon Ferry Lake, and then follow the railroad to
   Gallatin.
3--Follow the highway to Helena, then Boulder, and then fly
   direct to Gallatin on V343.
4--Follow the Missouri River to Three Forks, then the railroad to
   Gallatin.

35. (FAA 1319) What is the elevation of the tallest object in
the south portion of the city of St. Louis? (See Figure 13.6.)

1--1,049 ft.   2--1,156 ft.   3--1,214 ft.   4--1,649 ft.

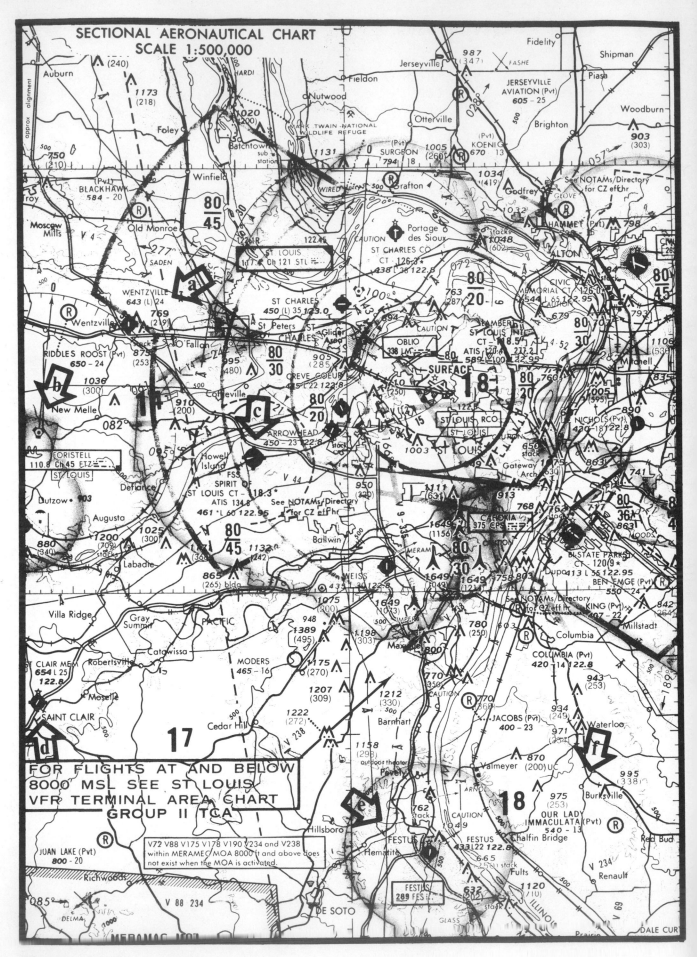

SECTIONAL AERONAUTICAL CHART
SCALE 1:500,000

Figure 13.6

36. (FAA 1318) Under what condition may a pilot operate from
the Spirit of St. Louis Airport (c) without an operable 4096
transponder? (See Figure 13.6.)

1--Request a clearance through the FSS on the airport.
2--Request a waiver from FAA to operate from the airport.
3--Advise Approach Control and the tower of the inoperative
     transponder and operate below the TCA floor.
4--Approach the airport above 8,000 ft. and remain within the
     control zone while descending to the traffic pattern altitude.

37. (FAA 1320) What danger is represented by the word CAUTION
printed on Figure 13.6. at various places along the Mississippi
River?

1--Radio or TV antennas.
2--Airports.
3--Powerline crossings.
4--Tall bridges.

38. (FAA 1321) What VFR restrictions exist in the St. Louis
area depicted in Figure 13.6?

1--Fixed-wing special VFR is prohibited in the Lambert-St. Louis
     control zone.
2--All aircraft must follow IFR in the TCA.
3--VFR traffic below 8,000 ft. receive minimum priority.
4--The airways in the immediate area may not be used by VFR
     traffic.

39. (FAA 1323) What is the maximum indicated airspeed within
and beneath the TCA depicted in Figure 13.6?

1--Both are 200 kts.
2--250 kts. within and 200 kts. beneath the TCA.
3--200 kts. within and 250 kts. beneath the TCA.
4--156 kts. within and 200 kts. beneath the TCA.

40. (FAA 1324) For a flight from over Festus Airport (e) to the
Foristell VORTAC (b) depicted in Figure 13.6, the omnibearing
selector should be set to

1--130°.   2--142°.   3--310°.   4--322°.

41. (FAA 1314) What is a procedure for an approach and landing
at Cornelia Fort Airport just north of Nashville Metro (g)?
(Figure 13.7.)

1--Request a clearance from Nashville Metro 10 NM from the
     airport.
2--Approach from the north at less than 2,000 ft. MSL and receive
     an airport advisory on 122.8.
3--Request a clearance from Nashville Approach on 122.0 MHz.
4--Remain below 2,000 ft. MSL and advise Memphis Flight Watch of
     your intentions.

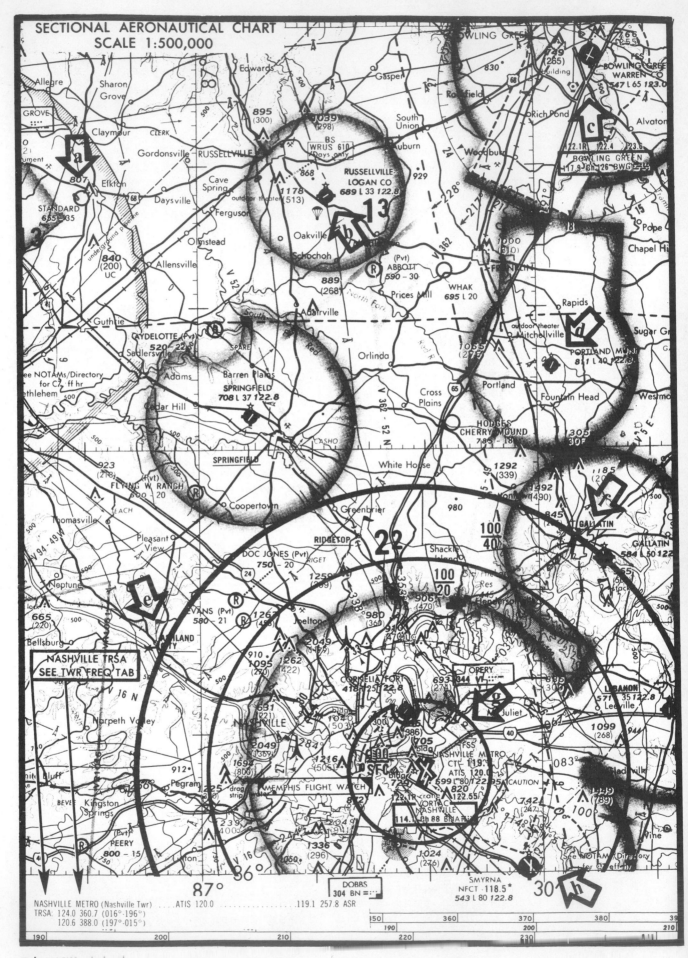

SECTIONAL AERONAUTICAL CHART
SCALE 1:500,000

Figure 13.7

42. (FAA 1307) The recommended frequency for an airport
advisory at Bowling Green Airport (c) in Figure 13.7 is

1--123.0 MHz.   3--123.6 MHz.
2--122.4 MHz.   4--121.1/117.9 MHz.

43. (FAA 1308) What is the status of Smyrna Airport (h)? (See
Figure 13.7.)

1--The airport has a non-Federal control tower which operates
   part time.
2--Night flight in the vicinity is prohibited.
3--FAA has no jurisdiction over operations at the airport.
4--Clearances for operations at the airport are made through
   Nashville Federal Control Tower.

44. (FAA 1309) The upper limits of the Nashville TRSA (Figure
13.7) is

1--up to, but not including 3,000 ft.
2--the base of the overlying continental control area.
3--18,000 ft.
4--10,000 ft.

45. (FAA 1310) Select two checkpoints that will provide the
most accurate time checks for a flight between Nashville VORTAC
(g) and Bowling Green VORTAC (c) on V5-49. (See Figure 13.7.)

1--The towns of Shackle Island and Mitchellville.
2--The road west out of Portland and the road east out of
   Franklin.
3--The powerline north of Shackle Island and the railroad between
   Portland and Mitchellville.
4--The powerline north of Shackle Island and the river between
   Franklin and Portland Municipal.

46. (FAA 1311) What type airspace does a flight penetrate from
over Standard Airport (a) to Portland Municipal (d) while
maintaining 3,500 ft. cruising altitude? (See Figure 13.7.)

1--Uncontrolled, then control area surrounding Portland.
2--Uncontrolled, then transition area surrounding Portland.
3--Control area, then control zone surrounding Portland.
4--Control area along the entire route.

47. (FAA 1312) The flag symbol at Ashland City (e) in Figure
13.7 is

1--a visual checkpoint to identify position for initial callup to
   the TRSA.
2--a compulsory reporting point for the TRSA.
3--marking one of the required approach lanes to the TRSA.
4--a visual boundary marker for the TRSA.

WORLD AERONAUTICAL CHART
SCALE 1:1,000,000

FOR (Magnetic)	N	30	60	E	120	150
STEER (Compass)	0	27	56	85	116	148
FOR (Magnetic)	S	210	240	W	300	330
STEER (Compass)	181	214	244	274	303	332

Figure 13.0

48. (FAA 1313) What is the approximate latitude and longitude of Gallatin Airport (f) in Figure 13.7?

1--36°22'N - 86°24'W.   3--36°45'N - 87°36'W.
2--36°27'N - 86°54'W.   4--36°52'N - 87°36'W.

49. (FAA 1317) Concerning the information about the parachute jumping area depicted near Russellville (b) in Figure 13.7, a pilot should refer to the

1--special use airspace listing on the aeronautical chart border.
2--special use airspace listed in AIM.
3--Airport/Facility Directory.
4--Notices to Airmen.

50. (FAA 1328) To navigate with the aid of the omni through the pass at Grass Lake (c) to Fort Jones VORTAC (b) as depicted in Figure 13.8, the omnibearing selector should be set to

1--050°.   2--070°.   3--230°.   4--250°.

51. (FAA 1329) What special precaution should a pilot take during a flight from Red Bluff (d) to Spaulding Airport (e)? (See Figure 13.8.)

1--File a DVFR flight plan prior to entering the MOA.
2--Exercise caution for possible military operations if cruising above 11,000 ft.
3--Secure permission for that part of the flight under or through the MOA from the FSS at Red Bluff.
4--Check with the FSS at Red Bluff for periods of operation and fly around the MOA if in operation.

52. (FAA 1330) What is the compass heading for the leg of the flight from Red Bluff VORTAC (d) to Fort Jones VORTAC (b)? The wind is from 040° at 25 kts. and the TAS is 135 kts. (See Figure 13.8.)

1--326°.   2--335°.   3--344°.   4--353°.

53. (FAA 1331) While en route on V25, the aircraft passes over Red Bluff VORTAC (d) at 1005 and over the road between Redding and Inwood at 1019. At what time should the aircraft pass over Klamath Falls VORTAC (a)? (See Figure 13.8.)

1--1110.   2--1114.   3--1120.   4--1128.

54. (FAA 1332) Determine the compass heading for a flight from Klamath Falls VORTAC (a) to Red Bluff Airport (d). The wind is from 270° at 18 kts. and the expected TAS is 118 kts. (See Figure 13.8.)

1--164°.   2--170°.   3--176°.   4--182°.

WORLD AERONAUTICAL CHART
SCALE 1:1,000,000

Figure 13.9

55. (FAA 1333) What is the pilot's position on V25 while
northbound from Red Bluff VORTAC (d)? One omni is tuned to the
Fort Jones VORTAC (b) with the omnibearing selector set on 105°
and the CDI is centered. (See Figure 13.8.)

1--Approximately 2 NM south of Big Bend.
2--Approximately 6 NM north of Big Bend.
3--Over the railroad 15 NM north of Big Bend.
4--Over the lake 23 NM north of Big Bend.

56. (FAA 1334) What is the estimated time en route from Tri
Cities Airport (c) to Lancaster Airport (h) via V149 to Allentown
VORTAC via direct to Lancaster? The average groundspeed is 115
kts. (See Figure 13.9.)

1--1 hr., 14 min.   3--1 hr., 26 min.
2--1 hr., 20 min.   4--1 hr., 32 min.

57. (FAA 1335) What is the expected groundspeed for a flight
from Hallstead Airport (d) to Marietta Airport (g) if the wind is
from 165° at 25 kts. and the TAS is 150 kts.? (See Figure 13.9.)

1--119 kts.   2--124 kts.   3--129 kts.   4--134 kts.

58. (FAA 1336) Determine the magnetic heading for a flight from
North Philadelphia Airport (i) direct to Towanda Airport (a).
The wind is from 025° at 18 kts. and the expected TAS is 125 kts.
(See Figure 13.9.)

1--324°.   2--334°.   3--344°.   4--350°.

59. (FAA 1337) What is the position of the aircraft if the omni
indicates the 080 radial of Williamsport VORTAC (e) and the 023
radial of the Wilkes Barre VORTAC (f)? (See Figure 13.9.)

1--1 mi. north of Thompson.
2--Over Carbondale-Clifford Airport.
3--6 mi. west of Carbondale-Clifford Airport.
4--2 mi. northeast of Harford.

60. (FAA 1338) While southbound on V499, an aircraft crosses
over the intersection with V36 (b) at 0852 and over the
intersection with V116 at 0901. At what time should the flight
pass over Lancaster Airport (h)? (See Figure 13.9.)

1--1000.   2--1005.   3--1010.   4--1015.

61. (FAA 1339) Which pilot action is most likely to eliminate
large fluctuations on the VOR course deviation indicator during
flight?

1--Recycle the ON-OFF switch.
2--Recycle the frequency selector.
3--Disconnect the microphone.
4--Change the engine RPM.

62. (FAA 1340) When performing a periodic VOR receiver
calibration check on a VOT, the OBS should be set to

1--360 or 180 and the omni should indicate FROM with the CDI
     centered.
2--the bearing from the station and the omni should indicate FROM
     with the CDI centered.
3--360 or 180 and the omni should indicate 360, FROM or 180, TO
     with the CDI centered.
4--360 or 180 and the omni should indicate 360, TO or 180, FROM
     with the CDI centered.

63. (FAA 1341) How can a pilot determine when a particular VOR
is unreliable?

1--A recorded voice stating "VOR shut down for maintenance."
2--A continuous series of dashes replacing the coded
     identification.
3--An absence of the coded identification.
4--A coded W added to the identification.

64. TACAN stations operate at _____ frequency.

1--low   3--very high
2--medium   4--ultra-high

65. Which of the following is **not** a frequency on which VORs
transmit?

1--107.8.   2--113.2.   3--116.0.   4--117.85.

66. For flying inbound to a VOR station, the magnetic course is
_____ to the VOR radial.

1--parallel to   3--the same as
2--perpendicular to   4--the reciprocal of

67. Which of the VOR cockpit displays is the first to indicate
passage over a VOR station?

1--An interruption in the auditory signal broadcast by the
     station.
2--Ccourse deviation indicator (CDI).
3--Omnibearing selector (OBS).
4--TO/FROM indicator.

68. Under which of the following conditions will the Off flag of
a VOR receiver appear?

1--When the TO indicator is used for an outbound radial.
2--When the FROM indicator is used for an inbound radial.
3--When you are 90° off the radial selected on the OBS.
4--Under all of the conditions described above

69. Which of the following is a disadvantage of VOR navigation?

1--Its inaccuracy.
2--Its omnidirectionality.
3--Its susceptibility to atmospheric interference.
4--Its signal operating on a line-of-sight basis.

70. To indicate that you have lost two-way radio communication during flight, you should set your transponder to

1--1200.   2--4096.   3--7500.   4--7600.

71. Which of the following is not a typical cockpit display on a transponder?

1--CDI.   2--Ident.   3--Reply.   4--Standby.

72. The use of phantom stations (waypoints) in which VOR and/or DME information is calculated for navigational purposes is part of a system referred to as

1--area navigation (RNAV).   3--TACAN.
2--ARTCC.   4--Victor Airways.

ANSWERS

Key Terms and Concepts, Part 1

1.	F	5.	B	9.	M	13.	E
2.	K	6.	C	10.	L	14.	H
3.	D	7.	G	11.	J	15.	P
4.	I	8.	A	12.	N	16.	O

Key Terms and Concepts, Part 2

1.	F	4.	L	7.	C	10.	J
2.	H	5.	E	8.	D	11.	B
3.	A	6.	G	9.	I	12.	K

Discussion Questions and Exercises

8.   T--H.
        U--C.
        V--F.
        W--A, E. Remember, it does matter where the nose of the airplane is pointed.
        X--G, I. Remember, you are 90° off the radial.
        Y--No match. You would have to be directly over the station in the zone of confusion to get this reading.
        Z--B, D.

11.   325°. The magnetic bearing to the station is equal to
           the heading plus the relative bearing.
13.   S--C, D.
           T--No match.
           U--A.
           V--B, E.
           W--G.
           X--No match.
           Y--F, H.
           Z--No match.

Review Questions

 1.   ?--This question has been designated as unusable by the FAA
         and removed from the FAA Question Selection Sheets.
 2.   3--The arrow points TO the station.
 3.   3--The arrow points TO the station.
 4.   3--Use 190° + 30° = 220° to intercept the 180° bearing.
 5.   2--It's the reciprocal of the direction the arrow indicates.
 6.   4--The arrow points TO the station. The aircraft is
         tracking to the station with a right crosswind.
 7.   1--It's the reciprocal of the direction the arrow indicates.
 8.   1--You must have a VHF transmitter and receiver.
 9.   4--Climb or descend on the right side of the airway center
         line. Once at altitude, fly in the center.
10.   4--For easterly magnetic courses (0° to 179°), fly at odd-
         thousand plus 500-ft. altitudes. For westerly magnetic
         courses (180° to 359°), fly at even-thousand plus 500-
         ft. altitudes.
11.   2--See answer 10.
12.   2--See answer 10.
13.   3--Maximum elevation figures (MELs) include terrain and
         obstructions shown in quadrangles bounded by ticked lines
         of latitude and longitude. The large numbers represent
         thousands of feet and the small numbers indicate hundreds
         of feet.
14.   1--Use your plotter or the compass rose of the Hopewell
         VORTAC. If you use the compass rose, you must correct
         the magnetic course for variation. Since it is westerly
         variation, you must **subtract** it from the magnetic course
         to determine the true course. Remember, the formula is
         MC = TC plus or minus VAR. Since the VAR is westerly, so
         you would **add** it to the TC to get the MC. But, in this
         example, you would **subtract** it from the MC to get TC.
15.   1--The total distance is 122 NM. At 185 kts., it will take
         0.66 hr. and consume 17.36 gal. At 165 kts., it will
         take 0.74 hr. and burn 15.91 gal. The difference is 1.45
         gal.
16.   1--See answer 14.
17.   3--Use your plotter and remember to **add** the westerly
         variation of 7°. TC = MC plus or minus VAR.
18.   3--The distance is 116 NM. At 145 kts., it will take 0.8
         hr. 0.8 hr. x 14.5 gal./hr. = 11.6 gal.

19. 2--The pilot should exercise extreme caution when passing
       through the MOA.
20. 1--The magnetic course (MC) is 205° (read from Hopewell
       VORTAC) from which you subtract the WCA of 10°.
21. 2--This is the line between the two VORTACs.
22. 1--Plot both lines. They intersect near the Virginia-North
       Carolina border. Fortunately, state borders are clearly
       visible from the sky; they are typically marked with a
       string of police cars with flashing lights.
23. 1--V194 is 246° FROM. Use the reciprocal, 66° TO.
24. 1--Full scale deflection to the right means you have not yet
       reached the 192° radial.
25. 4--The small dark box in the lower right corner of the
       AMSTERDAM rectangle indicates that TWEB is available.
26. 1--See answer 25.
27. 3--The triangles in the GREAT FALLS rectangle indicate that
       EFAS is available and the small dark box in the lower
       right corner means that TWEB is available. ATIS is also
       available, but that is a function of the control tower.
28. 4--For the first leg of the trip (Whitehall VORTAC to Helena
       VORTAC), the distance is 46 NM, TC is 11°, TAS is 135
       kts., GS= 116 kts., and time en route is 26 min. For the
       second leg (Helena VORTAC to Augusta), the distance is 55
       NM, TC is 342°, GS is 122 kts., and time en route is 27
       min.
29. 2--For the first leg of the trip (Gallatin Airport to Ox Bow
       Ranch), the distance is 80 NM, TC is 333°, GS is 85 kts.,
       and time en route is 56 min. For the second leg (Ox Bow
       Ranch to Great Falls VORTAC), the distance is 37 NM, the
       TC is 40°, GS is 92 kts., and the time en route is 24
       min.
30. 2--Note the turn on V21. The total distance is 102 NM. It
       takes 10 min. to go 20 mi., so it will take 51 min. to
       travel 102 NM.
31. 4--The time to travel 25 statute miles is 12 min. The time
       for 130 statute miles is 62 min.
32. 3--The distance is 112 NM, the TC is 340°, GS is 130 kts.,
       and the time en route is 52 min.
33. 1--Determine the magnetic course from the compass rose on
       the Great Falls VORTAC. Change wind direction to
       magnetic by subtracting 18°E VAR. Compute the WCA (-10°)
       and subtract from the MC (140°) to determine the MH
       (130°).
34. 4--Due to cloud clearances, you cannot go over 6,000 ft. and
       you must stay above 1,000 ft. AGL. The first three
       routes all violate the altitude restrictions.
35. 4--This may be difficult to find since the chart excerpt is
       hard to read. The elevation (MSL) of the highest
       obstruction is 1,649 ft. as indicated by the boldface
       numbers.
36. 3--Advise Approach Control and the tower of the inoperative
       transponder; they may authorize deviations from the
       transponder requirements for operations within the TCA.
37. 3--Study the chart.

38. 1--The line of TTTTTTs means that fixed-wing special VFR is
     prohibited.
39. 2--See FAR 91.71.
40. 3--For this one, you will need to use the reciprocal of the
     radial read from the compass face for Foristell.
41. 2--Stay under 2,000 ft. MSL and request an airport advisory
     from Cornelia Fort Airport on 122.8.
42. 3--Bowling Green has an FSS on the field. Contact FSS on
     123.6 MHz for airport advisories.
43. 1--The letters "NFCT" mean non-Federal control tower.
44. 4--The upper limit is 10,000 ft. as indicated by the numbers
     100/40. The two trailing zeros are dropped from this
     altitude designation.
45. 1--These two towns are located directly along the route and
     approximately equidistant along the route of flight.
     They will also be easy to identify from the air.
46. 4--There are control areas along the entire route.
47. 1--The flag is a visual checkpoint to identify position for
     initial contact with the TRSA.
48. 1--The lines of latitude increase as you go north and the
     lines of longitude increase as you go west.
49. 3--You find information about activities such as parachute
     jumping in the Airport/Facility Directory.
50. 3--Set it to the reciprocal of the line drawn from Grass
     Lake to the Fort Jones VORTAC.
51. 2--At and above cruising altitudes of 11,000 ft., the pilot
     should exercise extreme caution for possible military
     operations.
52. 2--The true course is 342°, the WCA is +9°, the VAR is -18,
     compass deviation is +2°. The compass heading is 335°.
53. 2--Using SM, it takes 14 min. to travel 30 mi. That
     translates to 129 MPH. The total distance is 147   mi.,
     which at 129 MPH, will take 1 hr., 8 min. 1005 + 60 min.
     + 8 min. yields an estimated arrival time of 1113.
54. 4--The TC is 189°, WCA is +9°, VAR is -18° and CD is +1°.
     Thus, the compass heading will be 181°. Another way to
     compute this is to add the WCA and CD to the magnetic
     course (171°) read from the Klamath Falls VORTAC compass
     rose.
55. 1--Draw the 105° line from the Fort Jones VORTAC (if you can
     read it) until it intersects V25. You will be about 2 NM
     south of Big Bend.
56. 1--The total distance is 144 NM (including the two legs on
     V149). Distance divided by 115 kts. yields a time en
     route of 1.25 hr., or 1 hr., 15 min.
57. 3--The TC is 198°. Using your flight computer or electronic
     calculator, determine the groundspeed to be 129 kts.
58. 3--The TC is 326°, the WCA is +6°, and the VAR is westerly.
     For the life of me I cannot find the VAR listed, but from
     an analysis of Figure 12.1., it will be close to +10°
     since Philadelpia, PA, is about +10°. Thus, the magnetic
     heading will be about 342°. Of course, you should not
     have to intuit VAR from Figure 12.1. It just happens to
     be shown there.

59. 2--Draw the two lines on the chart; they intersect over the
     Carbondale-Clifford Airport.
60. 2--It takes 9 min. to travel 14 statute miles (GS = 93.3
     MPH). The total trip is 114 mi., which will take 1.22
     hr., or 1 hr., 13 min. 0852 + 60 + 13 = 1005.
61. 4--Sometimes the engine RPM will interfere with the VOR CDI,
     so it frequently helps to make minor changes in RPM to
     stop the fluctuations.
62. 3--The omni should indicate 360, FROM or 180, TO with the
     CDI centered when the omnibearing selector is set to 360
     or 180.
63. 3--An absence of the coded identification tells the pilot
     that the VOR is unreliable.
64. 4--TACANs transmit UHF, or ultra-high frequency.
65. 1--VORs transmit on frequencies between 108.0 MHz and 117.95
     MHz.
66. 4--The radial is its magnetic course **from** the station; all
     radials are **from** the station.
67. 4--The TO/FROM indicator denotes station passage when it
     makes its first positive change from TO to FROM.
68. 3--It will also appear when you are passing directly over
     the station (the cone of confusion), when the station is
     off, when it is out of the line-of-sight (you are too far
     away or there is an obstruction), and when you are 90°
     off the radial selected on the OBS.
69. 3--VORs are characterized by accuracy, omnidirectionality,
     and freedom from atmospheric interference. The major
     drawback is its restriction to line-of-sight transmission
     and reception.
70. 4--7600 is the transponder code for loss of two-way radio
     communication.
71. 1--The course deviation indicator (CDI) is part of the VOR
     cockpit display.
72. 1--RNAV uses phantom stations in which VOR and/or DME data
     is calculated for navigational purposes.

14/COMPOSITE NAVIGATION: GOING CROSS-COUNTRY

MAIN POINTS

1.   This chapter, which describes the process of planning and executing a cross-country flight, draws on all the knowledge you have accumulated in the preceding chapters of this book, including preflight planning, dead reckoning, radio navigation, FARs, and weather.

2.   One of your tasks during this chapter will be to prepare a **flight log** for a proposed cross-country flight. The phases you will consider are general planning; preflight planning, including navigational computations; weather briefings and airport data; airplane preflight preparation; departure procedures; en route procedures; en route replanning; arrival procedures; and postflight activities, including such items as closing your flight plan.

CROSS-COUNTRY EXERCISE

For this exercise, you will need a flight computer or electronic calculator, plotter, navigation log, San Francisco sectional chart, and excerpts from the Airport/Facility Directory (provided in Figure 14.1). Routing and weather information will be provided.

Your aircraft is a black and blue Starship U2. You and your instructor will be going on a round-robin cross-country flight from Fresno Chandler Downtown (36°44'N, 119°49'W to Pine Mountain Lake near Groveland (37°52'N, 120°11'W), then on to Carmel Valley (36°29'N, 121°44'W), and finally back to Fresno. It is a Thursday in August; Pacific daylight time (PDT) is in effect.

```
-----------------------------------------------------------------------------
§ FRESNO-CHANDLER DOWNTOWN    °(FCH)    1.7 W    GMT −8( −7DT)          SAN FRANCISCO
   36°43'56"N 119°49'08"W                                                H-2F, L-2E
   278    B    S4    FUEL 80, 100    OX 3, 4    TPA −1078(800)               IAP
  RWY 12L-30R: H3475X75 (ASPH)    S-17
    RWY 12L: Thld dsplcd 460'. Road.           RWY 30R: Thld dsplcd 527'. Road. Rgt tfc.
  RWY 12R-30L: H3441X75 (ASPH)    S-17    MIRL
    RWY 12R: REIL. Thld dsplcd 474'. Pole. Rgt tfc.
    RWY 30L: REIL. VASI(V2L) — GA 3.42° TCH 21'. Thld dsplcd 536'. Road.
  AIRPORT REMARKS: Attended 1500-0630Z‡. Fee for acft over 12,500 pounds gross weight. Fuel avbl Mon-Sat
    1500Z‡-dusk & Sun 1500-0100Z‡, later thru ATCT. Control Zone effective 1500-0600Z‡.
  COMMUNICATIONS: UNICOM 123.0
    FRESNO FSS (FAT) LC 251-8269
  Ⓡ APP CON 132.35, 119.6    Ⓡ DEP CON 132.35
    CHANDLER TOWER 121.1 opr 1600-0400Z‡    GND CON 121.9
  RADIO AIDS TO NAVIGATION:
    (H) ABVORTAC 112.9    ■ FAT    Chan 76    36°53'12"N 119°48'11"W    167° 9.3 NM to fld. 361/17E.
    CHANDLER NDB (H-SAB) 344    ■ FCH    36°43'26.4"N 119°49'57.7"W    036° 0.8 NM to fld
      NDB unusable 095-120° beyond 40 NM    200-235° beyond 35 NM
-----------------------------------------------------------------------------
```

GROVELAND

```
  PINE MOUNTAIN LAKE    (Q68)    2.6 NE    GMT −8(−7DT)    37°51'45"N 120°10'40"W    SAN FRANCISCO
  2900    B    TPA — 3700(800)                                                         L-2F
  RWY 09-27: H3640X50 (ASPH)    S-2    MIRL
    RWY 09: Trees.           RWY 27: Tree. Rgt tfc.
  AIRPORT REMARKS: Unattended
  COMMUNICATIONS: UNICOM 123.0
    STOCKTON FSS (SCK)
  RADIO AIDS TO NAVIGATION:
    MODESTO (H) VOR/DME 114.6    MOD    Chan 93    37°37'39"N 120°57'25"W    052° 39.7 NM to fld.
      90/17E.
```

```
  CARMEL VALLEY    (O62)    0 NE    GMT −8(−7DT)    36°28'55"N 121°43'45"W    SAN FRANCISCO
  450    TPA — 1500(1050)
  RWY 11-29: 2475X35 (TRTD-GRVL)
    RWY 11: P-line. Rgt tfc.           RWY 29: Trash piles.
  AIRPORT REMARKS: Unattended. No touch & go lndgs. Straight out departure Rwy 29, no turns below 1000'. First
    600' runway 11-29 overgrown with 3' weeds.
  COMMUNICATIONS:
    SALINAS FSS (SNS) LC 372-6050
```

```
§ LOS BANOS MUNI    (LSN)    .9 W    GMT −8(−7DT)    37°03'43"N 120°52'05"W    SAN FRANCISCO
  119    B    S4    FUEL 80, 100    TPA — 919(800)                                     H-2F, L-2F
  RWY 14-32: H3000X75 (ASPH)    S-23    MIRL                                               IAP
    RWY 14: VASI(V4L) — GA 3.0° TCH 30'. Rgt tfc.           RWY 32: VASI(V4L). — GA 3.0° TCH 30'. Road.
  AIRPORT REMARKS: Attended 1600-0130Z‡. ACTIVATE VASIs Rwy 14/32-122.8.
  COMMUNICATIONS: UNICOM 122.8
    FRESNO FSS (FAT) Toll free dial 0, ask for ENTERPRISE 14598.
    PANOCHE LRCO 122.1R, 112.6T (FRESNO FSS)
  RADIO AIDS TO NAVIGATIONS:
    PANOCHE (L) VORTAC 112.6    PXN    Chan 73    36°42'56"N 120°46'40"W    333° 21 NM to fld.
      2060/16E.
      VOR unusable 230-280° beyond 7 NM below 9000'    280-290° beyond 6-10 NM below 10,000'
```

Figure 14.1

Given

Aircraft Identification	N7118Q
Transponder	4096 3/A
Two-way radio and VOR	Operable
Compass deviation	None
Proposed departure time--Fresno	0900 PDT
Fuel on board	23.0 gal.
Demonstrated crosswind component	17 kts.
IAS at 6,500 ft.	84 kts.
IAS at 7,500 ft.	82 kts.
TAS at 6,500 ft.	92 kts.
TAS at 7,500 ft.	91 kts.
Fuel flow at 6,500 ft.	5.4 gal./hr.
Fuel flow at 7,500 ft.	5.3 gal./hr.

Note: Add 1 mi. for each 1,000 ft. of climb to altitude to compensate for fuel burned and time spent to reach cruising altitude.

Routing

Leg 1: Fresno Chandler Downtown (FCH) direct at 6,500 ft. MSL to Pine Mountain Lake at Groveland (Q68); 1-hr. stopover; proposed departure time 1100 PDT.

Leg 2: Pine Mountain Lake (Q68) direct at 6,500 ft. MSL to Los Banos (LSN); fly over checkpoint.

Leg 3: Los Banos (LSN) direct at 6,500 ft. MSL to Carmel Valley (O62); 2-hr. stopover; proposed departure time 1430 PDT.

Leg 4: Carmel Valley (O62) direct at 7,500 ft. MSL to Panoche VOR (PXN); checkpoint.

Leg 5: Panoche VOR (PXN) at 7,500 ft. MSL direct to Fresno Chandler Downtown (FCH).

Weather Information (summarized)

Winds and temperature aloft:
      3,000 ft.:   240° at 15 kts.; 20° C
      6,000 ft.:   260° at 25 kts.; 15° C
      9,000 ft.:   260° at 25 kts.; 10° C

Fresno current: 8,000 broken and 10,000 overcast, visibility 10 mi., OAT 82°F, dewpoint 67°F, barometer 29.76, surface winds 240° at 10 kts.

Modesto current: 12,000 scattered, visibility 12 mi., OAT 84°F, dewpoint 67°F, barometer 29.86, surface winds 200° at 20 kts.

Modesto forecast: clear, visibility 12 mi., OAT 88°F, dewpoint 68°F, barometer 29.82, surface winds 240° at 20 kts.

Monterey current: field obscured, fog, visibility restricted, OAT 63°F, dewpoint 63°F, barometer 29.62, winds calm.

Monterey forecast: clear, visibility 6 mi. and haze, OAT 70°F, dewpoint 63°F, barometer 29.76, winds 210° at 10 kts.

Fresno forecast: 6,000 scattered, visibility 10 mi., OAT 84°F, dewpoint 67°F, barometer 29.86, winds 270° at 20 kts.

Discussion Questions and Exercises

1.   Complete the flight navigation log in Figure 14.2 for this cross-country trip.

    a.   Are the altitudes appropriate for the routing and weather conditions? Why or why not?

    b.   Calculate the headwind and crosswind components (Figure 6.1) for takeoffs and landings at:

        Fresno (FHC) current; assume you will use Rwy 30L or Rwy 30R.

        Pine Mountain Lake (Q68), both current and forecast; use Modesto for an approximation; assume you will use Rwy 27.

        Carmel Valley (O62); use the Monterey forecast; assume you will use Rwy 29.

Aircraft number: _____

Proposed takeoff: _____
Actual takeoff: _____

FLIGHT LOG

Route/Fix/Checkpoint	Alt. / IAS Temp. / TAS	TC	Wind Dir. / Vel.	WCA TH	Var. MH	Dev. CH	Ground Speed	Dist. Dist. Rem.	Time ETE / ATE	ETA / ATA	Fuel GPH: ___ Used	Rem.
Totals												

Remarks:

Airport Frequencies

Airport		
ATIS		
Cl. Del.		
Gnd.		
Twr.		
Dep.		
App.		
FSS		
UNICOM		

Airport/ATIS Advisories

	Time in	
Airport.	Time out	
Cig.	Total Time	
Vsby.		
Temp.	Field Elevation MSL	
Wind	Departure	
Alt.	Destination	
Rwy.		
NOTAMs		

REMEMBER:
CLOSE YOUR FLIGHT PLAN
Freq. _____
Phone no. _____

Figure 14.2

Fresno (FCH) forecast; assume you will use Rwy 30L or
Rwy 30R.

c. Have you exceeded the demonstrated crosswind component
in any of the above? Which, if any?

d. Pine Mountain Lake, as you discovered above, may present
a problem in landing since the current Modesto conditions
suggest that the demonstrated crosswind component might be
exceeded. Should you change your flight plan at this point?
Analyze the situation.

2.   Your flight plan:

a. With whom will you file the plan? How will you contact
them? Where are they located?

b. Complete a flight plan for this trip using Figure 14.3.

c. How will you activate your flight plan on your departure
from FCH?

3.   Departure and Leg 1:

       a. Given your proposed route of flight, to which runway
       will you probably be directed for takeoff? Why?

       b. After your runup, you are ready for takeoff. It is 0857
       PDT. What do you do next?

       c. How would you use your VOR to obtain a fix exactly 21 NM
       from Fresno Chandler (FCH) along your intended route of
       flight? What visual landmarks are also available at this
       point?

Form Approved: OMB No. 04-R0072

| DEPARTMENT OF TRANSPORTATION FEDERAL AVIATION ADMINISTRATION **FLIGHT PLAN** | **CIVIL AIRCRAFT PILOTS**. FAR Part 91 requires you file an IFR flight plan to operate under instrument flight rules in controlled airspace. Failure to file could result in a civil penalty not to exceed $1,000 for each violation (Section 901 of the Federal Aviation Act of 1958, as amended). Filing of a VFR flight plan is recommended as a good operating practice. See also Part 99 for requirements concerning DVFR flight plans. |

1. TYPE	2. AIRCRAFT IDENTIFICATION	3. AIRCRAFT TYPE/ SPECIAL EQUIPMENT	4. TRUE AIRSPEED	5. DEPARTURE POINT	6. DEPARTURE TIME		7. CRUISING ALTITUDE
VFR / IFR / DVFR			KTS		PROPOSED (Z)	ACTUAL (Z)	

8. ROUTE OF FLIGHT

9. DESTINATION (Name of airport and city)	10. EST. TIME ENROUTE		11. REMARKS
	HOURS	MINUTES	

12. FUEL ON BOARD		13. ALTERNATE AIRPORT(S)	14. PILOT'S NAME, ADDRESS & TELEPHONE NUMBER & AIRCRAFT HOME BASE	15. NUMBER ABOARD
HOURS	MINUTES			

16. COLOR OF AIRCRAFT	**CLOSE VFR FLIGHT PLAN WITH_____FSS ON ARRIVAL**

FAA Form 7233-1 (5-77)

Figure 14.3

d. Use the road heading northwest out of Mariposa-Yosemite Airport as another checkpoint. A crosscheck is the 025° radial from the Merced (MCE) VOR. If it takes you 20 min. to get from the checkpoint in question 3c to the Mariposa-Yosemite checkpoint, what is your actual groundspeed? Why do you suppose it is different from your estimated groundspeed?

e. How would you call Pine Mountain to obtain an airport advisory?

f. Suppose Pine Mountain UNICOM reports that winds are 230° at 20 kts. Are you going to land?

g. Describe how you would enter and fly the pattern at Pine Mountain, including directions on the various legs of the pattern and compensation for the surface winds to keep the airplane in a rectangular pattern relative to the runway.

4. Leg 2:

a. What should you know about the **alert area** that lies between Pine Mountain and Los Banos?

b. Since you will be flying over Castle Air Force Base, are
you required to report your position to them? If so, where
should you report?

c. Shortly after passing over Castle Air Force Base, you
notice, to your amazement, that the oil temperature gauge is
near red line. What should you do?

d. Suppose that, as you approach Los Banos, the oil
temperature has risen further and is now on the red line.
You and your instructor decide to land and have it checked.
As you begin your descent, you need to check Los Banos. The
winds are reported as 180° at 15 and Rwy 14 is the active
runway. Describe how you will enter and fly the pattern,
the direction on the various legs of the pattern, and what
crabbing you will have to do to fly a rectangular pattern.

e. Suppose you decide to use VASI to aid you on your final
approach. How would you get it turned on?

f. As your instructor and a mechanic analyze the problem (a
faulty gauge), you decide to call FSS and file an amended
flight plan that will allow an extra hour for lunch at
Carmel Valley. How would you contact FSS and where is the
station located?

g. After departing Los Banos, you climb to 6,500 ft. and proceed to Carmel Valley. Assume it takes you 12 min. to travel from V107 to V485. What is your groundspeed?

h. Briefly explain how you will make your approach to Carmel Valley, including a description of whom you will call and how you will fly the pattern.

5. Leg 3:

a. From studying the sectional chart, with what should you be concerned shortly after departing Carmel Valley? What do you plan to do?

b. How would you set your VOR to fly from Carmel Valley to Panoche VOR?

c. How would you set your VOR to fly from Panoche to Fresno Chandler?

d. Assume you decide to contact Fresno Chandler over Kerman. Whom would you call, what frequency would you use, and what would you say?

e. With whom would you close your flight plan? How would you contact them?

REVIEW QUESTIONS

1.   (FAA 1514 DI) The information that should be entered in
Block 9 (Figure 14.3), for a VFR day flight is the

1--name of the airport of the first intended landing.
2--name of the airport of the last intended landing for the
    flight.
3--name of the airport where the aircraft is based.
4--names of all the airports where landings are to be made.

2.   (FAA 1515) What information should be entered in Block 12,
(Figure 14.3) for a VFR day flight?

1--The estimated time en route plus 30 min.
2--The estimated time en route plus 45 min.
3--The maximum endurance time as shown in the Pilot's Operating
    Handbook.
4--The amount of usable fuel on board expressed in time.

3.   (FAA 1575) Prior to each flight, the pilot-in-command must

1--check the personal logbook for appropriate recent experience.
2--become familiar with all available information concerning that
    flight.
3--calculate the weight and balance of the aircraft to determine
    if the CG is within limits.
4--check with ATC for the latest traffic advisories and any
    possible delays.

4.   (FAA 1573) Preflight action, as required by regulations for
all flights away from the vicinity of an airport, shall include a
study of the weather, taking into consideration fuel requirements
and

1--an operational check of the navigation radios.
2--the designation of an alternate airport.
3--the filing of a flight plan.
4--an alternate course of action if the flight cannot be
    completed as planned.

5.   (FAA 1572) Which preflight action is required for every
flight?

1--Check weather reports and forecasts.
2--Determine runway lengths at airports of intended use.
3--Determine alternatives if the flight cannot be completed.
4--Check for any known traffic delays.

ANSWERS

Cross-Country Exercise

Do not use the data and charts provided in this chapter for
flight planning purposes! They are outdated. Only updated
charts and information should be used. Furthermore, should you
ever want to fly this particular cross-country trip, be sure to
check with your local FBO, or others familiar with the airports,
especially Carmel Valley.

1.   See Figure 14.4. Note that distances have been modified
        to account for time to altitude by adding 1 mi. for each
        1,000-ft. increment.
    a. Yes, they meet FAA standards as far as the intended
       routes are concerned and they keep you away from adverse
       weather, assuming that Monterey clears as forecast.
    b. Your headwind and crosswind components for takeoff at
       Fresno current: headwind, 5 kts.; crosswind, 8.5 kts.
       Pine Mountain Lake current: assume you will use Rwy 27;
       headwind, 7 kts.; crosswind, almost 19 kts.
       Pine Mountain Lake forecast: headwind, 17 kts;
       crosswind, 10 kts.
       Carmel Valley forecast: assume you will use Rwy 29;
       headwind, 2 kts.; crosswind, almost 10 kts.
       Fresno forecast: assume you will use Rwy 30L and Rwy
       30R; headwind, 17 kts.; crosswind, 10 kts.
    c. Yes. Pine Mountain may have winds that exceed the
       demonstrated crosswind component, assuming Modesto is
       representative.
    d. According to the Modesto forecast, the winds should
       become more favorable throughout the morning. It might
       be wise to call Pine Mountain, but the airport is
       unattended. You might also ask FSS for any pilot
       reports. The winds are a factor to monitor as your
       flight progresses but at present are not serious enough
       to warrant canceling that portion of the flight.
2. a. Fresno FSS, located at Fresno Air Terminal, can be
       reached by calling 251-8269 (local call). See Figure
       14.1, the Airport/Facility Directory excerpt.
    b. See Figure 14.5.
    c. Call Fresno Radio on 122.55 or 123.65, or transmit on
       122.1 and listen on the VOR (112.9). Refer to the
       sectional chart for frequencies. The small shaded box in
       the lower right corner of the FRESNO rectangle indicates
       TWEB is available.
3.   Departure and Leg 1:
    a. Since you will be departing to the northwest, you will
       probably be assigned to Rwy 30R to take into account the
       right traffic pattern. Rwy 30L has left traffic, which
       would take you away from your proposed flight route.
    b. Call the tower on 121.1 and tell them you are ready to
       take off. Note: The sectional chart does not indicate
       that Chandler has an active control tower. However, the

Aircraft number: 7118Q

FLIGHT LOG

Proposed takeoff: 0900/1600 Z
Actual takeoff: _____

Fuel GPH: 23.0

Route/Fix/Checkpoint	Alt. / Temp.	IAS / TAS	TC	Wind Dir. / Vel.	WCA / TH	Var. / MH	Dev. / CH	Ground Speed	Dist. / Rem	Time ETE/ATE	ETA / ATA	Fuel Used / Rem
FCH → Q68	6,500 / +15	84 / 92	346	260 / 25	-16 / 330	-16E / 314	0 / 314	87	70+6 / 207	52	0949 / 0952	4.7 / 18.3
Q68 → LSN	6,500 / +15	84 / 92	215	260 / 25	+11 / 226	-16E / 210	0 / 210	73	58+6 / 149	53	1137 / 1152	4.7 / 13.6
LSN → 062	6,500 / +15	84 / 92	230	260 / 25	+8 / 238	-16E / 222	0 / 222	69	54 / 95	47	1224 / 1239	4.2 / 9.4
062 → PXN(112.6)	7,500 / +12	82 / 91	73	260 / 25	-2 / 71	-16E / 55	0 / 55	116	48+7 / 47	28	1449 / 1458	2.5 / 6.9
PXN → FCH	7,500 / +12	82 / 91	89	260 / 25	+2 / 91	-16E / 75	0 / 75	116	47 / 0	24	1513 / 1522	2.1 / 4.8
Totals									296	3+24		18.2 / 4.8

Airport Frequencies

	FCH	Q68	062	MRY
Airport	FCH	Q68	062	MRY
ATIS				119.25
Cl. Del.				
Gnd.	121.9		121.9	
Twr.	121.1		118.4	
Dep.	132.35		ATIS 119.25	
App.				120.8
FSS	123.65		SNS 122.0	122.95
UNICOM	123.0	122.8		

Airport/ATIS Advisories

Airport	
Cig.	
Vsby.	
Temp.	
Wind	
Alt.	
Rwy.	
NOTAMs	

Time in	
Time out	
Total Time	

Field Elevation MSL

Departure	Destination
FCH 278	Q68 2900
	062 450

Remarks:

Stopover Q68: 1 hour
ETD 1100
Stopover 062: 2 hours
ETD 1430

REMEMBER:
CLOSE YOUR FLIGHT PLAN
Freq. 123.65 or 122.55
Phone no. 251-8269

Figure 14.4

Airport/Facility Directory excerpt (Figure 14.1) states that there is a control tower and that it is operated from 1600-0400Z.

c. There are several options here. You can use the Panoche VOR (PXN, 112.6) and the 49° radial, or you can use the Merced VOR (MCE, 114.2) and the 94 radial, or you can use the Friant VOR (FRA, 115.6) and the 245° radial. The best one is the Friant VOR since it is approximately perpendicular to your line of flight in that region. A radial that is perpendicular to your route of flight is more accurate than one that runs at an obtuse or acute angle.

Form Approved: OMB No. 04-R0072

DEPARTMENT OF TRANSPORTATION FEDERAL AVIATION ADMINISTRATION **FLIGHT PLAN**	CIVIL AIRCRAFT PILOTS. FAR Part 91 requires you file an IFR flight plan to operate under instrument flight rules in controlled airspace. Failure to file could result in a civil penalty not to exceed $1,000 for each violation (Section 901 of the Federal Aviation Act of 1958, as amended). Filing of a VFR flight plan is recommended as a good operating practice. See also Part 99 for requirements concerning DVFR flight plans.

1. TYPE	2. AIRCRAFT IDENTIFICATION	3. AIRCRAFT TYPE/ SPECIAL EQUIPMENT	4. TRUE AIRSPEED	5. DEPARTURE POINT	6. DEPARTURE TIME		7. CRUISING ALTITUDE
					PROPOSED (Z)	ACTUAL (Z)	
☒ VFR ☐ IFR ☐ DVFR	N 7118Q	Starship U-2	92 KTS	FCH	1600		6500

8. ROUTE OF FLIGHT

D→ Q68 D→ LSN D→ 062 D→ PXN D→ FCH

9. DESTINATION (Name of airport and city)	10. EST. TIME ENROUTE		11. REMARKS
Fresno Chandler Downtown Fresno	HOURS 3	MINUTES 13	Stopovers at Q68 (1t), 062 (2t)

12. FUEL ON BOARD		13. ALTERNATE AIRPORT(S)	14. PILOT'S NAME, ADDRESS & TELEPHONE NUMBER & AIRCRAFT HOME BASE	15. NUMBER ABOARD
HOURS 4	MINUTES 00	N/A	V. Fraser 309 South Willard Fresno 555-0324	2

16. COLOR OF AIRCRAFT	
Black and blue	CLOSE VFR FLIGHT PLAN WITH ___Fresno___ FSS ON ARRIVAL

FAA Form 7233-1 (5-77)

Figure 14.5

d. 27.5 NM in 20 min. (.33 hr.) yields a groundspeed of 82.5 kts., considerably less than you estimated. Perhaps the winds are not as forecasted, or perhaps you are not following your course precisely. Finally, you should realize that, over relatively short distances, rounding errors may also account for some variance in your calculations.

e. The sectional chart and Airport/Facility Directory both say it is 123.0 MHz.

    f. You will be able to land, since the crosswind component
       is 13 kts., which is below the demonstrated crosswind
       component of 17 kts. Since this component is close to
       the airplane's limit, you must be sure to consider your
       own proficiency before deciding to attempt the landing.
    g. You should plan a pattern altitude of 3,700 ft. MSL (800
       ft. AGL), as described in the <u>Airport/Facility Directory.</u>
       Since you will be landing on Rwy 27, you should plan for
       a right-hand traffic pattern. Fly over the field, make a
       circle to the west, and enter downwind at a 45° angle
       midway along the runway. On downwind, you will be flying
       090°, keeping in mind that you will have to crab to the
       south to maintain a straight downwind path. When you
       turn base (180°), the wind will retard you somewhat and
       also push you to the east, which means you will have to
       crab to the right. On final (270°), you will have to
       crab to the left to maintain a straight final approach.

4.   Leg 2:
    a. The alert area is in effect from 0700 to 0200 Monday
       through Friday from 1,000 to 4,000 ft. MSL. You will
       overfly the area, so there is no need for alarm. If you
       were flying through it, you would exercise extreme
       caution.
    b. You are not required to contact them since the ATA
       extends from ground level up to but not including 3,000
       ft. AGL. You are well above the ATA.
    c. You might be tempted to land at Merced or Atwater, but
       since it is a hot day the high oil temperature might not
       be abnormal. You should check frequently to see if the
       temperature continues to rise, or if it is stable.
    d. You are northeast of the field and will land on Rwy 14,
       which has a right-hand traffic pattern. As at Pine
       Mountain, you should fly over the field so you can enter
       the pattern on the southwest side of the field. Pattern
       altitude is 800 ft. AGL, so you will fly the pattern at
       919 ft. MSL. The downwind leg will be 320° with a crab
       to the left; base will be 50° with crab to the right; and
       final will be 140° with a crab to the right.
    e. Call on 122.8 and ask for it to be activated.
    f. Use the telephone; ask the operator for Enterprise 14598,
       which will connect you with Fresno FSS (FAT).
    g. It is 13 NM and it took you 0.2 hr., for a groundspeed
       of 65 kts., only slightly less than you estimated.

5.   Leg 3:
    a. Note that the elevation rises from 450 ft. MSL to 3,560
       ft. MSL about 7 NM from Carmel Valley along your proposed
       route of flight. Make certain you check that you will be
       well above that altitude when you go over the mountains.
    b. Set the VOR to 112.6 MHz, check for the identification
       (PXN), and fly TO 057°.
    c. Fly FROM 072°.
    d. Contact Chandler Tower on 121.1 and tell them where you
       are (over Kerman) and what your intentions are. Again,
       note that the sectional chart does not show a control
       tower at Chandler; but, according to the <u>Airport/Facility</u>

<u>Directory</u>, there is one that operates from 1600-0400Z.
e. Fresno FSS.

Review Questions

1. ?--This question has been designated as unusable by the FAA
    and removed from the FAA Question Selection Sheets.
2. 4--Block 12 should contain the amount of fuel on board
    expressed in time.
3. 2--See FAR 91.5.
4. 4--See FAR 91.5.
5. 2--See FAR 91.5.

15/THE PHYSIOLOGY OF FLIGHT

MAIN POINTS

1. Flight has many effects on the human body; some are easy to understand and cope with, and others can produce serious consequences in a short period of time. This chapter explores both.

2. **Respiration** is the exchange of gases between your body and the environment. In particular, your body derives oxygen from the environment and expels carbon dioxide. As the amount of carbon dioxide increases, breathing becomes deeper and more frequent and vice versa. Furthermore, as altitude increases and atmospheric pressure decreases, less oxygen is available to the body. Gases inside the body also expand as altitude increases since they exert more pressure than is being exerted from the outside atmosphere. Thus it is sometimes necessary to clear your ears by yawning or blowing gently through your nose while pinching your nostrils (the **Valsalva technique**). Pilots should also remember that because underwater diving reverses the pressure gradient, diving should not be mixed with piloting--the body takes time to readjust.

3. Two oxygen-related disorders of particular interest to pilots are **hypoxia** (too little oxygen) and **hyperventilation** (too little carbon dioxide). **Hypoxia,** or oxygen deficiency, is a progressive condition characterized by impaired vision, euphoria, dizziness, hot and cold flashes, breathlessness, repeated thought patterns, headaches, slowed reaction time, tingling sensation, and perspiration. The time available from the onset of hypoxia to the point of inability to act is called the **time of useful consciousness (TUC).** As exposure to oxygen deficiency continues, the effects become more pronounced and may include increased respiration, inability to perform even simple computations, bluing of the skin (cyanosis), and finally, unconsciousness. The **rate of** onset increases dramatically with altitude. From sea level to 10,000 ft. (indifferent stage), vision may be affected,

particularly at night. From 10,000 to 15,000 ft. (compensatory stage), pulse, respiration, and blood pressure increase. From 15,000 to 20,000 ft. (disturbance stage), TUC is 20 to 30 min., and from 20,000 to 25,000 ft. (critical stage), TUC can be as short as 3 to 5 min. Rapid decompression in a pressurized cabin can produce extremely rapid (less than a minute) TUC. Furthermore, anything that affects the body's ability to get oxygen to the brain, such as hypertension or smoking, can increase the rate of onset. One way to counteract the effects of hypoxia is to descend at once to a lower altitude since recovery is rapid in an oxygen-rich environment. Oxygen should be used above 10,000 ft. during the day and above 5,000 ft. at night. FAR oxygen restrictions were discussed in Chapter 11.

4.   **Hyperventilation**, or overbreathing, is an excessively fast rate of respiration that often occurs unknowingly in response to stress. Hyperventilation can cause the individual to pass out due to a severe shortage of carbon dioxide available to the voluntary respiration center. Symptoms include dizziness, tingling sensation, nausea, muscle tightness (tetany), cooling, and fainting. Hyperventilation can begin rapidly, particularly in anxiety-producing situations. Countermeasures are to reduce the rate of breathing or to breathe into an enclosed space, such as a bag, to increase the carbon dioxide content in the blood.

5.   Because the symptoms of hypoxia and hyperventilation are similar, it is sometimes difficult to tell which one is occurring. One way to tell is to breathe oxygen (if it is available). If you are experiencing hypoxia, the symptoms should begin to go away almost immediately. If they do not, suspect hyperventilation as the culprit.

6.   **Carbon monoxide poisoning** can be particularly insidious because the gas is odorless. Carbon monoxide (CO) deprives the brain of oxygen. Symptoms include a vague uneasy feeling, inability to concentrate, and headaches; if uncorrected, it will lead to unconsciousness and death. Effects can last for days because the blood's oxygen-carrying ability has been altered. Carbon monoxide usually enters the cabin through the heating system. If CO poisoning is suspected, close all heating vents, flood the cabin with air, and land at the nearest facility available.

7.   The motion produced in an airplane also affects the body, in particular the motion associated with **acceleration** (changes in velocity). These changes are measured in gravity-level equivalents, or Gs. Forward- and aft-working Gs are called **transverse Gs**; side-force Gs are called **lateral Gs**; and up-and-down Gs are referred to as **positive** (those that push you down) or **negative** (those that push you up). Gs produced in varying combinations are called **asymmetric Gs**. Positive Gs tend to deprive blood from the head and upper extremities, while negative Gs do just the opposite. Excessive positive Gs can lead to grayout or blackout due to the extreme lack of blood. The effect of positive Gs, pushing blood to the lower extremities, can be

counteracted by tightening the calf, thigh, and abdominal muscles. The effects of negative Gs, pushing blood to your head, are rare in general aviation and have no easy remedy.

8.   The dynamics of flight can have a dramatic effect on spatial orientation. We receive special cues from three sources: eyes, motion sensors (the vestibular system) in the semicircular canals of the inner ear, and proprioceptive feedback from the postural system of touch, pressure, and tension. False cues may begin to develop, however, due to the motion produced in the cockpit, especially under conditions of low visibility. For the most part, we have learned to trust our eyes; and, as long as VMC conditions prevail, they provide the quickest way to correct any false cues from the vestibular or postural senses.

9.   **Spatial disorientation**, which is frequently referred to as **vertigo**, occurs when a pilot cannot orient the aircraft to the natural horizon. The condition is physiological, not learned, but understanding it will help you cope with it. By all means, if you suspect vertigo, **trust your aircraft instruments**. There are several motion-related disorientations. The **graveyard spiral** is a result of illusions in which the vestibular and postural systems adjust and stabilize to a spiral (power-on descending turn). Without reference to the appropriate instruments, you may be led to aggravate the situation by adding power and increasing the angle of bank. The **Coriolis illusion** occurs when you move your head quickly once the vestibular system has adjusted to constant conditions, such as a constant turn. It is your head that has moved, however, not the airplane. The **acceleration illusion** may occur when you experience transverse Gs due to a change in velocity. The sensation is that you are climbing, which may lead you to lower the nose of the airplane. The **oculogyral illusion** may occur when you perceive the visual field as moving, a condition that can be overcome by holding your head still and concentrating on the instrument panel. **Indefinite horizons** can be a problem at night when you confuse stars and ground lights or when angled cloud decks suggest a horizon. Staring at a stationary light may lead to **autokinesis**, in which the light source appears to move. This effect is why many aviation lights such as beacons rotate, flash, or are arranged in patterns. Finally, **postural illusions** ("seat of the pants") can result from the various motions and G forces that act on the body. To prevent spatial disorientation, trust your eyes when you can see the horizon and your flight instruments when you cannot see it.

10.   **Airsickness**, a sympathetic nervous system reaction to conflicting sensations, may have both physical components (for example, a warm, stuffy cockpit or changes in velocity) and psychological components (anxiety or anticipation of the possibility of becoming sick). The best way to cure airsickness is to prevent it. If it occurs or begins to occur, it is wise to scan the horizon or administer fresh air (or oxygen if it is available).

11. **Night vision** is less acute than daytime vision. It takes the eyes approximately 30 min. to fully adapt to a dark environment. Furthermore, the eyes are less sensitive to objects straight ahead at night, which means you may have to glance at cockpit instruments often and out of the side of your eye at night (peripheral vision). **Noise and vibration** also affect pilots, particularly over a long period of time. Sound, measured in **decibels**, can also become so intense that it produces pain or, in extreme cases, physical damage to the middle and inner ears. Persons who fly more than eight hours a week may want to wear ear plugs to help avoid hearing loss.

12. Stress (psychological and bodily tensions) is a natural consequence of living. Not all stress is bad, nor is it necessarily produced from unpleasant things. Stress compels us to resolve certain situations and pushes us on to new endeavors. Extremely high levels of stress (such as panic) and extremely low levels of stress (such as sleep) both lead to low or ineffective performance. As stress increases (as in landing an airplane), our ability to perform increases up to a maximal level, after which performance capability decreases rapidly.

13. Stress has three sources: the environment, the body, and psychological processes. As you have probably experienced, the three sources frequently overlap and interact with one another. **Environmental stresses** of importance to pilots are apprehension, anxiety, and frustration. **Body stress** includes fatigue, both **acute** (short-lived) and **chronic** (continued). Chronic fatigue can have serious physiological effects on the body, such as slowed reactions and an ambivalent attitude. Diet also affects body stress, as do drugs and alcohol. Drugs, even readily available cold remedies and pain relievers, should not be taken until cleared by an aviation medical examiner; many contain substances that dramatically affect bodily functions, and these effects may be exaggerated at high altitudes. Alcohol, which is a **depressant**, should be avoided 24 hr. before flight time, even though FARs only prohibit its use 8 hr. before flight. In addition to reduced judgment and a false sense of capability, the effects of alcohol increase rapidly with altitude. The third type of stress, **psychological stress**, includes worry, job or marital difficulty, and anxiety. Reactions to stress, regardless of its source, include increases in pulse rate, blood pressure, perspiration, respiration, and muscle tension, all of which produce deleterious outcomes if they occur over a long period of time. One way to deal with stress is to learn to accommodate through relaxation.

14. **Panic**, a sudden overpowering and unreasoning fright, can seriously impede your ability to function and can be avoided, in part, by limiting your susceptibility to stress. This may be accomplished, for example, by relaxation exercises, by gaining more knowledge about your airplane, and by expanding your aviation experience.

15. Keeping your body in a general state of good health, knowing you are under stress or in a state of fatigue, and observing regulations about medication, alcohol, and drugs all will contribute to being a safe pilot.

KEY TERMS AND CONCEPTS, PART 1

Match each term or concept (1-16) with the appropriate description (A-P) below. Each item has only one match.

___ 1. hyperventilation
___ 2. negative Gs
___ 3. carbon dioxide
___ 4. graveyard spiral
___ 5. positive Gs
___ 6. aviation physiology
___ 7. acceleration illusion
___ 8. stress

___ 9.   TUC
___ 10. eustachian tube
___ 11. airsickness
___ 12. Valsalva technique
___ 13. hypoxia
___ 14. respiration
___ 15. tetany
___ 16. decibel

A. tube that connects the space behind the eardrum to the throat
B. time from the onset of oxygen deficiency to unconsciousness
C. blood begins to accumulate in the upper extremities as a result of these Gs
D. blood begins to accumulate in the lower extremities as a result of these Gs
E. science dealing with body functions in the flying environment
F. an excessively fast rate of respiration
G. used to measure your body's need for oxygen
H. illusion caused by transverse Gs
I. sympathetic reaction of the stomach to conflicting vestibular, visual, and postural sensations
J. pinching your nostrils and blowing gently through your nose
K. psychological, environmental, or body tension
L. constant descending turn that is the result of disorientation
M. condition in which the brain and body tissues receive too little oxygen
N. exchange of gas between you and the environment
O. muscle tightness
P. measure of sound intensity

KEY TERMS AND CONCEPTS, PART 2

Match each term or concept (1-16) with the appropriate description (A-P) below. Each item has only one match.

___ 1. vertigo
___ 2. transverse Gs
___ 3. acceleration
___ 4. oculogyral illusion
___ 5. alcohol
___ 6. fatigue
___ 7. vestibular
___ 8. oxygen

___ 9.   panic
___ 10. Coriolis illusion
___ 11. carbon monoxide
___ 12. aviation medicine
___ 13. autokinesis
___ 14. lateral Gs
___ 15. alveoli
___ 16. cyanosis

A. branch of medicine that deals with body functions in the flying environment
B. perceptual system located in the semicircular canals
C. believing that your field of vision is moving
D. a depressant drug that may lead to a false sense of confidence
E. the belief that the airplane is spinning in space
F. forward- and aft-working Gs
G. sudden, overpowering, and unreasoning fright
H. bluing of the skin
I. apparent movement of a stationary light after you have been staring at it
J. colorless, odorless, poisonous gas given off by internal combustion engines
K. air sacs in the lungs where oxygen and carbon dioxide are exchanged
L. gas your body needs to convert food to energy
M. any change in velocity
N. side-force Gs
O. illusion caused by a rapid head movement when in a constant-rate turn
P. caused by lack of sleep, too much stress, too much activity, or improper diet

DISCUSSION QUESTIONS AND EXERCISES

1.   What is respiration? How does the blood's chemistry regulate normal respiration?

2.   Shortly after leveling off at 9,500 ft., a passenger with a mild cold begins to complain of a severe headache. What physiological disorder might you suspect and what action should you take?

3.   Outline the effects of positive and negative Gs on the body's circulatory system?

4.   What three body systems provide cues for spatial orientation? Which is most reliable in VMC conditions?

5.   Name three sources of stress. What special types of flight-related stress may student pilots encounter? What effects do very high and very low levels of stress have on our ability to perform?

6.   Briefly explain what effects each of the following may have on your ability to perform as a pilot:

    a.   fatigue

    b.   diet

    c.   alcohol

7.   Explain the difference between acute fatigue and chronic fatigue. Which is more dangerous? Why?

8.   What is the FAA regulation concerning the time between consuming alcohol and acting as a pilot-in-command?

9.   Name at least five symptoms associated with hypoxia. How does the rate of onset vary with increasing altitude? What can you do to counteract it?

10. What is hyperventilation? Name at least four symptoms. How is it related to stress and how can it be counteracted?

11. What is carbon monoxide poisoning? Why is it so dangerous? What should you do if you suspect it while in flight?

12. What is airsickness? What action should you take if a passenger begins to complain of nausea during flight?

13. Briefly characterize each of the following motion-related disorientations and state what can be done to overcome them:

    a. vertigo

    b. graveyard spiral

c.   Coriolis illusion

d.   oculogyral illusion

e.   acceleration illusion

f.   autokinesis

g.   postural illusion

14.   What are two general rules for minimizing the occurrence and effects of spatial disorientation?

15.   What is panic?   How can it be avoided?

16.   How is night vision different from daytime vision?   Why is this difference of particular importance to pilots?

REVIEW QUESTIONS

1.   (FAA 1905) To preclude the effects of hypoxia, you should

1--avoid flying above 10,000 ft. MSL for prolonged periods without
    breathing supplemental oxygen.
2--rely on your body's built-in alarm system to warn when you are
    not getting enough oxygen.
3--try swallowing, yawning, or holding the nose and mouth shut and
    forcibly try to exhale.
4--avoid hyperventilation which is caused by rapid heavy breathing
    and results in excessive carbon dioxide in the bloodstream.

2.   (FAA 1906 DI) Hypoxia is caused by

1--nitrogen bubbles forming in the blood at high altitudes.
2--trapped gases in the body.
3--reduced atmospheric pressure.
4--toxic substances in the blood.

3.   (FAA 1907) Rapid or extra deep breathing while using oxygen
can cause a condition known as

1--hypoxia.   3--aerotitis.
2--aerosinusitis.   4--hyperventilation.

4.   (FAA 1908) A pilot should be able to overcome the symptoms or
avoid future occurrence of hyperventilation by

1--closely monitoring the flight instruments to control the
    airplane.
2--slowing the breathing rate, breathing into a bag, or talking
    aloud.
3--increasing the breathing rate in order to increase lung
    ventilation.
4--refraining from the use of over-the-counter remedies and drugs
    such as antihistamines, cold tablets, tranquilizers, etc.

5.   (FAA 1909 DI) What action should be taken if hyperventilation
is suspected?

1--Breathe at a slower rate by taking very deep breaths.
2--Consciously breathe at a slower rate than normal.
3--Consciously force yourself to take deep breaths and breathe at
    a faster rate than normal.
4--If oxygen is available go on 100 percent oxygen.

6.   (FAA 1910) Which would most likely lead to hyperventilation?

1--Emotional tension, anxiety, or fear.
2--The excessive consumption of alcohol.
3--An extremely slow rate of breathing and insufficient oxygen.
4--An extreme case of relaxation or sense of well-being.

7.   (FAA 1911) Large accumulations of carbon monoxide in the
human body result in

1--tightness across the forehead.
2--loss of muscular power.
3--an increased sense of well-being.
4--being too warm.

8.   (FAA 1912) Susceptibility to carbon monoxide poisoning
increases as

1--altitude increases.
2--altitude decreases.
3--air pressure increases.
4--humidity of the air decreases.

9.   (FAA 1913) A pilot is more subject to spatial disorientation
if

1--ignoring or overcoming the sensations of muscles and inner ear.
2--kinesthetic senses are ignored.
3--eyes are moved often in the process of cross-checking the
    flight instruments.
4--body signals are used to interpret flight attitude.

10. (FAA 1914) A pilot experiences spatial disorientation during
flight in a restricted visibility condition. The best way to
overcome the effect is to

1--depend on sensations received from the fluid in the
    semicircular canals of the inner ear.
2--concentrate on any yaw, pitch, and roll sensations.
3--consciously slow your breathing rate until symptoms clear and
    then resume normal breathing rate.
4--rely upon the aircraft instrument indications.

11. (FAA 1915) A state of temporary confusion resulting from
misleading information being sent to the brain by various sensory
organs is defined as

1--spatial disorientation.   3--hypoxia.
2--hyperventilation.   4--motion sickness.

12. (FAA 1916) What preparation should a pilot make to adapt the
eyes for night flying?

1--Wear sunglasses after sunset until ready for flight.
2--Avoid red lights at least 30 min. before the flight.
3--Wear amber colored glasses at least 30 min. before the flight.
4--Avoid bright white lights at least 30 min. before the flight.

13. (FAA 1917) What is the most effective way to use the eyes during night flight?

1--Look only at far away, dim lights.
2--Scan slowly to permit off center viewing.
3--Blink the eyes rapidly when concentrating on an object.
4--Concentrate directly on each object for a few seconds.

14. (FAA 1103) What type of oxygen should be used to replenish an aircraft's oxygen system for high-altitude flights?

1--Medical oxygen.   3--Welder's oxygen.
2--Therapeutic oxygen.   4--Aviation breathing oxygen.

15. The Valsalva technique is used to counteract

1--differential pressure in the eustachian tube.
2--hyperventilation.
3--hypoxia.
4--spatial disorientation.

16. A _____ positive G force will push blood toward your head while you are flying an airplane.

1--lateral   2--negative   3--positive   4--transverse

17. The motion-sensing semicircular canals in the middle ear are part of the _____ perceptual system.

1--auditory
2--proprioceptive (postural)
3--vestibular
4--visual

18. Continued exposure to fatigue is called _____ fatigue.

1--acute   2--active   3--chronic   4--delayed

19. Which of the following is **not** a common reaction to alcohol?

1--Improved reaction time.
2--Increased self-confidence.
3--Loss of visual acuity.
4--Poor reasoning capability.

20. Which statement is true regarding alcohol in the human system?

1--A common misconception is that coffee alters the rate at which the body metabolizes alcohol.
2--An increase in altitude decreases its adverse effects.
3--Alcohol increases judgment and decision-making abilities.
4--Aspirin increases the rate the body metabolizes alcohol.

21. Which of the following techniques is a common way to combat the effects of positive Gs?

1--Administer fresh air.
2--Breathe gently through the nose while pinching the nostrils.
3--Tighten your abdominal, thigh, and calf muscles.
4--All of these are common, effective techniques.

22. Which of the following statements about hypoxia is correct?

1--Hypoxia is a singular condition that typically strikes all at once.
2--Hypoxia symptoms occur more rapidly as altitude increases, particularly above 10,000 ft. MSL.
3--One of the most effective countermeasures for the effects of hypoxia is the Valsalva technique.
4--All of the above are correct.

ANSWERS

Key Terms and Concepts, Part 1

1.	F	5.	D	9.	B	13.	M
2.	C	6.	E	10.	A	14.	N
3.	G	7.	H	11.	I	15.	O
4.	L	8.	K	12.	J	16.	P

Key Terms and Concepts, Part 2

1.	E	5.	D	9.	G	13.	I
2.	F	6.	P	10.	O	14.	N
3.	M	7.	B	11.	J	15.	K
4.	C	8.	L	12.	A	16.	H

Review Questions

1. 1--Although FAR regulations do not require it until you reach 12,500 ft., it is generally good practice to avoid flying above 10,000 ft. for prolonged periods without supplemental oxygen.
2. ?--This question has been designated as unusable by the FAA and removed from the FAA Question Selection Sheets.
3. 4--Hyperventilation leads to rapid or extra breathing.
4. 2--The effects of hyperventilation can be attenuated by decreasing the rate of breathing or, if necessary, breathing into a bag.
5. ?--This question has been designated as unusable by the FAA and removed from the FAA Question Selection Sheets.
6. 1--Emotional tension, anxiety, and fear are all factors associated with hyperventilation.

7. 3--Euphoria, a sense of well-being, is one symptom of carbon
    monoxide poisoning.
8. 1--Increases in altitude increase susceptibility to carbon
    monoxide poisoning.
9. 4--Using body signals ("flying by the seat of your pants") to
    interpret flight attitudes is a good way to become
    spatially disoriented (vertigo).
10. 4--**Always** rely on the aircraft instruments if you suspect
    that you have become spatially disoriented, a condition
    that is more likely to develop when visibility is poor.
11. 1--Confusion resulting from misleading information being sent
    to the brain by various sensory organs is defined as
    spatial disorientation, or vertigo.
12. 4--For night flight, you should avoid bright lights for at
    least 30 min. prior to flight as it takes that long for
    the eyes to adapt to low levels of illumination such as
    those experienced in the cockpit environment.
13. 2--When flying at night, scan slowly to permit off center
    viewing (peripheral vision).
14. 4--Only FAA-approved oxygen should be used.
15. 1--The Valsalva technique helps to reduce the pressure
    difference; it is used when yawning does not work.
16. 2--Negative Gs are most commonly encountered in general
    aviation during turbulent conditions.
17. 3--They are filled with fluids that respond to acceleration.
18. 3--Short exposure to fatigue is referred to as acute.
19. 1--Reaction time increases--that is, your reactions are
    slower.
20. 1--Your body metabolizes (burns) alcohol at a constant rate;
    there are no known ways to increase how fast it burns.
21. 3--This helps prevent blood from collecting in the lower
    extremities.
22. 2--Hypoxia is a progressive condition (not a single event);
    symptoms occur more rapidly at high altitudes.

16/HANDLING AIRBORNE EMERGENCIES

MAIN POINTS

1. Emergency procedures are no different from normal operating procedures. They are simply encountered less frequently.

2. Three general rules apply to airborne emergencies: (1) maintain control of the aircraft; (2) attempt to locate the cause, analyze the situation, and take proper action; and (3) land as soon as conditions permit. Specific procedures are found in your airplane's Pilot's Operating Handbook (POH). Learn them as if your life depended on it. It does.

3. Three factors cited most frequently in general aviation accidents are: (1) inadequate preflight procedures and/or planning; (2) failure to maintain flying speed; and (3) improper in-flight decision making.

4. Unless you are confronted with a real emergency, such as an engine fire, you should first confirm a suspected emergency with at least one other verifying indication. Pilot experience and confidence also play a factor in identifying and declaring an emergency. When more than one emergency exists at once, such as a radio/navigation failure in marginal weather conditions, it is referred to as a compound emergency.

5. Equipment malfunctions can be diagnosed more easily if you are familiar with each instrument and how it is powered. For example, if you suspect an electrical malfunction, you should check the proper circuit breaker before taking further action.

6. Safety begins at home during preflight planning and inspections. Two important aspects of flight preparation are filing a flight plan and using flight following services. Once you file a flight plan, follow it, or advise the nearest FSS if you make changes. When the flight is complete, close it!

7.   Emergency landing procedures are grouped into four categories: (1) land as soon as practicable means to land at the nearest suitable airport; (2) land as soon as possible means to land at the nearest airport regardless of facilities; (3) land immediately means that continued flight is inadvisable, as for example when there is an engine fire; and (4) forced landing, when the airplane will land very soon whether the pilot wishes it or not, as for example when there is a total engine failure.

8.   There are two types of forced landings, those that occur immediately after takeoff and those that occur after the airplane has gained a reasonable amount of altitude. When forced to land on the takeoff leg, land straight ahead! It is acceptable to make turns to avoid obstacles, but do not attempt to turn back to the runway because maintaining airplane control is essential.

9.   Once the airplane has gained altitude, forced landing procedures include five phases. First, as outlined in your POH, there are several immediate actions you should take, including restarting procedures in case of engine malfunction. In any event, it is absolutely essential to maintain altitude and airspeed. Establish a glide path, if necessary, aimed at maintaining as much altitude as possible. If restart fails, set most light trainers as follows: mixture, full lean; fuel selector, off; ignition system, off; master switch, on (for radios and flaps). Second, select a landing site, preferably one with a hard surface that will allow you to land into the wind. Third, report your position and situation to the nearest facility, or use the standard emergency frequency. Fourth, in preparation for landing, secure seatbelts and shoulder harnesses (if provided) and stow loose items in the cabin. Fifth, maintain airspeed, avoid improvising at the last minute, and hold back pressure as long as possible after touchdown.

10.   Know and use your **emergency locator transmitter (ELT)**. When activated, an ELT transmits on 121.5 or 243.0 MHz. If the ELT has been activated, you can monitor it on 121.5 MHz prior to engine shut down. Operational tests of the ELT should be made only during the first 5 min. of the hour. Finally, the non-rechargeable batteries of an ELT need to be replaced when 50 percent of their useful life expires or they have been in use for a cumulative period of one hour.

11.   You should be familiar with ground-to-air signals in the event you land in a remote location. Stay near or in your airplane until you are located, unless you see a source of assistance. If you regularly fly in remote areas, you should carry a survival kit.

12.   Being lost **is** an emergency. Remember the **"four Cs"**: confess, communicate, climb, comply. **Confess** your problem. **Communicate** your situation over the appropriate radio channel. **Climb** when possible to improve radio communication, but maintain VFR conditions. **Comply** with controller instructions, again staying within VFR conditions. Radio assistance may include ATC

directions if you have a transponder, or a direction-finding (DF)
steer if you have a VHF transmitter and receiver.

13.   The VHF emergency frequency, 121.5 MHz, is to be used
only by an aircraft in distress. In an emergency, communicate
your situation in plain English. State **mayday** three times (or **pan**
if you are uncertain about the situation). Next, repeat your
airplane identification number three times. Give your type of
aircraft, position, heading, TAS, altitude, fuel remaining, nature
of distress, intentions, and request. Finally, hold your mike
button open for two consecutive ten-second intervals and then
repeat your airplane ID and say "over." If you have a
transponder, squawk 7700 to indicate an emergency.

14.   If your communications receiver fails, you may still be
able to transmit and you may be able to receive over your
navigation radio. If neither appears to solve the problem, use
the information you have already learned about responding to
controller light signals. If you have a transponder, squawk 7600
to indicate two-way radio communication.

15.   The FAA has several guidelines for safe piloting: know
your limits, both physiological and technical; use a checklist,
particularly for routine tasks; preplan your flight: prepare,
file, open, fly, and close a flight plan; preflight your airplane;
know your airplane and its performance limits.

KEY TERMS AND CONCEPTS

Match each term and concept (1-12) with the appropriate
description (A-L) below. Each term has only one match.

___ 1. forced landing   ___ 7. 7600
___ 2. POH   ___ 8. maintain airplane control
___ 3. 7700   ___ 9. pan
___ 4. compound emergency   ___ 10. as soon as practicable
___ 5. 121.5 MHz   ___ 11. confess your problem
___ 6. mayday   ___ 12. as soon as possible

A. landing at the nearest airport with suitable facilities
B. first thng you should do in an emergency situation
C. words used to indicate an actual airborne emergency
D. landing at once whether the pilot wishes it or not
E. more than one emergency condition occurs at the same time
F. transponder code to indicate a failure in two-way radio
    communication capability
G. VHF emergency frequency
H. where you find specific emergency procedures for your airplane
I. transponder code to squawk to declare an airborne emergency
J. what you should radio first when you are lost
K. words to use to notify the ground of a probable airborne
    emergency
L. landing in which the pilot lands at the nearest airport at
    once regardless of available facilities

DISCUSSION QUESTIONS AND EXERCISES

1.   Outline the three general rules for handling any airborne emergency.

2.   What are three factors cited most frequently in general aviation accidents?

3.   What is the primary way to distinguish between a suspected malfunction and an actual malfunction?

4.   What are compound emergencies? Give an example illustrating how a pilot might help to create a compound emergency.

5.   Suppose you have tried unsuccessfully several times to contact a nearby FSS station on your radio. Do you have a malfunctioning radio? How would you troubleshoot the situation?

6.   Identify, describe, and give an example of each of the four types of emergency landings:

        a.   land as soon as practicable –

        b.   land as soon as possible –

    c.   land immediately –

    d.   forced landing –

7.   Outline three immediate actions you should take in case of a sudden engine failure.

8.   Name three criteria you should use in selecting a forced landing site.

9.   What are the four Cs and how do they apply to being lost?

10. Identify the appropriate frequencies, code, or word for each of the following:

    a.   VHF guard frequency

    b.   ATC transponder code for an airborne emergency

    c.   ATC transponder code for two-way radio failure

    d.   international code for an airplane in distress

    e.   international code word for a probable emergency

11. Refer to your San Francisco sectional chart and the cross-country exercise in Chapter 14. Suppose you are en route from Los Banos direct to Carmel Valley at 6,500 ft. You are 30 NM from Los Banos on course when your oil pressure skyrockets and the engine begins to run exceedingly rough. Use the data from the cross-country exercise in Chapter 14 to aid in your planning. Your airplane has a glide speed of 63 kts., which translates to about 1.3 NM per 1,000 ft. above the terrain.

    a. Where would you attempt to land? Why?

    b. What characteristics would you look for in selecting a landing site?

    c. Describe in detail what you would communicate over your radio.

12. Once activated, on what frequency does an ELT transmit? On what frequency can you monitor it? During what times can you make operational tests of an ELT? When should non-rechargeable batteries be replaced?

13. Briefly outline the six steps in the FAA Accident Prevention Program.

REVIEW QUESTIONS

1.   (FAA 1497) When activated, an ELT transmits on

1--122.3 and 122.8 MHz.   3--121.5 and 243.0 MHz.
2--123.0 and 119.0 MHz.   4--118.0 and 118.8 MHz.

2.   (FAA 1498) Operational tests of the ELT should be made only

1--during the first 5 min. of the hour.
2--during annual inspection.
3--after one-half the shelf life of the battery.
4--upon replacing the battery.

3.   (FAA 1499) Which procedure is recommended to ensure that the
ELT has not been activated?

1--Turn off the aircraft ELT after landing.
2--Ask the airport tower if they are receiving an ELT signal.
3--Monitor 121.5 before engine shut down.
4--Have a certified repair station inspect the ELT.

4.   (FAA 1500) When making routine transponder code changes
pilots should avoid inadvertent selection of which code?

1--3100, 7600, 7700.   3--7000, 7600, 7700.
2--7500, 7600, 7700.   4--4000, 7600, 7700.

5.   (FAA 1503) To use VHF/DF facilities for assistance in
locating an aircraft's position, the aircraft must have

1--a VHF transmitter and receiver.
2--an IFF transponder.
3--a VOR receiver and DME.
4--an ELT.

6.   (FAA 1502) The letters VHF/DF appearing in the
Airport/Facility Directory for a certain airport, indicate that

1--this airport is designated as an airport of entry.
2--the flight service station has equipment with which to
   determine your direction from the station.
3--this airport has a direct-line phone to the flight service
   station.
4--this airport is a defense facility.

7.   (FAA 1621) Operational tests of the ELT (emergency locator
transmitter) should be made only

1--during the first 5 min. of an hour.
2--during the annual inspection.
3--after one-half the shelf life of the battery.
4--upon replacing the battery.

8.   (FAA 1622) When are non-rechargeable batteries of an ELT (emergency locator trasmitter) required to be replaced?

1--Every 24 mo.
2--When 50 percent of their useful life expires or they were in use for a cumulative period of 1 hr.
3--At the time of each 100-hr. or annual inspection.
4--Annually.

9.   Which of the following things should you do **first** in any airborne emergency?

1--Contact air traffic control (ATC) and confess your problem.
2--Contact the nearest FSS facility.
3--Maintain aircraft control.
4--Stay in VMC conditions.

10. Which of the following factors is cited **least** in general aviation accidents?

1--Engine failure.
2--Failure to obtain or maintain flying speed.
3--Improper in-flight decision making or planning.
4--Inadequate preflight preparation and/or planning.

11. Where do you find specific guidance and procedures for emergency situations?

1--AIM.
2--Airport/Facility Directory.
3--Pilot's Operating Handbook (POH).
4--Rescue Coordination Handbook.

12. _____ means to locate the nearest airport and land at once regardless of available facilities.

1--Forced landing
2--Land as soon as practicable
3--Land as soon as possible
4--Land immediately

13. What is the transponder code for an airborne emergency?

1--1200.   2--4096.   3--7600.   4--7700.

14. What is the international word to notify a ground agency of an urgent condition?

1--Help.   2--Mayday.   3--Pan.   4--Ouch.

15.   The red line on an airspeed indicator means a maximum
airspeed that

1--may be exceeded only in an emergency situation.
2--may be exceeded only if gear and flaps are retracted.
3--may be exceeded only in smooth air.
4--should not be exceeded.

16.   The **most** important rule to remember in the event of a power
failure after becoming airborne is

1--determining wind direction to plan for a forced landing.
2--maintaining safe airspeed.
3--quickly checking the fuel supply for possible fuel exhaustion.
4--turning back immediately to the takeoff runway.

ANSWERS

Key Terms and Concepts

1.	D	4.	E	7.	F	10.	A
2.	H	5.	G	8.	B	11.	J
3.	I	6.	C	9.	K	12.	L

Discussion Questions and Exercises

11.   a. Given the winds aloft (240° at 25), it makes sense to turn
         around and glide to either Christensen ranch (7.5 NM to
         the north) or Hollister (9.0 NM to the north). The glide
         ratio gives you 1.3 x 5.5, or about 7 NM in a no-wind
         situation with 1,000 ft. to spare when you reach your
         destination. Given the moderately favorable winds, you
         should be able to glide to Christensen ranch. Further-
         more, the topography between your present position and
         Christensen ranch is relatively flat, which should give
         you even more options.
      b. If you do not make it to Christensen ranch, select the
         best landing site available.
      c. mayday, mayday, mayday.
         Starship 7118Q, Starship 7118Q, Starship 7118Q.
         Starship U2.
         30 NM southwest of Los Banos, 7 south of Christensen.
         steering 51° true.
         true airspeed 97 kts.
         6,500 ft. and descending.
         with over 2 hr. fuel.
         engine out.
         proceeding direct to Christensen ranch.
         notify Christensen ranch.
         hold mike button for 10 seconds.
         Starship 7118Q.
         over.

If you have recently contacted FSS, call on that
frequency. Otherwise use 121.5 MHz, the VHF guard
frequency.

Review Questions

1. 3--When activated, an ELT transmits on 121.5 and 243.0 MHz.
2. 1--Operational tests should be made only during the first 5
    min. of the hour.
3. 3--You can test to see if the ELT has been activated by
    monitoring 121.5 before shutting down the engine.
4. 2--7500 is the code for hijacking, 7600 is the code for a
    loss of two-way radio communication, and 7700 is the code
    to indicate an airborne emergency.
5. 1--To receive a VHF/DF steer, the aircraft must have a VHF
    transmitter and receiver.
6. 2--The VHF/DF designation means the FSS has equipment with
    which to determine your direction from the station.
7. 1--See answer 2.
8. 2--Non-rechargeable batteries must be replaced when 50
    percent of their useful life expires or they were in use
    for a cumulative period of 1 hr.
9. 3--First and foremost, maintain aircraft control.
10. 1--Equipment malfunctions (for example, engine failure) play
    a minor role in general aviation accidents.
11. 3--Your POH contains **specific** emergency procedures.
12. 3--Land as soon as possible means that safety requires that
    the airplane be landed at once, without regard to
    available facilities.
13. 4--7700 is the emergency code.
14. 3--"Pan" is the word to use to indicate a probable emergency
    or urgent situation. "Mayday" is the word to use to
    indicate an airborne emergency.
15. 4--Red line indicates the speed that should not be exceeded.
16. 2--The most important rule is to maintain aircraft control.

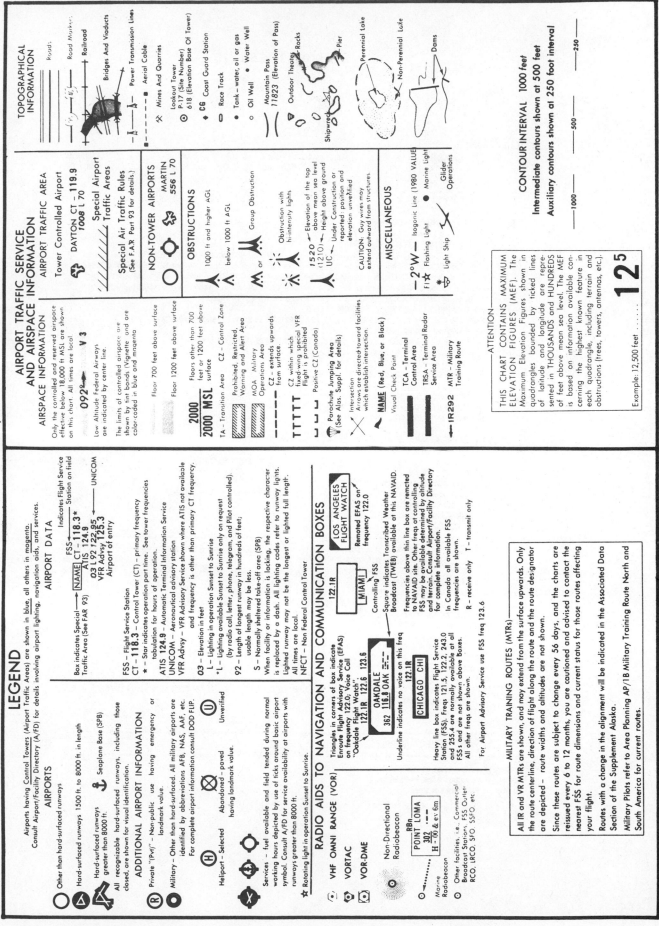

DIRECTORY LEGEND
SAMPLE

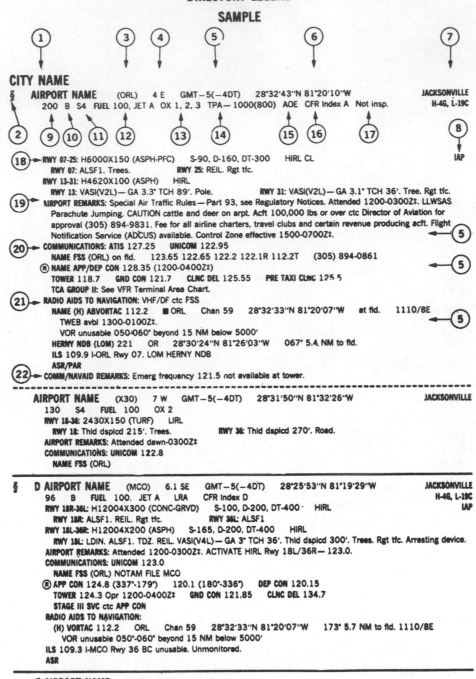

CITY NAME

§ AIRPORT NAME   (ORL)   4 E   GMT−5(−4DT)   28°32'43"N 81°20'10"W   JACKSONVILLE
    200 B S4 FUEL 100, JET A OX 1, 2, 3   TPA—1000(800)   AOE CFR Index A Not insp.   H-4G, L-19C

(18)➤ RWY 07-25: H6000X150 (ASPH-PFC)   S-90, D-160, DT-300   HIRL CL
      RWY 07: ALSF1. Trees.   RWY 25: REIL. Rgt tfc.
      RWY 13-31: H4620X100 (ASPH)   HIRL
        RWY 13: VASI(V2L)— GA 3.3° TCH 89'. Pole.   RWY 31: VASI(V2L)— GA 3.1° TCH 36'. Tree. Rgt tfc.
(19)➤ AIRPORT REMARKS: Special Air Traffic Rules—Part 93, see Regulatory Notices. Attended 1200-0300Z‡. LLWSAS.
      Parachute Jumping. CAUTION cattle and deer on arpt. Acft 100,000 lbs or over ctc Director of Aviation for
      approval (305) 894-9831. Fee for all airline charters, travel clubs and certain revenue producing acft. Flight
      Notification Service (ADCUS) available. Control Zone effective 1500-0700Z‡.
(20)➤ COMMUNICATIONS: ATIS 127.25   UNICOM 122.95
      NAME FSS (ORL) on fld.   123.65 122.65 122.2 122.1R 112.2T   (305) 894-0861
     ® NAME APP/DEP CON 128.35 (1200-0400Z‡)
      TOWER 118.7   GND CON 121.7   CLNC DEL 125.55   PRE TAXI CLNC 125.5
      TCA GROUP II: See VFR Terminal Area Chart.
(21)➤ RADIO AIDS TO NAVIGATION: VHF/DF ctc FSS
      NAME (H) ABVORTAC 112.2   ■ ORL.   Chan 59   28°32'33"N 81°20'07"W   at fld.   1110/8E
        TWEB avbl 1300-0100Z‡.
        VOR unusable 050-060° beyond 15 NM below 5000'
      HERNY NDB (LOM) 221   OR   28°30'24"N 81°26'03"W   067° 5.4 NM to fld.
      ILS 109.9 I-ORL Rwy 07. LOM HERNY NDB
      ASR/PAR
(22)➤ COMM/NAVAID REMARKS: Emerg frequency 121.5 not available at tower.

    AIRPORT NAME   (X30)   7 W   GMT−5(−4DT)   28°31'50"N 81°32'26"W   JACKSONVILLE
    130   S4   FUEL 100   OX 2
    RWY 18-36: 2430X150 (TURF)   LIRL
      RWY 18: Thld dsplcd 215'. Trees.   RWY 36: Thld dsplcd 270'. Road.
    AIRPORT REMARKS: Attended dawn-0300Z‡
    COMMUNICATIONS: UNICOM 122.8
      NAME FSS (ORL)

---

§ D AIRPORT NAME   (MCO)   6.1 SE   GMT−5(−4DT)   28°25'53"N 81°19'29"W   JACKSONVILLE
    96   B   FUEL 100, JET A   LRA   CFR Index D   H-4G, L-19C
    RWY 18R-36L: H12004X300 (CONC-GRVD)   S-100, D-200, DT-400   HIRL   IAP
      RWY 18R: ALSF1. REIL. Rgt tfc.   RWY 36L: ALSF1
    RWY 18L-36R: H12004X200 (ASPH)   S-165, D-200, DT-400   HIRL
      RWY 18L: LDIN. ALSF1. TDZ. REIL. VASI(V4L)— GA 3° TCH 36'. Thld dsplcd 300'. Trees. Rgt tfc. Arresting device.
    AIRPORT REMARKS: Attended 1200-0300Z‡. ACTIVATE HIRL Rwy 18L/36R— 123.0.
    COMMUNICATIONS: UNICOM 123.0
      NAME FSS (ORL) NOTAM FILE MCO
     ® APP CON 124.8 (337°-179°)   120.1 (180°-336°)   DEP CON 120.15
      TOWER 124.3 Opr 1200-0400Z‡   GND CON 121.85   CLNC DEL 134.7
      STAGE III SVC ctc APP CON
    RADIO AIDS TO NAVIGATION:
      (H) VORTAC 112.2   ORL   Chan 59   28°32'33"N 81°20'07"W   173° 5.7 NM to fld. 1110/8E
        VOR unusable 050°-060° beyond 15 NM below 5000'
    ILS 109.3 I-MCO Rwy 36 BC unusable. Unmonitored.
    ASR

    E AIRPORT NAME   (See PLYMOUTH)

All Bearings and Radials are Magnetic unless otherwise specified.
All mileages are nautical unless otherwise noted.
All times are GMT except as noted.

DIRECTORY LEGEND
LEGEND

This Directory is an alphabetical listing of data on record with the FAA on all airports that are open to the public, associated terminal control facilities, air route traffic control centers and radio aids to navigation within the conterminous United States, Puerto Rico and the Virgin Islands. Airports are listed alphabetically by associated city name and cross referenced by airport name. Facilities associated with an airport, but with a different name, are listed individually under their own name, as well as under the airport with which they are associated.

The listing of an airport in this directory merely indicates the airport operator's willingness to accommodate transient aircraft, and does not represent that the facility conforms with any Federal or local standards, or that it has been approved for use on the part of the general public.

The information on obstructions is taken from reports submitted to the FAA. It has not been verified in all cases. Pilots are cautioned that objects not indicated in this tabulation (or on charts) may exist which can create a hazard to flight operation.

Detailed specifics concerning services and facilities tabulated within this directory are contained in Airman's Information Manual, Basic Flight Information and ATC Procedures.

The legend items that follow explain in detail the contents of this Directory and are keyed to the circled numbers on the sample on the preceding page.

(1) CITY/AIRPORT NAME

Airports and facilities in this directory are listed alphabetically by associated city and state. Where the city name is different from the airport name the city name will appear on the line above the airport name. Airports with the same associated city name will be listed alphabetically by airport name and will be separated by a dashed rule line. All others will be separated by a solid rule line.

(2) NOTAM SERVICE

§—NOTAM "D" (Distant teletype dissemination) and NOTAM "L" (Local dissemination) service is provided for airport. Absence of annotation § indicates NOTAM "L" (Local dissemination) only is provided for airport. See AIM. Basic Flight Information and ATC Procedures for detailed description of NOTAM.

(3) LOCATION IDENTIFIER

A three or four character code assigned to airports. These identifiers are used by ATC in lieu of the airport name in flight plans, flight strips and other written records and computer operations.

(4) AIRPORT LOCATION

Airport location is expressed as distance and direction from the center of the associated city in nautical miles and cardinal points, i.e., 4 NE.

(5) TIME CONVERSION

Hours of operation of all facilities are expressed in Greenwich Mean Time (GMT) and shown as "Z" time. The directory indicates the number of hours to be subtracted from GMT to obtain local standard time and local daylight saving time GMT−5(−4DT). The symbol ‡ indicates that during periods of Daylight Saving Time effective hours will be one hour earlier than shown. In those areas where daylight saving time is not observed that (−4DT) and ‡ will not be shown. All states observe daylight savings time except Arizona and that portion of Indiana in the Eastern Time Zone and Puerto Rico and the Virgin Islands.

(6) GEOGRAPHIC POSITION OF AIRPORT

(7) CHARTS

The Sectional Chart and Low and High Altitude Enroute Chart and panel on which the airport or facility is located.

(8) INSTRUMENT APPROACH PROCEDURES

IAP indicates an airport for which a prescribed (Public Use) FAA Instrument Approach Procedure has been published.

(9) ELEVATION

Elevation is given in feet above mean sea level and is the highest point on the landing surface. When elevation is sea level it will be indicated as (00). When elevation is below sea level a minus (−) sign will precede the figure.

(10) ROTATING LIGHT BEACON

B indicates rotating beacon is available. Rotating beacons operate dusk to dawn unless otherwise indicated in AIRPORT REMARKS.

(11) SERVICING

S1:   Minor airframe repairs.
S2:   Minor airframe and minor powerplant repairs.
S3:   Major airframe and minor powerplant repairs.
S4:   Major airframe and major powerplant repairs.

DIRECTORY LEGEND

(12) FUEL

CODE	FUEL
80	Grade 80 gasoline (Red)
100	Grade 100 gasoline (Green)
100LL	Grade 100LL gasoline (low lead) (Blue)
115	Grade 115 gasoline
A	Jet A—Kerosene freeze point—40° C.
A1	Jet A-1—Kerosene, freeze point—50° C.
A1+	Jet A-1—Kerosene with icing inhibitor, freeze point—50° C.
B	Jet B—Wide-cut turbine fuel, freeze point—50° C.
B+	Jet B—Wide-cut turbine fuel with icing inhibitor, freeze point—50° C.

(13) OXYGEN

OX 1 High Pressure
OX 2 Low Pressure
OX 3 High Pressure—Replacement Bottles
OX 4 Low Pressure—Replacement Bottles

(14) TRAFFIC PATTERN ALTITUDE

Traffic Pattern Altitude (TPA)—The first figure shown is TPA above mean sea level. The second figure in parentheses is TPA above airport elevation.

(15) AIRPORT OF ENTRY AND LANDING RIGHTS AIRPORTS

AOE—Airport of Entry—A customs Airport of Entry where permission from U.S. Customs is not required, however, at least one hour advance notice of arrival must be furnished.

LRA—Landing Rights Airport—Application for permission to land must be submitted in advance to U.S. Customs. At least one hour advance notice of arrival must be furnished.

NOTE: Advance notice of arrival at both an AOE and LRA airport may be included in the flight plan when filed in Canada or Mexico, where Flight Notification Service (ADCUS) is available the airport remark will indicate this service. This notice will also be treated as an application for permission to land in the case of an LRA. Although advance notice of arrival may be relayed to Customs through Mexico, Canadian, and U.S. Communications facilities by flight plan, the aircraft operator is solely responsible for insuring that Customs receives the notification. (See Customs, Immigration and Naturalization, Public Health and Agriculture Department requirements in the International Flight Information Manual for further details.)

(16) CERTIFICATED AIRPORT (FAR 139)

Airports serving Civil Aeronautics Board certified carriers and certified under FAR, Part 139, are indicated by the CFR index; i.e., CFR Index A, which relates to the availability of crash, fire, rescue equipment.

FAR—PART 139 CERTIFICATED AIRPORTS

INDICES AND FIRE FIGHTING AND RESCUE EQUIPMENT REQUIREMENTS

Airport Index	Required No. Vehicles	Aircraft Length	Scheduled Departures	Agent + Water for Foam
A	1	≤90′	≥1	500#DC or 450#DC + 50 gal H_2O
AA	1	>90′, ≤126′	<5	300#DC + 500 gal H_2O
B	2	>90′, ≤126′	≥5	Index A + 1500 gal H_2O
		>126′, ≤160′	<5	
C	3	>126′, ≤160′	≥5	Index A + 3000 gal H_2O
		>160′, ≤200′	<5	
D	3	>160′, ≤200′	≥5	Index A + 4000 gal H_2O
		>200′	<5	
E	3	>200′	≥5	Index A + 6000 gal H_2O

> Greater Than; < Less Than; ≥ Equal or Greater Than; ≤ Equal or Less Than; H_2O–Water; DC–Dry Chemical.

NOTE: If AFFF (Aqueous Film Forming Foam) is used in lieu of Protein Foam, the water quantities listed for Indices AA thru E can be reduced 33 1/3 %. See FAR Part 139.49 for full details. The listing of CFR index does not necessarily assure coverage for non-air carrier operations or at other than prescribed times for air carrier. CFR index Ltd.—indicates CFR coverage may or may not be available. for information contact airport manager prior to flight.

DIRECTORY LEGEND

FAA INSPECTION

All airports not inspected by FAA will be identified by the note: Not insp. This indicates that the airport information has been provided by the owner or operator of the field.

18 RUNWAY DATA

Runway information is shown on two lines. That information common to the entire runway is shown on the first line while information concerning the runway ends are shown on the second or following line. Lengthy information will be placed in the Airport Remarks.

Runway direction, surface, length, width, weight bearing capacity, lighting, gradient (when gradient exceeds 0.3 percent) and appropriate remarks are shown for each runway. Direction, length, width, lighting and remarks are shown for sealanes. The full dimensions of helipads are shown, i.e., 50X150.

RUNWAY SURFACE AND LENGTH

Runway lengths prefixed by the letter "H" indicate that the runways are hard surfaced (concrete, asphalt). If the runway length is not prefixed, the surface is sod, clay, etc. The runway surface composition is indicated in parentheses after runway length as follows:

(AFSC)—Aggregate friction seal coat	(GRVD)—Grooved	(TURF)—Turf
(ASPH)—Asphalt	(GRVL)—Gravel, or cinders	(TRTD)—Treated
(CONC)—Concrete	(PFC)—Porous friction courses	(WC)—Wire combed
(DIRT)—Dirt	(RFSC)—Rubberized friction seal coat	

RUNWAY WEIGHT BEARING CAPACITY

Runway strength data shown in this publication is derived from available information and is a realistic estimate of capability at an average level of activity. It is not intended as a maximum allowable weight or as an operating limitation. Many airport pavements are capable of supporting limited operations with gross weights of 25-50% in excess of the published figures. Permissible operating weights, insofar as runway strengths are concerned, are a matter of agreement between the owner and user. When desiring to operate into any airport at weights in excess of those published in the publication, users should contact the airport management for permission. Add 000 to figure following S, D, DT, DDT and MAX for gross weight capacity:

S—Runway weight bearing capacity for aircraft with single-wheel type landing gear, (DC-3), etc.
D—Runway weight bearing capacity for aircraft with dual-wheel type landing gear, (DC-6), etc.
DT—Runway weight bearing capacity for aircraft with dual-tandem type landing gear, (707), etc.
DDT—Runway weight bearing capacity for aircraft with double dual-tandem type landing gear, (747), etc.

Quadricycle and dual-tandem are considered virtually equal for runway weight bearing consideration, as are single-tandem and dual-wheel.

Omission of weight bearing capacity indicates information unknown.

RUNWAY LIGHTING

Lights are in operation sunset to sunrise. Lighting available by prior arrangement only or operating part of the night only and/or pilot controlled and with specific operating hours, are indicated under airport remarks. Since obstructions are usually lighted, obstruction lighting is not included in this code. Unlighted obstructions on or surrounding an airport will be noted in airport remarks.

Temporary, emergency or limited runway edge lighting such as flares, smudge pots, lanterns or portable runway lights will also be shown in airport remarks.

Types of lighting are shown with the runway or runway end they serve.

LIRL—Low Intensity Runway Lights
MIRL—Medium Intensity Runway Lights
HIRL—High Intensity Runway Lights
REIL—Runway End Identifier Lights
CL—Centerline Lights
TDZ—Touchdown Zone Lights
ODALS—Omni Directional Approach Lighting System.
AF OVRN—Air Force Overrun 1000' Standard Approach Lighting System.
LDIN—Lead-In Lighting System.
MALS—Medium Intensity Approach Lighting System.
MALSF—Medium Intensity Approach Lighting System with Sequenced Flashing Lights.
MALSR—Medium Intensity Approach Lighting System with Runway Alignment Indicator Lights.

SALS—Short Approach Lighting System.
SALSF—Short Approach Lighting System with Sequenced Flashing Lights.
SSALS—Simplified Short Approach Lighting System.
SSALF—Simplified Short Approach Lighting System with Sequenced Flashing Lights.
SSALR—Simplified Short Approach Lighting System with Runway Alignment Indicator Lights.
ALSAF—High Intensity Approach Lighting System with Sequenced Flashing Lights
ALSFI—High Intensity Approach Lighting System with Sequenced Flashing Lights, Category I, Configuration.
ALSF2—High Intensity Approach Lighting System with Sequenced Flashing Lights, Category II, Configuration.
VASI—Visual Approach Slope Indicator System.

VISUAL APPROACH SLOPE INDICATOR SYSTEMS

VASI—Visual Approach Slope Indicator
SAVASI—Simplified Abbreviated Visual Approach Slope Indicator

DIRECTORY LEGEND

S2L	2-box SAVASI on left side of runway
S2R	2-box SAVASI on right side of runway
V2R	2-box VASI on right side of runway
V2L	2-box VASI on left side of runway
V4R	4-box VASI on right side of runway
V4L	4-box VASI on left side of runway
V6R	6-box VASI on right side of runway
V6L	6-box VASI on left side of runway
V12	12-box VASI on both sides of runway
V16	16-box VASI on both sides of runway
*NSTD	Nonstandard VASI, VAPI, or any other system not listed above

VASI approach slope angle and threshold crossing height will be shown when available; i.e., GA 3.5° TCH 37.0'.

PILOT CONTROL OF AIRPORT LIGHTING

Key Mike	Function
7 times within 5 seconds	Highest intensity available
5 times within 5 seconds	Medium or lower intensity (Lower REIL or REIL-Off)
3 times within 5 seconds	Lowest intensity available (Lower REIL or REIL-Off)

Available systems will be indicated in the Airport Remarks, as follows:

ACTIVATE MALSR Rwy 7, HIRL Rwy 7/25-122.8.
or
ACTIVATE MIRL Rwy 18/36-122.8.
or
ACTIVATE VASI and REIL, Rwy 7-122.8.

Where the airport is not served by an instrument approach procedure and/or has an independent type system of different specification installed by the airport sponsor, descriptions of the type lights, method of control, and operating frequency will be explained in clear text. See AIM, "Basic Flight Information and ATC Procedures," for detailed description of pilot control of airport lighting.

RUNWAY GRADIENT

Runway gradient will be shown only when it is 0.3 percent or more. When available the direction of slope upward will be indicated, i.e., 0.5% up NW.

RUNWAY END DATA

Lighting systems such as VASI, MALSR, REIL; obstructions; displaced thresholds will be shown on the specific runway end. "Rgt tfc"—Right traffic indicates right turns should be made on landing and takeoff for specified runway end.

⑲ AIRPORT REMARKS

LLWSAS—Indicates a Low Level Wind Shear Alert System consisting of a centerfield and several field perimeter anemometers is installed.

SAWRS—Identifies airports that have a Supplemental Aviation Weather Reporting Station available to pilots for current weather information.

Landing Fee indicates landing charges for private or non-revenue producing aircraft, in addition, fees may be charged for planes that remain over a couple of hours and buy no services, or at major airline terminals for all aircraft.

Remarks—Data is confined to operational items affecting the status and usability of the airport.

⑳ COMMUNICATIONS

Communications will be listed in sequence in the order shown below:

Automatic Terminal Information Service (ATIS) and Private Aeronautical Stations (UNICOM) along with their frequency is shown, where available, on the line following the heading "COMMUNICATIONS". Whenever a second UNICOM frequency is shown in parentheses, it represents a proposed change to the current frequency with an uncertain date of change. If unable to contact the ground station on the regularly published UNICOM frequency, attempt to establish communications on the frequency shown in parenthesis.

Flight Service Station (FSS) information. The associated FSS will be shown followed by the identifier and information concerning availablity of telephone service, e.g. Direct Line (DL), Local Call (LC), etc. Where the airport NOTAM File identifier is different than the associated FSS it will be shown as "NOTAM FILE IAD." Where the FSS is located on the field it will be indicated as "on arpt" following the identifier. Frequencies available will follow. The FSS telephone number will follow along with any significant operational information. FSS's whose name is not the same as the airport on which located will also be listed in the normal alphabetical name listing for the state in which located. Limited Remote Communication Outlet (LRCO) or Remote Communications Outlet (RCO) providing service to the airport followed by the frequency and name of the Controlling FSS.

FSS's provide information on airport conditions, radio aids and other facilities, and process flight plans. Airport Advisory Service is provided at the pilot's request on 123.6 or 123.65 by FSS's located at non-tower airports or when the tower is not in operation. (See AIM, ADVISORIES AT NON TOWER AIRPORTS.)

DIRECTORY LEGEND

Aviation weather briefing service is provided by FSS specialists. Flight and weather briefing services are also available by calling the telephone numbers listed.

Limited Remote Communications Outlet (LRCO)—Unmanned air/ground communications facility, which may be associated with a VOR. These outlets have receive-only capability and rely on a VOR or a remote transmitter for full capability.

Remote Communications Outlet (RCO)—An unmanned air/ground communications facility, remotely controlled and providing UHF or VHF communications capability to extend the service range of an FSS.

Civil Communications Frequencies—Civil communications frequencies used in the FSS air/ground system are now operated simplex on 122.0, 122.2, 122.3, 122.4, 122.6, 123.6; emergency 121.5; plus receive-only on 122.05, 122.1, 122.15, and 123.6.

- a. 122.0 is assigned as the Enroute Flight Advisory Service channel at selected FSS's.
- b. 122.2 is assigned to all FSS's as a common enroute simplex service.
- c. 123.6 is assigned as the airport advisory channel at non-tower FSS locations, however, it is still in commission at some FSS's collocated with towers to provide part time Airport Advisory Service.
- d. 122.1 is the primary receive-only frequency at VOR's. 122.05, 122.15 and 123.6 are assigned at selected VOR's meeting certain criteria.
- e. Some FSS's are assigned 50 kHz channels for simplex operation in the 122-123 MHz band (e.g. 122.35). Pilots using the FSS A/G system should refer to this directory or appropriate charts to determine frequencies available at the FSS or remoted facility through which they wish to communicate.

Part time FSS hours of operation are shown in remarks under facility name.

Emergency frequency 121.5 is available at all Flight Service Stations, Towers, Approach Control and RADAR facilities, unless indicated as not available.

Frequencies published followed by the letter "T" or "R", indicate that the facility will only transmit or receive respectively on that frequency. All radio aids to navigation frequencies are transmit only.

TERMINAL SERVICES

ATIS—A continuous broadcast of recorded non-control information in selected areas of high activity.

UNICOM—A non-government air/ground radio communications facility utilized to provide general airport advisory service.

APP CON—Approach Control. The symbol ⓡ indicates radar approach control.

TOWER—Control tower

GND CON—Ground Control

DEP CON—Departure Control. The symbol ⓡ indicates radar departure control.

CLNC DEL—Clearance Delivery.

PRE TAXI CLNC—Pre taxi clearance

VFR ADVSY SVC—VFR Advisory Service. Service provided by Non-Radar Approach Control.

Advisory Service for VFR aircraft (upon a workload basis) ctc APP CON.

STAGE II SVC—Radar Advisory and Sequencing Service for VFR aircraft

STAGE III SVC—Radar Sequencing and Separation Service for participating VFR Aircraft within a Terminal Radar Service Area (TRSA)

TCA—Radar Sequencing and Separation Service for all aircraft in a Terminal Control Area (TCA)

TOWER, APP CON and DEP CON RADIO CALL will be the same as the airport name unless indicated otherwise.

㉑ RADIO AIDS TO NAVIGATION

The Airport Facility Directory lists by facility name all Radio Aids to Navigation, except Military TACANS, that appear on National Ocean Survey Visual or IFR Aeronautical Charts and those upon which the FAA has approved an Instrument Approach Procedure. All VOR, VORTAC and ILS equipment in the National Airspace System has an automatic monitoring and shutdown feature in the event of malfunction. Unmonitored, as used in this publication for any navigational aid, means that FSS or tower personnel cannot observe the malfunction or shutdown signal.

NAVAID information is tabulated as indicated in the following sample:

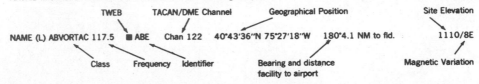

| TWEB | TACAN/DME Channel | Geographical Position | Site Elevation |

NAME (L) ABVORTAC 117.5 ▪ ABE   Chan 122   40°43'36"N 75°27'18"W   180°4.1 NM to fld.   1110/8E

Class   Frequency   Identifier   Bearing and distance facility to airport   Magnetic Variation

VOR unusable 020°-060° beyond 26 NM below 3500°

Restriction within the normal altitude/range of the navigational aid (See primary alphabetical listing for restrictions on VORTAC and VOR/DME).

ASR/PAR—Indicates that Surveillance (ASR) or Precision (PAR) radar instrument approach minimums are published in U.S. Government Instrument Approach Procedures.

DIRECTORY LEGEND

RADIO CLASS DESIGNATIONS

Identification of VOR/VORTAC/TACAN Stations by Class (Operational Limitations):

Normal Usable Altitudes and Radius Distances

Class	Altitudes	Distance (miles)
(T)	12,000' and below	25
(L)	Below 18,000'	40
(H)	Below 18,000'	40
(H)	Within the Conterminous 48 States only, between 14,500' and 17,999'	100
(H)	18,000' FL 450	130
(H)	Above FL 450	100

(H) = High   (L) = Low   (T) = Terminal

NOTE: An (H) facility is capable of providing (L) and (T) service volume and an (L) facility additionally provides (T) service volume.

The term VOR is, operationally, a general term covering the VHF omnidirectional bearing type of facility without regard to the fact that the power, the frequency protected service volume, the equipment configuration, and operational requirements may vary between facilities at different locations.

AB — Automatic Weather Broadcast (also shown with ▓ following frequency.)

DF — Direction Finding Service.

DME — UHF standard (TACAN compatible) distance measuring equipment.

H — Non-directional radio beacon (homing), power 50 watts to less than 2,000 watts (50 NM at all altitudes).

HH — Non-directional radio beacon (homing), power 2,000 watts or more (75 NM at all altitudes).

H-SAB — Non-directional radio beacons providing automatic transcribed weather service.

ILS — Instrument Landing System (voice, where available, on localizer channel).

ISMLS — Interim Standard Microwave Landing System.

LDA — Localizer Directional Aid.

LMM — Compass locator station when installed at middle marker site (15 NM at all altitudes).

LOM — Compass locator station when installed at outer marker site (15 NM at all altitudes).

MH — Non-directional radio beacon (homing) power less than 50 watts (25 NM at all altitudes).

S — Simultaneous range homing signal and/or voice.

SABH — Non-directional radio beacon not authorized for IFR or ATC. Provides automatic weather broadcasts.

SDF — Simplified Direction Facility.

TACAN — UHF navigational facility-omnidirectional course and distance information.

VOR — VHF navigational facility-omnidirectional course only.

VOR/DME — Collocated VOR navigational facility and UHF standard distance measuring equipment.

VORTAC — Collocated VOR and TACAN navigational facilities.

W — Without voice on radio facility frequency.

Z — VHF station location marker at a LF radio facility.

APPENDIX B

FAA PRIVATE PILOT EXAM CROSS-REFERENCE GUIDE

Appendix B lists 651 questions that appear in the Private Pilot Question Book (FAA-T-8080-1). The questions that appear apply to all aircraft, airplanes, and powered aircraft. These are the **only** questions that may be assigned on the Private Pilot Written Exam (Airplane). Each potential FAA exam question is cross-referenced to the appropriate chapter in both the text and the study guide. Finally, each potential FAA exam question is cross-referenced to the specific chapter and item number in this study guide.

It is important to remember that each item in this study guide manual that may also appear on the FAA Private Pilot Written Exam (Airplane) is denoted in **boldface** as follows:

6. (FAA 1015)

Thus, item six in this chapter is FAA <u>Private Pilot Question Book</u> Item Number 1015.

Appendix B works in the opposite direction. Each item in the <u>Private Pilot Question Book</u> that may appear on the FAA Private Pilot Written Exam (Airplane) below is cross-referenced to the appropriate chapter and item number in this study guide. For example, consider the following:

1031   6   15

In this example, FAA **Private Pilot** Question Book Item Number 15 appears in Chapter 6 of this study guide as Item Number 15.

As of August 1984, the FAA published a list of items that are unusable for test purposes and that have been removed from the FAA Question Selection Sheets. These items are indicated in **boldface** in the following list.

FAA	Ch.	Item	FAA	Ch.	Item	FAA	Ch.	Item
1001	6	14	1002	6	15	1003	4	1
1004	4	2	1005	6	16	1006	4	3
1007	4	4	1008	6	17	1010	6	1
1011	6	2	1012	6	3	1013	6	4
1014	6	5	1015	6	6	1017	6	7
1018	6	8	1019	6	9	1020	6	10
1031	6	11	1032	6	12	1033	6	13
1034	6	19	1035	6	20	1036	6	21
1037	6	22	1038	6	23	1039	6	24
1045	6	18	1046	6	29	1047	6	30
1048	6	31	1049	6	32	1050	6	25
1054	2	17	1055	3	1	1056	3	2

FAA	Ch.	Item	FAA	Ch.	Item	FAA	Ch.	Item
1057	3	3	1058	3	4	1059	3	5
1060	3	6	1061	3	7	1062	3	8
1063	3	9	1064	3	10	1065	3	11
1066	3	12	1067	3	13	1070	3	14
1071	3	15	1072	3	16	1073	3	17
1074	3	18	1075	3	19	1076	3	20
1077	3	21	1081	3	22	1082	3	23
1083	3	24	1087	4	5	1088	4	6
1089	4	7	1090	4	8	1091	4	9
1092	4	10	1095	4	11	1096	4	12
1098	4	13	1100	4	14	1101	4	15
1102	4	16	1103	15	14	1104	4	17
1105	4	18	1106	4	19	1107	4	20
1108	4	21	1109	4	22	1110	4	23
1111	4	24	1112	4	25	1113	4	26
1114	4	27	1115	4	28	1116	4	29
1117	4	30	1118	4	31	1119	4	32
1120	4	33	1121	4	34	1122	4	35
1123	4	36	1124	4	37	1125	4	38
1126	4	39	1127	13	1	1150	2	1
1151	2	2	1152	2	3	1153	2	4
1154	2	5	1155	2	6	1157	2	15
1158	2	16	1159	2	13	1160	2	14
1161	2	7	1162	2	8	1163	2	18
1164	2	19	1165	2	20	1166	2	21
1167	2	10	1168	2	17	1169	2	30
1170	2	9	1171	5	1	1172	5	2
1173	5	3	1174	2	22	1175	2	23
1176	2	11	1177	2	24	1178	2	25
1179	2	12	1180	2	26	1181	2	27
1182	2	28	1183	2	29	1198	5	4
1199	5	5	1200	5	6	1207	5	7
1208	5	8	1209	5	9	1210	5	10
1211	5	11	1222	5	12	1223	5	13
1224	5	14	1225	5	15	1226	5	16
1227	5	17	1228	5	18	1229	5	19
1244	12	1	1245	12	2	1246	12	3
1247	12	4	1248	12	5	1249	12	6
1250	12	7	1251	12	8	1252	12	9
1253	12	10	1254	12	11	1255	12	12
1256	12	13	1257	12	14	1258	12	15
1259	12	16	1260	12	17	1261	12	18
1262	12	19	1264	12	20	1265	12	21
1269	12	22	1270	12	23	1271	12	24
1273	12	25	1274	12	26	1275	12	27
1276	12	28	1277	12	29	1278	12	30
1279	12	31	1285	13	14	1286	13	15
1287	13	16	1288	13	17	1289	13	18
1290	13	19	1291	13	20	1292	13	22
1293	13	23	1294	13	21	1295	13	24
1296	13	13	1297	13	25	1298	13	26
1299	13	27	1300	13	28	1301	13	29
1302	13	30	1303	13	31	1304	13	32

FAA	Ch.	Item	FAA	Ch.	Item	FAA	Ch.	Item
1305	13	33	1306	13	34	1307	13	42
1308	13	43	1309	13	44	1310	13	45
1311	13	46	1312	13	47	1313	13	48
1314	13	41	1317	13	49	1318	13	36
1319	13	35	1320	13	37	1321	13	38
1323	13	39	1324	13	40	1328	13	50
1329	13	51	1330	13	52	1331	13	53
1332	13	54	1333	13	55	1334	13	56
1335	13	57	1336	13	58	1337	13	59
1338	13	60	1339	13	61	1340	13	62
1341	13	63	1342	13	2	1343	13	3
1344	13	4	1345	13	5	1346	13	6
1347	13	7	1348	3	25	1353	7	1
1354	7	2	1355	7	3	1356	7	4
1357	7	5	1358	7	6	1359	7	7
1361	3	26	1362	3	27	1363	3	28
1364	3	29	1369	3	30	1370	3	31
1371	3	32	1373	4	40	1419	7	8
1420	7	9	1421	7	10	1422	7	11
1423	3	33	1424	2	33	1425	2	31
1426	2	32	1427	3	34	1428	3	35
1429	3	36	1433	7	12	1434	7	13
1435	7	14	1436	7	15	1437	7	16
1438	7	17	1439	7	18	1441	7	19
1442	7	20	1443	7	21	1444	7	22
1445	7	23	1446	7	24	1447	7	25
1448	7	26	1449	7	27	1450	7	28
1451	7	29	1452	7	30	1453	7	31
1454	7	32	1455	7	33	1456	7	34
1458	7	35	1459	7	36	1460	7	37
1461	7	38	1462	7	39	1463	7	40
1464	7	41	1465	7	42	1466	7	43
1467	7	44	1468	7	45	1469	7	46
1470	10	1	1471	10	2	1472	10	3
1473	10	4	1474	10	5	1475	10	6
1476	10	7	1477	7	47	1478	7	48
1479	7	49	1480	7	50	1481	7	51
1483	7	52	1484	7	53	1485	7	54
1486	7	55	1487	7	56	1488	7	57
1489	7	58	1490	7	59	1491	7	60
1492	7	61	1493	7	62	1494	9	1
1495	9	2	1496	7	63	1497	16	1
1498	16	2	1499	16	3	1500	16	4
1501	7	112	1502	10	8	*1502	16	6
1503	13	8	*1503	16	5	1504	9	3
1505	9	4	1506	9	5	1507	10	9
1508	10	10	1509	10	11	1510	10	12
1511	10	13	1512	10	14	1513	10	15
1514	14	1	1515	14	2	1516	13	9
1517	7	64	1518	7	65	1519	11	1
1520	11	2	1521	11	3	1522	11	4
1523	11	5	1524	11	6	1525	11	7
1526	10	16	1527	7	66	1528	7	67

FAA	Ch.	Item	FAA	Ch.	Item	FAA	Ch.	Item
1529	7	68	1530	7	69	1531	9	6
1534	1	3	1535	1	4	1536	1	1
1537	1	6	1538	1	2	1539	1	5
1540	11	8	1541	11	9	1547	11	10
1548	11	11	1549	11	17	1554	11	13
1555	11	14	1557	11	15	1558	11	16
1561	11	12	1562	11	18	1563	11	19
1564	7	70	1565	7	71	1566	7	72
1567	7	73	1568	7	74	1569	11	20
1571	11	21	1572	11	22	1573	11	23
1574	11	24	1575	11	25	*1575	14	3
1576	11	26	1577	11	27	1578	11	32
1582	11	29	1584	11	30	1585	11	31
1586	11	28	1594	11	33	1595	11	34
1596	11	35	1598	7	114	1601	7	113
1605	11	36	1606	11	37	1607	11	38
1608	11	39	1612	11	40	1613	11	41
1616	11	42	1620	11	43	1621	16	7
1622	16	8	1624	11	44	1625	11	46
1626	11	47	1627	11	48	1628	11	49
1629	11	50	1630	11	45	1631	11	51
1632	11	52	1633	11	53	1634	11	54
1635	11	55	1636	11	56	1637	11	57
1638	11	58	1640	11	59	1641	11	60
1642	11	61	1643	11	62	1644	11	63
1647	7	78	1648	7	79	1649	7	80
1650	7	81	1651	7	82	1652	11	64
1654	11	65	1657	11	66	1658	11	67
1659	11	68	1660	7	109	1661	7	110
1662	7	11	1663	7	83	1664	7	84
1665	7	85	1667	7	86	1669	7	87
1670	7	88	1671	7	89	1672	7	90
1673	7	91	1674	7	92	1679	7	105
1686	7	93	1687	7	94	1688	7	95
1689	7	96	1690	7	97	1691	7	98
1692	7	99	1693	7	100	1694	7	101
1695	7	102	1696	7	103	1697	7	104
1698	7	106	1699	7	107	1700	7	108
1701	7	75	1703	13	10	1704	13	11
1705	13	12	1707	11	69	1708	11	70
1710	11	71	1714	11	72	1715	11	73
1716	11	74	1717	11	75	1722	11	76
1723	8	1	1724	8	2	1725	8	3
1726	8	4	1727	8	5	1728	8	6
1729	6	26	1730	6	27	1731	6	28
1732	6	33	1733	6	37	1734	6	38
1735	6	39	1736	6	40	1737	6	41
1738	6	42	1739	6	43	1740	6	34
1741	6	35	1742	6	36	1743	6	44
1744	6	45	1745	6	46	1746	6	47
1747	6	48	1748	6	49	1749	6	50
1750	6	51	1751	8	7	1752	8	8
1754	8	9	1755	8	10	1756	8	11

FAA	Ch.	Item	FAA	Ch.	Item	FAA	Ch.	Item
1757	8	12	1758	8	13	1759	8	14
1760	8	15	1761	8	16	1762	8	17
1763	8	18	1764	8	19	1765	8	20
1766	8	21	1767	8	22	1768	8	23
1770	8	24	1771	8	25	1772	8	26
1773	8	27	1774	8	28	1775	8	29
1776	8	30	1777	8	31	1778	8	32
1779	8	33	1780	8	34	1781	8	35
1782	9	7	1783	8	36	1784	8	37
1785	8	39	1786	8	40	1787	8	38
1788	8	41	1789	8	42	1790	3	37
1791	3	38	1792	8	44	*1792	5	21
1793	8	43	1794	5	20	1795	8	45
1796	8	46	1797	8	47	1798	8	48
1799	8	49	1800	8	50	1801	8	51
1802	8	52	1803	8	53	1804	8	54
1805	8	55	1806	8	56	1807	8	57
1808	8	58	1809	8	59	1826	8	60
1827	9	10	1828	9	11	1829	9	12
1830	9	13	1831	9	8	1832	9	9
1833	9	14	1834	9	15	1835	9	16
1836	9	17	1837	9	18	1838	9	19
1839	9	20	1840	9	21	1841	9	22
1842	9	23	**1843**	**9**	**24**	1844	9	25
1845	9	26	1846	9	27	1847	9	28
1848	9	29	1849	9	30	1850	9	31
1851	9	32	1852	9	33	1853	9	34
1854	9	35	1855	9	36	1856	9	37
1857	9	38	1858	9	39	1859	9	40
1860	9	41	1861	9	42	1862	9	43
1863	9	44	1864	9	45	1865	9	46
1866	9	47	1867	9	48	1868	9	49
1869	9	50	1870	9	51	1871	9	52
1872	9	66	1873	9	59	1874	9	53
1875	9	54	1876	9	55	1877	9	56
1878	9	57	1879	9	58	1884	9	60
1885	**9**	**61**	1886	9	62	1887	9	63
1888	9	64	1889	9	65	1890	9	67
1891	9	68	1892	9	69	1893	9	70
1894	9	71	1895	9	73	1896	9	74
1897	9	75	1898	9	76	1899	9	77
1900	9	78	1901	9	79	1902	9	72
1903	9	80	1904	9	81	1905	15	1
1906	**15**	**2**	1907	15	3	1908	15	4
1909	**15**	**5**	1910	15	6	1911	15	7
1912	15	8	1913	15	9	1914	15	10
1915	15	11	1916	15	12	1917	15	13

* Denotes an item that appears in more than one chapter of this study guide.

Boldface items have been designated as unusable by the FAA and have been removed from the FAA Question Selection Sheets.